ENVIRONMENTAL CHANGE
IN SOUTH–EAST ASIA

In the wake of the Rio Earth Summit, how have political leaders sought to rec-oncile the quest for economic development with the new world-wide concern about environmental conservation? Do policy changes denote real political change or mere rhetoric designed to placate Western aid donors? How have non-state groups reacted to environmental change and government policies in a post-Rio world? These questions illustrate the need to situate the current interest in sustainable development in the context of broader questions per-taining to the political economy of environmental change.

Environmental Change in South-East Asia brings together scholars, journalists, consultants and NGO activists in order to explore how people, politics and the quest for sustainable development are interrelated in South-East Asia. As a region characterized by explosive economic growth, grave socio-economic inequities and pervasive environmental degradation, South-East Asia epito-mizes the dilemmas facing policy-makers as they seek to implement sustain-able development policies. It illustrates the centrality of politics to environmental change, and the human response to that change. Key economic and technical elements of the quest for sustainable development – ecotourism, plantation forestry, remote sensing and GIS – are set in a context that is sen-sitive to the political dimensions of that quest. Highlighting the practical polit-ical obstacles to the attainment of sustainable development in South-East Asia, the authors present an important and essential corrective to a literature for too long dominated by economists and ecologists. The authors assume that nei-ther the quest for sustainable development nor the process of environmental change itself can be understood without reference to political processes.

Environmental Change in South-East Asia will be of interest to all those con-cerned with understanding the interaction of politics, sustainable development and environmental change in the developing world.

Michael J. G. Parnwell is Senior Lecturer at the Centre for South-East Asian Studies, University of Hull; **Raymond L. Bryant** is Lecturer in Geography at King's College, London.

GLOBAL ENVIRONMENTAL CHANGE SERIES
Edited by Michael Redclift, Wye College, University of
London, Martin Parry, University of Oxford, Timothy
O'Riordan, University of East Anglia, Robin Grove-White,
University of Lancaster and Brian Robson, University of
Manchester.

The *Global Environmental Change Series*, published in association with the
ESRC Global Environmental Change Programme, emphasizes the way that
human aspirations, choices and everyday behaviour influence changes in the
global environment. In the aftermath of UNCED and Agenda 21, this series
helps crystallize the contribution of social science thinking to global change
and explores the impact of global changes on the development of social
sciences.

ENVIRONMENTAL CHANGE IN SOUTH-EAST ASIA

People, Politics and Sustainable Development

Edited by
Michael J. G. Parnwell and
Raymond L. Bryant

Global Environmental Change Programme

London and New York

First published 1996
by Routledge
11 New Fetter Lane, London EC4P 4EE

Simultaneously published in the USA and Canada
by Routledge
29 West 35th Street, New York, NY 10001

Routledge is an International Thomson Publishing company I(T)P

Typeset in Garamond by
Florencetype Limited, Stoodleigh, Devon

Printed and bound in Great Britain by
Biddles Ltd, Guildford and King's Lynn

British Library Cataloguing in Publication Data
A catalogue record for this book is available from the British Library

Library of Congress Cataloguing in Publication Data
Environmental change in South-East Asia : people, politics and sustainable
development / edited by Michael Parnwell and Raymond Bryant
p.   cm. – (Global environmental change series)
Includes bibliographical references and index.
ISBN 0–415–12932–X. – ISBN 0–415–12933–8 (pbk.)
1. Environmental policy–Asia, Southeastern.   2. Sustainable
development –Asia, Southeastern.   3. Conservation of natural resources–
Economic aspects–Asia, Southeastern.   I. Parnwell, Mike.   II. Bryant,
Raymond, 1953–   .   III. Series.
HC441.Z9E5287   1996
333.7′0959–dc20   95–26276
                                                                    CIP

ISBN 0–415–12932–X (hbk)
ISBN 0–415–12933–8 (pbk)

CONTENTS

PLATES

FIGURES

TABLES AND BOXES

TABLES

x

BOXES

CONTRIBUTORS

Dr Gilbert Braganza is a research associate in the Environmental Research Division, Manila Observatory, the Philippines.

Dr Raymond L. Bryant is a lecturer in Geography at King's College, University of London.

Dr Owen Cameron has recently completed his PhD in the Department of Geography, University of Cambridge.

Janet Cochrane is an ecotourism consultant and a PhD student, Centre for South-East Asian Studies, University of Hull.

Rili Hawari Djohani works for the World Wildlife Fund in Indonesia.

Dr Alan P. Dykes is a lecturer at the University of Huddersfield.

Dr Bernard Eccleston is at the Open University.

Randi Jerndal is at the Centre for Southern Asian Studies, Gothenburg University.

Professor Victor T. King is Professor of South-East Asian Development Sociology, University of Hull.

Chris R. Lang works for Earth Action Resource Centre, Oxford.

Helen Lawes is at the University of Manchester.

Dr Mark Lloyd is a lecturer in the Department of Geography, Royal Holloway, University of London.

Dr Larry Lohmann is the assistant editor of *The Ecologist*.

Dr Duncan McGregor is Senior Lecturer in the Department of Geography, Royal Holloway, University of London.

Julia McMorrow is a lecturer in the Department of Geography, University of Manchester.

Dr Michael J.G. Parnwell is Senior Lecturer in South-East Asian Geography, University of Hull.

Professor David Potter is at the Open University.

Dr Jonathan Rigg is Reader in South-East Asian Geography, University of Durham.

Dr Colin Sage is a lecturer at Wye College, University of London.

Dr David M. Taylor is a lecturer in Biogeography, in the Department of Geography, University of Hull.

Ann Danaiya Usher is a former journalist with *The Nation*, Bangkok.

Dr John Wills is lecturer at the University of Huddersfield.

ACKNOWLEDGEMENTS

This volume owes its genesis to the 1994 annual conference of the Association of South-East Asian Studies in the United Kingdom (ASEASUK), held at Royal Holloway, University of London. The conference subject was 'The Environment in South-East Asia'. The co-editors, who were also the co-convenors of the conference, would like to express their gratitude to Dr Tony Stockwell (Royal Holloway) for his considerable input into the organization of the conference, and the Committee of ASEASUK for their encouragement and support. This is the second book to have emerged from the annual conferences of ASEASUK, following *Tourism in South-East Asia* (edited by Mike Hitchcock, Terry King and Mike Parnwell: Routledge, 1993).

It is very appropriate that this volume should appear in the Routledge Global Environmental Change series, as seven of the contributors have received funding support under the Economic and Social Research Council's (ESRC) Global Environmental Change Programme. The contributions by Victor (Terry) King, Mike Parnwell, David Taylor, Julia McMorrow and Helen Lawes were made possible by funding within Phase I of this programme for a collaborative research programme entitled 'Tropical Rainforests, Communities and Global Environmental Change in Borneo'. In addition to the ESRC, the team wishes to express its gratitude to the following institutions and individuals. In Sarawak: Dr Abdul Rashid Abdullah and Dr Peter Songan of Universiti Malaysia Sarawak and formerly of Universiti Pertanian, Kampus Bintulu, who provided local support for the research team; Miss Idan Lebit, Dr Jean Morrison and David Hortin for their work as research assistants; the Sarawak State Planning Unit, for granting research permission and for otherwise facilitating the research; and several state departments for their willing and generous support. In Sabah: Dr Mustapa Abd. Talip, and other members of the Sabah team; Universiti Kebangsaan Malaysia Sabah campus; the Institute of Development Studies, Sabah; Rakjat Berjaya Sdn. Bhd.; Boustead Estates Sdn. Bhd. (Kota Kinabalu office); the Lahad Datu District Officer; the Sabah Departments of Forestry, Agriculture and Lands and Surveys; and the Malaysian Centre for Remote Sensing.

The Hull branch of the team would also like to take this opportunity to acknowledge the help, friendship and insight which we were privileged to receive from Henry Gana Ngadi, who recently passed away. Henry had dedicated his life and career to recording the oral history and traditions of his people, the Ibans. These were also the focus of his Ph.D. research in the Centre for South-East Asian Studies, University of Hull. Sadly, Henry was taken from us before he was able to complete either task – but his memory and contribution to the cause live on.

The research by Bernard Eccleston and David Potter was supported by an award to the Open University Research Group under Phase II of the ESRC GEC Programme for a project entitled 'Setting Environmental Agendas: NGOs, Democracy and Global Politics'.

Duncan McGregor would like to express his thanks to the British Academy Committee for South-East Asian Studies for the award of a grant which enabled him to undertake a reconnaissance visit to Thailand and Malaysia; the Central Research Fund of the University of London, which provided funds for the digitization of aerial photographic materials; Dr Elizabeth Moore (Department of Art and Archaeology, School of Oriental and African Studies, University of London), for freely giving of her time in introducing him to the Williams-Hunt Collection; and Andrew Kennedy and Justin Jacyno for their considerable input into digitizing images and producing maps.

John Wills and Alan Dykes would like to acknowledge the assistance of the following: Land Survey Department of Brunei Darussalam; the members of the Brunei Rainforest Project; Universiti Brunei Darussalam and the Royal Geographical Society; the Earl of Cranbrook and Dr David Edwards; and the sponsors of the RGS/UBD Project, Royal Brunei Airlines, Greencard Trust, the Hong Kong Bank, and Brunei Shell.

Larry Lohmann would like to acknowledge the support of Heinrich-Böll-Stiftung and the International Development Research Centre. Rili Djohani's research was conducted under the auspices of the World Wide Fund for Nature Indonesia Programme, and the Directorate General of Forest Protection and Nature Conservation of the Department of Forestry in Indonesia, in co-operation with the Foundation for Development of Indonesian Village Settlements. Colin Sage's research was funded by the Population and Environment Research Programme of the Overseas Development Administration. Field-work in North Lampung would not have been possible without the efforts of Ir Edi Dwi Cahyono of Brawijaya University, as well as the contribution of Ir Nita Ariastuti and Ir Herna of Lampung University.

The above clearly shows that individuals and institutions within South-East Asia have played an essential role in the research undertaken by the contributors to this volume. Of equal importance are the many hundreds of local people who, perhaps unknowingly, have helped in so many ways in the creation of the ideas, ideologies and insights upon which this volume has been built. It is our genuine hope that this volume will help in some way, however modest, to improve their prospects, and those of subsequent generations, for a sustainable future.

1

INTRODUCTION

Politics, sustainable development and environmental change in South-East Asia

Raymond L. Bryant and Michael J.G. Parnwell

In so far as environmental change has become an important preoccupation of our times, 'sustainable development' has become the leitmotif of the environment and development literature. With its promise to set all environmental problems right (thereby averting a feared ecological Armageddon), it is not surprising that this concept has been embraced by policy-makers, business leaders, grassroots activists and scholars alike with, at times, almost religious fervour.

However, uncritical acceptance of sustainable development as the 'solution' to the world's environmental problems is problematic. To begin with, the meaning of 'sustainable development' remains elusive – a chameleon-like concept, it means many things to many people. Such flexibility helps to explain its popularity, but simultaneously raises serious questions about its utility as a concept capable of uniting widely disparate objectives and interests (Lele, 1991; Redclift, 1992).

Further, sustainable development is increasingly used as a means to classify a wide variety of economic activities according to their apparent 'greenness' or lack thereof. Thus, certain activities such as ecotourism and plantation forestry are 'sustainable', while other activities, notably manufacturing and clear-cut logging, are 'unsustainable'. This classificatory exercise, however, fails to take into account the location-specific nature of economic activities – what might be sustainable in one context might be unsustainable in another. More seriously, and as various chapters in this book illustrate, such an approach neglects to situate discrete activities in a wider political and economic context. How 'sustainable', after all, are economic activities that form part of a global capitalist economy which is seen by some to be incompatible with environmental conservation? (Redclift, 1987). If 'a capitalist society based on competition and growth for its own sake must ultimately devour the natural world, just like an untreated cancer must ultimately devour its host' (Bookchin: cited in Cutter, 1994, p. 217), then a reformist approach that promotes 'greener' activities is inevitably doomed to failure – unless the social context within which it is applied changes.

If the debate still rages as to whether sustainable capitalist development is an oxymoron, what is becoming increasingly clear is the intensely political nature of sustainable development – from its initial definition to its attempted implementation. However, the politics of sustainable development, as with the politics of environmental change generally, has received surprisingly little attention in the literature. Rather, attention hitherto has largely focused on specifying the economic measures needed for sustainable development – that is, on seeking to integrate ecological factors into the economic calculations that are believed to underpin the decision-making process (Schramm and Warford, 1987; Pearce *et al.*, 1990). Such work has always had a quality of the surreal about it. How, after all, could it be possible for policies to be devised and implemented as if in a political vacuum? As scholars begin to explore systematically the ways in which discursive and material practices are politically constructed and mediated, it is becoming impossible not to consider politics in addressing the issues of environmental change and sustainable development (see Adams, 1990; Bryant, 1992; Harvey, 1993; Peet and Watts, 1993; Silva, 1994). Such 'political-ecology' research stands in sharp contrast to the economistic work that remains even today the predominant element in the environment and development literature.

In order to make sense of the inter-relatedness of politics, sustainable development and environmental change, it is important to locate general debates in an empirical or 'grounded' setting. Such work may be undertaken at various scales, but it is at the regional level that the interplay between political and ecological forces is most fruitfully analysed and understood (Blaikie and Brookfield, 1987). This book thus explores the interaction of politics and ecology in the South-East Asian setting. South-East Asia (see Figure 1.1), with its explosive mix of rapid but uneven economic growth and pervasive environmental degradation, is a region in which many of the political issues and problems associated with sustainable development and environmental change can be clearly seen. Here, perhaps more than anywhere else in the developing world, the contradictions between environment and development, economic growth and environmental conservation, are visible, and inform the political process. This is particularly the case with the exploitation of the region's forest resources: accordingly, and because of its current prominence, the ecological and social consequences of forestry provide an important focus for discussion in this volume.

There is now a rapidly growing literature on environment and development issues in South-East Asia (e.g. Poffenberger, 1990; Brookfield and Byron, 1993; Bryant *et al.*, 1993; Howard, 1993). Scholars have also begun to emphasize the ways in which regional environmental degradation is linked to the political process. Thus, one avenue of enquiry has been to describe how state policies provide economic incentives for large-scale logging, mining and other destructive activities (Repetto and Gillis, 1988; Barbier, 1993). An alternative approach has been to emphasize the manner in which the

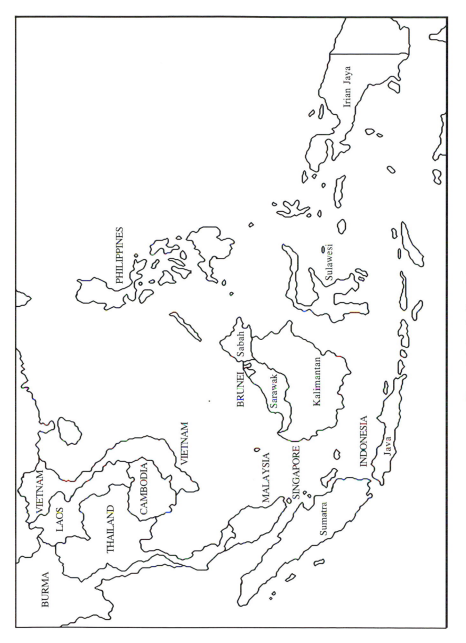

Figure 1.1 South-East Asia

empowerment of political and economic élites is linked directly to profit-making from environmentally destructive practices (Hurst, 1990; Rush, 1991; Broad and Cavanagh, 1993; Colchester and Lohmann, 1993; Dauvergne, 1994). However, what this literature has scarcely begun to address is the question of the politics of sustainable development – that is, the response of state leaders and private citizens in South-East Asia to environmental change. In the wake of the Rio Earth Summit, how have political leaders sought to reconcile the quest for economic development with the new world-wide concern about environmental conservation? Do policy changes denote real political change or mere rhetoric designed to placate Western aid donors? How have non-state groups reacted to environmental change and government policies in a post-Rio world? These questions illustrate the need for research that situates the current interest in sustainable development in the context of broader questions pertaining to the political economy of environmental change in South-East Asia.

This book is a preliminary exploration of this research terrain. It encompasses research setting out some of the key economic and technical elements of the quest for sustainable development (i.e. ecotourism, plantation forestry, remote sensing and Geographical Information Systems (GIS)), but in a context that is sensitive to the political dimensions of that quest. This book assumes, therefore, that neither the quest for sustainable development nor the process of environmental change itself can be understood without reference to political processes. The following discussion considers the grounds on which this assumption is based.

THE POLITICS OF ENVIRONMENTAL CHANGE

The past two centuries have witnessed human-induced environmental change on an unprecedented scale in South-East Asia. At the heart of this process has been the integration of the region into a globalizing capitalist economy, initially during the colonial era, but with greater momentum in post-colonial times. Yet such integration has not taken place 'naturally' but, rather, has been linked to political processes that have prompted South-East Asia's emergence as one of the key natural resource regions in the world.

In the late nineteenth and early twentieth centuries, the colonial powers reorganized and expanded pre-colonial patterns of resource exploitation in such a way that by the end of colonial rule export-oriented commercial resource extraction was central to economic life in the region. Certainly, this process occurred unevenly in South-East Asia depending on the vicissitudes of local political and ecological conditions as well as market demand. However, by the beginning of the Second World War, only the most remote territories were not integrated into national and international markets. The social and environmental implications of this process were immense. The growing resource dependency of South-East Asian economies was reflected in export and revenue figures as well as in social indicators such as occupational status. The most vivid indication of this dependency, however, was the identification of places – and even entire countries – with the large-scale production of selected natural resources and plantation crops which exploited land resources and prevailing natural conditions: Burma and Siam (Thailand) with rice, teak and minerals; Java with coffee and sugar; the Philippines with sugar, abaca and coconuts; and Malaya with tin, palm oil and rubber. Resource exploitation put South-East Asia on the (colonial) world map, thereby creating national and local identities that are only today beginning to break down with the uneven spread of industrialization through the region.

The development of South-East Asia's resource-based identity was intimately associated with large-scale environmental change. Social and economic transformation was accompanied by environmental mutation: changes in forest cover and type, the extension of agricultural production, deteriorating soil conditions, and increasing levels of pollution. Prior to 1850, much of South-East Asia was covered in forests, but one hundred years later, large swathes of low-lying forest had already been cleared. This process was partly a response to the rapidly growing imperial and indigenous demand for timber for housing, government buildings, bridges, boats, fuel, railway sleepers, and so on.

However, by far the main impetus for widespread deforestation was permanent agriculture, with cleared land being used to produce such cash crops as coffee, tea, rubber, sago, palm oil, rice, abaca and sugar cane. The sheer scale and rapidity of such environmental change were breathtaking. In British Burma, for example, the rice-growing area in the Irrawaddy and Sittang deltas expanded from between 700,000 to 800,000 acres in 1852 to nearly 6,000,000 acres in 1906, while during roughly the same period the local population climbed from about one million to over four million (Adas, 1974). In the early twentieth century, this area was the largest rice-exporting territory in the world. Yet such growth necessitated massive forest clearance with at least three million hectares of *kanazo* forest alone eliminated for this purpose (Adas, 1983). A comparable process of large-scale forest clearance occurred in other parts of South-East Asia around the same time (Tucker and Richards, 1983; Richards and Tucker, 1988; Rush, 1991).

Politics played a crucial role in the environmental transformation of South-East Asia. To take the Burmese case noted above, for example, the widespread conversion of forest to field in southern Burma would not have occurred without of a package of incentives offered to peasants by the colonial state designed to facilitate this process. Thus, peasants who undertook permanent cultivation in hitherto forested areas were entitled to tax holidays and legal title to the land. Further, the colonial state also funded the construction of canals and embankments, and improved river and land transport networks, 'to facilitate the movement of labour and export products and to make cultivation of empty lands possible' (Adas, 1974, p. 35).

However, the role of colonial rule in promoting resource exploitation and environmental change was even more far-reaching than this example would indicate. Indeed, colonialism resulted in a series of political and administrative transformations that have conditioned resource exploitation and environmental change to the present day. To begin with, colonial rule ushered in systematic changes in the way in which states were organized and run. Thus, administration was reorganized along functional lines with the result that in the field of resource management, departments responsible for the management of 'forests', 'agriculture' and 'mining' were created, beginning in the mid-nineteenth century. Although the specific remit of these departments varied, they nevertheless shared a common objective – namely, to conceptualize and manage resources in a functionally defined manner in order to maximize commercial production.

Four important implications followed from this colonial functionalist approach to resource management. First, this approach greatly enhanced the 'efficiency' of resource extraction in South-East Asia. Such efficiency usually encompassed the selective conservation of especially valuable renewable resources (teak being the classic example), but in a context of expanded overall production. The goal was the attainment of maximum extraction levels consonant with long-term commercial exploitation. Second, specialized knowledge and professional training became pre-requisites for entry to service in the state. This process enabled the rapid accumulation of knowledge about the resource in question, but also encouraged a parochial outlook among staff. As J.S. Furnivall (1956, p. 77) noted, 'none of these officials saw life whole and, by reason of frequent transfers, none of them saw it steadily.' Third, conflict between specialist departments often ensued as a result of a basic disjuncture between the 'political and administrative world' and the 'real resource world' that it sought to administer. The latter did not neatly conform to official resource categories (i.e. 'forests', 'agriculture'), but overlapped categories in complicated ways thereby virtually guaranteeing bureaucratic conflict. Perhaps the most common type of bureaucratic conflict concerned agricultural and forestry officials, with the former keen to clear suitable low-lying forest for permanent agriculture and the latter often equally keen to protect such forest if it contained valuable commercial species. Fourth, the functionalist approach often exacerbated conflict between officials and private citizens. The advent of functionally defined departments signalled a growing effort by the state to regulate the activities of the citizens under its political control. Such regulation took several forms. Thus, while forest departments sought to deny or limit popular access to commercial forests, agriculture departments insisted on fixed and inflexible cultivation taxes which, during economic downturns, were the source of much peasant hardship (Scott, 1976; Peluso, 1992).

Colonial rule was also associated with the assertion of territorial political control. It was thus partly about the definition of political control in terms of 'inside/outside' (Walker, 1993). There were two inter-related elements to this process. There was first 'external territoriality' – that is, the attempt to define state control clearly and permanently in terms of what was within its legal jurisdiction. Based on European notions of political power and of the state itself, the quest for fixed frontiers in South-East Asia went against the pre-colonial pattern in which a state was 'typically defined not by its perimeter, but by its center' (Anderson, 1990, p. 41).

As with the functionally defined state, the advent of fixed frontiers in South-East Asia had important implications for resource exploitation and environmental change in the region. Fixed borders thus permitted a state to act with greater confidence within 'its' territory than was hitherto the case. Especially in 'peripheral' areas – often rich in forest and mineral resources – a state no longer need fear that its resource exploitation policies might precipitate inter-state conflict due to contested ownership. Fixed borders also reinforced the power of ethnic majorities over ethnic minorities in the region in so far as peripheral lands traditionally used by the latter and relatively free from outside control were incorporated into territories controlled by the dominant ethnic group. As the latter has sought to assert control over peripheral areas of the nation–state, ethnic conflict has been the almost inevitable result – conflict in part about who is to control resource exploitation and environmental change in the contested area. Finally, fixed borders have not always served to eliminate inter-state conflict over resources. As contemporary fishing disputes highlight, not all resources fit neatly within politically defined borders – migratory patterns of fish, for example, defy conventional notions of aquatic territoriality (Innes-Brown and Valencia, 1993). Once again, the disjuncture between the 'political and administrative world' and the 'real resource world' has been a fertile source of conflict in South-East Asia.

'Internal territoriality' was a second means by which the colonial state sought to assert political control over people and resources. Here, the objective was to develop a national profile or inventory of all people and resources within a given nation–state as part of a broader attempt to promote economic activity and central political control. Using such tools as the map and census, the colonial state was able to achieve these objectives to an extent that pre-colonial states were never in a position to do. In some cases, the specific objective was to obtain an accurate picture of private land-holdings and agricultural production in order to facilitate taxation (Adas, 1974; Scott, 1976). In the case of forest management, the goal was to differentiate between state- and non-state-owned territory – hence, the creation of 'reserved' state forests in which non-state access and rights were severely curtailed, if not eliminated altogether (Peluso, 1992; Bryant, 1994). In diverse resource sectors, therefore, the colonial state measured and mapped to enhance central control over resource use and management (a similar process occurred in nominally independent Siam/Thailand: see Vandergeest and Peluso, 1993; Winichakul, 1994).

Finally, colonial rule was linked to the systematic introduction and dissemination of European science and technology in South-East Asia. To some extent, this point has been implicit in the discussion so far. Thus, the census and map which helped to define internal and external territoriality (thereby permitting increased resource exploitation) formed part of a broader set of statistical techniques that was applied to increasingly powerful effect in the quest to map, measure and manipulate. The functional organization of the state itself reflected new principles of 'scientific' administrative organization and efficiency in the nineteenth century. An array of powerful new technologies – the railway, the telegraph, the steamship – facilitated the flow of resources, people and information both within the region and between South-East Asia and other parts of the world. Equally significant but less evident was the application of science and technology to enhance resource production levels and productivity. Thus, the use of mechanical means to extract minerals (including petroleum), the scientific estimation of tree growth rates, and the genetic manipulation of cash crops (e.g. rubber) were means to maximize the physical output of the natural resources themselves.

As European science and technology were introduced to facilitate resource exploitation and environmental change, the state became ever more critical of non-state and 'traditional' approaches to resource use and environmental management. Shifting cultivators in particular were singled out as practising a 'primitive' and 'destructive' land use, but peasants too were often condemned for their 'ignorant' and 'backward' ways (Peluso, 1992; Bryant, 1994). As human-induced environmental change intensified in the twentieth century, the propensity of the state to differentiate between 'scientific' and 'unscientific' resource use increased, thereby exacerbating conflict over the environment.

The political and administrative changes just described with reference to colonial times have conditioned resource exploitation and environmental change in post-colonial times as well. Indeed, what is striking when comparing the colonial and post-colonial eras in terms of resource exploitation and environmental change is the essential continuity of processes and practices between the two eras. Certainly, the magnitude of environmental change and resource exploitation has increased enormously since countries in the region obtained independence. Thus, the proportion of the national territory under forest cover has declined precipitously in most countries: for example, in Burma from 75 per cent to 39 per cent, and in Thailand from 69 per cent to 15 per cent, between the late 1940s and the late 1980s. Yet, the political and economic processes associated with such rapid change do not differ substantially from those elaborated during the colonial era. The scale of the problem has become more apparent in the context of a more populous and economically prosperous South-East Asia, but the underlying logic is similar.

Nevertheless, there have been political and economic developments in the region since independence that have influenced patterns of resource exploitation and environmental change. A key development in this regard is associated with the emergence of close links between political and economic élites in many South-East Asian countries. The emergence of 'crony capitalism' has been an especially important process in patterns of resource exploitation in the region. The allocation of timber leases by politicians to allies and family members in order to advance personal or party political interests has been a regular phenomenon in contemporary South-East Asia. Thus, during the Marcos era, most of the Philippines' commercial forests were given out to friends and allies of the President who felled forests indiscriminately and in clear violation of the rules governing forest exploitation (Remigio, 1993). In Indonesia, the advent of the New Order government of President Suharto in 1967 was associated with a massive expansion in the logging industry subject to few restrictions on its operations. Key timber concessions have been given invariably to those closely linked to the President, including members of Suharto's immediate family (Dauvergne, 1994). Control over resource use has thus been an important source of political patronage designed to award supporters and punish opponents in the broader struggle for political power. Widespread environmental degradation has been a central outcome of this process.

A further change relates to the growing importance of trans-national corporations (TNCs) in South-East Asia's contemporary economic development. Although there were a few TNCs in colonial times (e.g. the Bombay Burmah Trading Corporation Limited with timber operations in Burma, Siam and Java), it is only since the end of the Second World War that these organizations have become prominent. An early focus of TNC activity in the region centred on the natural resource sector. Thus, firms such as Weyerhauser, Georgia Pacific, Mitsui, C. Itoh, Mitsubishi, and Amaco have played an important role in the creation of large-scale timber and mining operations in the Philippines, Malaysia, Indonesia and Papua New Guinea (Hurst, 1990). Such firms are also active in the drive to plant eucalyptus and other fast-growing species as part of a globalizing pulp and paper industry (Lohmann, this volume). More recently, TNCs – many of Japanese provenance (see Cameron, this volume) – have been instrumental in South-East Asia's nascent industrialization. The important role of TNCs in the natural resource and industrial sectors illustrates that it is not only states that are having an important effect on environmental conditions in South-East Asia.

Since the mid-1980s industrialization has been an additional factor in understanding environmental change in the region. Although industrial development has been the goal of all states in the region since independence (including the mainland socialist states), it is only in recent years as Japanese, South Korean and Taiwanese firms have transferred operations to South-East Asia that this goal has been within their grasp. However, such industrialization has been concentrated largely in Malaysia, Indonesia and Thailand (the 'new NICs') and Singapore (the original South-East Asian 'Tiger' economy), and has yet to have an appreciable effect in other countries. The environmental effects of this change are none the less increasingly evident in the form of increased air, land and water pollution (McDowell, 1989). Such pollution exacerbates regional environmental degradation which is already quite severe in many places as a result of unsustainable natural resource extraction practices.

A final change that needs to be noted relates to the uneven spread of democratization in South-East Asia since the 1980s which has enabled non-governmental organizations (NGOs) and grassroots organizations to lobby government and business for changes in resource use practices and policies. Yet as Eccleston and Potter (this volume) make plain with reference to environmental NGOs, the record of 'civil society' activism is not straightforward, but rather depends on local political conditions including the nature and degree of civil liberties, the relative tolerance of the state towards popular protest, the extent of civil society 'cohesion' and organization, and so on.

The role of the state is crucial in this process. Thus, the ability of NGOs and grassroots organizations to protest against resource practices that lead to environmental degradation is contingent upon the state's willingness to allow such protest in the first place. In Burma, for example, the ruling junta has crushed all manifestations of popular discontent in the wake of the 1988 uprising. In contrast, in the Philippines, the demise of the Marcos regime as a result of popular action ('people's power') in the mid-1980s has facilitated a political climate in which popular protest is the norm, and in which there is scope for community-based environmental management initiatives (Braganza, this volume). Other countries fall between these two extremes. Thus, in Indonesia, Thailand and Malaysia there is room for resistance, but in each country civil society activism is constrained through formal and informal means including legal restrictions on action, threats to close an organization for 'inappropriate conduct', and the threat (or use) of military force. Further, the ability to protest does not imply successful change on the ground. Here again, the role of the state in introducing reform is crucial. For these reasons, it is premature to equate the rise of civil society activism with a political 'sea-change' in South-East Asia. As various chapters in this book highlight, NGOs and grassroots organizations can play a significant role in altering patterns of resource exploitation and environmental change. Yet the persistence today of practices such as large-scale logging and dam construction in the face of local, national and international popular protest serves to emphasize that, however important they may be, these actors do not determine the course of events. Rather, the latter are an outcome of a complex political process in which diverse and often competing groups seek to influence how the environment is changed, and in whose interest.

THE POLITICS OF SUSTAINABLE DEVELOPMENT

During the 1990s the politics of environmental change has been given a new twist. Following the Rio Earth Summit in June 1992, the need to promote sustainable development has been almost universally accepted by state and non-state groups in South-East Asia. Yet, as noted earlier, such virtual unanimity belies serious differences over the meaning of that concept – differences which have come to the fore as governments in the region introduce policies they suggest are in keeping with sustainable development. In effect, political conflict over the environment persists but in a new guise as different groups struggle over the meaning of the concept, and even over who is to be responsible for its implementation.

An initial source of conflict has centred on the question of whether a given practice constitutes sustainable development or not. Two types of conflict merit attention. The first type of conflict is concerned with the modification of 'traditional' state-sponsored activities such as logging, mining, transmigration and dam construction that are widely blamed for much of the environmental degradation that has occurred in South-East Asia. The response of the state (and in some cases business) has been to minimize the destructiveness of these activities in the past while at the same time introducing new measures to render such activities less detrimental to long-term environmental well-being in the future. Thus, various states have sought to reform logging practices in keeping with 'forestry action plans' with the assistance of Western aid agencies and consultants (see the chapters by Lang, Rigg and Jerndal in this volume). In Sarawak (Malaysia), for example, the government is developing an integrated and multi-sectoral approach to forest land use management and has also established 'the largest wet tropical forest reserve in the world' along the Indonesian border as one response to criticism of unsustainable logging practices in the state (Bruenig, 1993; Tasker and Ai, 1994). Yet such efforts have not stopped the criticism of state-sponsored activities that are seen to be inherently in conflict with sustainable social and environmental conditions. There is a strong degree of two-handedness in state activities: on the one hand facilitating or even encouraging the continued rapacious exploitation of the resource base, whilst on the other seeking to create a virtuous external image by engaging in various forms of ecological 'window-dressing' or 'façadism'.

The second type of conflict is concerned with 'new' state-sponsored activities such as ecotourism and plantation forestry that are hailed as being the epitome of 'sustainable development'. States and businesses in South-East Asia have used the 'green' image of these activities in order to promote their rapid growth, and to overcome local opposition to them. Indeed, these 'green' activities have been somewhat of a boon to political and economic élites who have co-operated closely in the establishment of the timber plantation and ecotourism industries. Yet these activities have often had adverse local social and environmental ramifications, and have generated considerable opposition from grassroots groups. 'Eucalyptus politics' has been especially pronounced in largely deforested Thailand as the government – with the assistance of Thai and foreign firms (notably Shell) – undertakes a massive reforestation campaign. In doing so, however, it has prompted the creation of a vocal anti-plantation movement that condemns plantations as being ecologically destructive and detrimental to local villager interests (Hirsch and Lohmann, 1989; Puntasen *et al.*, 1992; Lohmann, this volume). Further, the official claim that such plantations relieve the pressure on existing tropical forests is contested by these groups who point to growing evidence that companies are felling old-growth forests to make way for eucalyptus plantations (Sargent and Bass, 1992). A similar process of conflict is developing around the booming ecotourist industry, raising comparable issues about the ecological viability of an ostensibly 'sustainable' activity (Cochrane, this volume).

The development of conflict over plantation forestry and ecotourism is symptomatic of a wider material and discursive struggle as sustainable development policies affect power relations in society generally. Just as human-induced environmental degradation reflects, and in turn, often reinforces power relations, so too the quest for sustainable development has a socially differentiated impact that is ripe with political meaning. Thus, the spread of eucalyptus plantations, golf courses and ecotourism facilities has prompted opposition through much of South-East Asia not so much because of the dubiousness of the green credentials of these activities, but because their spread has often resulted in widespread land dispossession and financial hardship among local poor people. In contrast, these 'growth' industries are among the top money-spinners for business in the region (see the chapters by Lohmann, Cameron in this volume). Thus, as with 'traditional' logging and mining activities, the new 'green' industries tend to reinforce political and economic inequality wherever they are introduced.

However, an important difference between the traditional and new industries is that political and economic élites have been able to use the latter (in a way not possible with the former) to support their general claim that they are promoting activities consonant with sustainable development. That general claim is used, in turn, by states to proclaim their 'responsible' stewardship of the environment, solicit donor assistance, and neutralize popular criticism. Thus, the Thai, Malaysian, Indonesian and Filipino states use green discourse associated with the promotion of the new 'green' industries to enhance their position in society, while the former socialist countries of mainland South-East Asia are now following a similar route. Even in the case of the region's 'pariah' state, Burma (Myanmar), a national campaign to 'green the central dry zone' is being used to promote the ruling junta's commitment to sustainable development (Government of Myanmar: Forest Department, 1994). Just as states use the quest for sustainable development to promote their own interests, so too businesses use their participation in the new 'green' industries as proof that they are good corporate citizens fulfilling the leading role in the fight against environmental degradation that was envisioned for them at the Rio Earth Summit in 1992 (Middleton *et al.*, 1993). The new 'green' industries thus generate important discursive benefits for both states and businesses in South-East Asia.

It is not surprising, therefore, that the issue of who is to control the formulation and implementation of sustainable development policies is of growing popular concern. At one level, it would appear self-evident that it is the role of the state to undertake this task. Yet, as noted above, the state acting often in conjunction with private business has played a prominent part in generating the environmental problems that now confront South-East Asia. How can the state be trusted in such circumstances to implement policies that would resolve those problems?

Beyond the question of the environmental credibility of the state, there is also the question as to whether the state even has the capacity to implement and enforce sustainable development policies that would, inevitably, entail significant changes to the political and economic *status quo*. While the region has its share of what might be termed 'strong states' (i.e. Burma, Indonesia), it also includes a number of 'weak states' (i.e. Laos and the Philippines) which may not be in a position to co-ordinate centrally the reform process – even if the political will to act were there in the first place (Migdal, 1988: see also the chapter by Usher in this volume). Further, the prevailing pro-business intellectual climate among the multilateral institutions and Western governments is such that 'less government' is frequently seen as the preferred development option – a perspective given practical significance through the mechanism of 'structural adjustment' programmes.

At the same time, the growth of civil society activism has prompted the growth of what has been termed 'civic politics' – politics between different groups in society that falls outside the realm of state-centred politics (Wapner, 1995). Thus, for example, NGOs and grassroots groups lobby business directly to alter environmentally damaging practices through the mechanism of media publicity and boycott campaigns. Yet such civic politics can, and in places such as the Philippines and Thailand today increasingly do, take the form of community environmental management initiatives designed to pursue 'sustainable development' independent of state control and direction (Leungaramsri and Rajesh, 1992; Braganza, this volume). In this manner, the politics of sustainable development is becoming intertwined in a much broader process of political, economic and social change in South-East Asia.

OVERVIEW

The objective of this book, then, is to contribute to the understanding of human-induced environmental change and the quest for sustainable development in South-East Asia. In this volume, people (indeed, actors) are placed at centre stage, politics provides the back-drop, and the script follows the process of resource (especially forest) exploitation. The production in the main takes the form of a rich and diverse narrative drawn from a broad cross-section of loci, foci, scales of analysis, and ideological perspectives. The latter are contributed by scholars, journalists, consultants, and representatives of NGOs and advocacy groups, who offer views both from within and outside the region. We have attempted to present an holistic view of the context within which the process of sustainable development can, or might, occur. We avoid the superficial notion that sustainability simply concerns the nexus of production and environment, but instead seek to emphasize the complex interplay of politics, history, society, culture, regionalization and globalization, in creating a formidable barrier to the promotion and adoption of sustainable economic practices.

The analyses are informed by in-depth and/or field-based research, and are on occasion quite hard-hitting when drawing attention to the human consequences of poorly conceived development projects, the misuse of political power, and the prioritization of resource exploitation and economic growth over conservation and redistribution. Criticism is fairly evenly apportioned between domestic and international actors, with politicians and planners in the former, and Japan and the Nordic countries in the latter, the subject of particularly close scrutiny.

The volume is structured in four main parts. The first examines the political context of ecological change in South-East Asia. Larry Lohmann leads in with an incisive assessment of the political ecology of resource exploitation, using the fascinating illustration of the burgeoning pulp and paper industry to draw out some of the sinister processes whereby land is being transformed on a dramatic scale from natural or degraded forest into a monoculture of commercial plantation forest. The discussion emphasizes the extreme vulnerability of local people and local environments in face of the immense power of business, military and political élites in Thailand and Indonesia who are driven by an urgent desire to capitalize on this industry's short-term potential, either for themselves or as conduits for international capital. Protest is 'dealt with', not responded to. The two-handedness of government to which we referred earlier is clearly manifest in this context: heavy subsidies are provided to promote an industry which either directly or indirectly (by displacing people to the forested margins) places pressure on the region's few remaining stands of moist forest, whilst governments seek to improve their 'green' credentials in the public eye by claiming to be 'reforesting' the countryside. Meanwhile, a complex social, cultural and political struggle is mitigating prospects for sustainable development.

Bernie Eccleston and David Potter also focus on the political context, in this case the role it plays in defining the conditions within which the growing corpus of non-governmental organizations must operate in South-East Asia. Case studies of Indonesia, Malaysia and Vietnam show how powerful government and business organizations, buttressed by global economic and political structures, are able to influence the pace and extent of democratization, and with it the role, freedom and effectiveness of NGOs. Symptomatic of their juxtapositioning in this respect, environmental NGOs typically point to the lack of democracy as constituting a significant impediment to the advancement of their fellow citizens' human rights, whilst politicians have often used the power of the media to project a countervailing image of NGOs as 'anti-democratic subversives and saboteurs'. Paradoxically, when governments were rooting around for respectability ahead of the 1992 Rio Earth Summit, NGOs suddenly became flavour of the month – a situation which has subsequently led, in some cases, to an increased domestic tolerance of their activities, a closer dialogue, and the narrowing of formerly diametrically opposed positions.

The chapter by Owen Cameron shows how South-East Asia's increasing integration into regional and global systems has added to the pressures which are being placed on the region's natural resources and environment and which, in turn, serve to compromise the potential for sustainable development. Japan, as an important global and, especially, regional power, has played a significant role in the environmental transformation of South-East Asia, both through its voracious market for the region's forest, marine and other resources, and by helping to intensify the process of industrialization therein. Cameron asks whether South-East Asia can learn anything from Japan's domestic experience of environmental conservation, but in truth the latter has largely been made possible by exporting degradation and the externalities of growth to her South-East Asian neighbours. Unless the dynamo economies of South-East Asia can do likewise to their late-starting neighbours, it is difficult to see how the experience can be repeated. Meanwhile, commentators see a tension between Japan's professed aim of becoming a world leader in the field of international environmental policy, and its continuing and over-riding commitment to the promotion of its own economic development.

The second section of the volume illustrates some of the processes and forms of human-induced environmental change which are occurring in South-East Asia today. Colin Sage uses the example of Indonesia's transmigration programme to highlight the considerable difficulties that migrants and indigenous peoples alike have experienced in sustaining their livelihoods after resettlement. The use and abuse of land resources are described in some instances as 'mining' – treating potentially renewable resources in much the same way as non-renewable resources such as coal and petroleum – driven by the urgency of short-term needs, a lack of appropriate knowledge on the part of transmigrants, and inadequate administrative support and funding. Sage's detailed research has highlighted the complexity and diversity of livelihood strategies and local settings, and he uses this to argue the need for a people-centred approach to sustainable development which is sensitive to the heterogeneity of local needs and potential.

The remaining chapters in this section deal with the little-known, landlocked country of Laos. Jonathan Rigg and Randi Jerndal ask whether the Lao PDR's relatively recent arrival on the fringes of the global capitalist system, following the partial dismantling of its command economy, might allow the opportunity to learn some lessons from the mistakes made by her neighbours in the field of resource and environmental management. The early signs are sadly unpromising: the country's resource base, especially the extensive forests, are already being targeted as a source of much-needed foreign exchange. International agencies and foreign businesses have become powerful agents of forest exploitation, whilst the Lao government struggles to keep up, making policy decisions 'on the run' and prompting questions about who is effectively in control – a crucial issue in the context of sustainable development. A similar situation prevails in relation to the exploitation of the country's massive hydro-power potential. Anne Usher shows very clearly how northern interests and agencies, especially from the Nordic countries, have provided both the means and the incentive to harness this 'white gold'. Through the 'pervasive appraisal optimism' which exudes from the proponents and evaluators of dam schemes (and especially aid donors and financiers), the negative environmental and social consequences are significantly downplayed, leaving a widely held perception of a 'win-win' situation: hard currency and an environmentally benign, or even 'green', source of energy. Local people who face the brunt of these activities have very little opportunity to express their preferences or feelings, and there are few NGOs to work as advocates. Meanwhile, the close link between aid provision and the pressures for resource mobilization greatly restricts the Lao government's room for manoeuvre.

The third section integrates discussion of the various methods which strengthen our understanding of human-induced environmental change in South-East Asia with further illustrations of its process and context. Victor King highlights the role that the academic community can play in researching the human consequences of environmental degradation, drawing illustrations from collaborative research programmes which have explored human–environment interaction in South-East Asia from a variety of perspectives. He then presents findings from a collaborative research project on Borneo in which he was recently involved, under the aegis of the UK Global Environmental Change Programme which has sought to wrest global environmental research from the domain of the natural sciences. King's chapter deals with the very topical issue of drought and forest fires, and attempts to identify the respective roles played by commercial logging, shifting cultivation and global climatic change in the apparently growing incidence of these devastating phenomena. His findings point to the value of locally informed research and the dangers both of generalized views and those informed by stereotypes and scapegoating, not least of native shifting cultivators. The study also shows how environmental change arising from human actions is having a major effect on human livelihoods.

Duncan McGregor *et al.* provide, in admirable detail, evidence of the value (and also some of the shortcomings) of various methodological tools –

16

Geographical Information Systems, aerial photography and remote sensing – in providing the informational basis for identifying and evaluating environmental change, and for the construction of predictive models with which to inform future policy. They argue that the growing awareness of the nature and consequences of environmental change in South-East Asia has not been matched by appropriate efforts and mechanisms to quantify and map the progress and processes of change. Their worked examples of land conversion in Sabah, Thailand and Brunei emphasize the value of being able to make comparisons across time and space, synthesizing large banks of environmental data and presenting them in an accessible format. Armed with such information and methodological tools, practitioners are better able to identify the nature and extent of resource depletion and environmental degradation, and to lay the foundation for more sustainable ecological practices.

In contrast, Chris Lang emphasizes the weakness of the Tropical Forestry Action Plan (TFAP) as a largely theoretical tool for regulating forest exploitation, especially where the process of its compilation and implementation is inherently flawed. Using the example of the TFAP for Vietnam, Lang shows how a reliance on foreign consultants and a very limited degree of consultation with local people and community organizations yielded a Plan which contained numerous internal contradictions and which, when implemented, would do little to preserve the country's rapidly diminishing forests. Popular misconceptions of the role played by shifting cultivators in rainforest destruction run unquestioned throughout the Plan, whilst some of the main causes of deforestation – commercial logging (legal and illegal), large-scale development projects, land encroachment – are addressed only tangentially. The complexity of both the process and impact of deforestation is also largely overlooked, rendering almost impossible the introduction of small-scale and locally based initiatives which are seen by many to hold the key to sustainable development.

Finally, the volume explores some of the options for change which are necessary if sustainable development is to be turned from rhetoric into reality. As one of the region's booming industries, tourism provides an excellent basis for exploring the potential for sustainable development. Ecotourism is often cited as an ideal means of promoting 'sustainable tourism' in South-East Asia, but when Janet Cochrane probes beneath the surface of this phenomenon she finds little more than mass tourism re-packaged in a 'greener' guise. She finds that the opening up of national parks in Indonesia to the tourists' gaze has had a significant, and seldom beneficial, impact on ecosystems, wildlife and local people. We should not be greatly surprised by this: 'sustainable development' is an oxymoron – environment and exploitation cannot be rationalised, and thus any form of tourism will have a range of impacts, however 'green' it may be made to appear. 'Ecotourism' has come to describe any holiday that is centred around a natural attraction; the label has become both fashionable and marketable.

Just as Cochrane uses her intimate knowledge of ecotourism development and the Indonesian setting to excellent effect, so Rili Djohani's involvement with WWF Indonesia is used to inform her assessment of the potential stewardship role of the formerly nomadic Bajau people in managing the country's recently designated marine parks. As such, the case study informs the general concept of sustainable development by exploring the principle and potential of popular participation. The Bajau have an intimate familiarity with the marine environment, and an inherent understanding of the limits to, and consequences of, its exploitation. Their stewardship of marine parks, were there the political will and foresight to incorporate a marginal minority people in this way, has the added advantage – crucial to the 'development' component of the sustainability issue – of creating a role and economic function for the Bajau in a situation where their traditional way of life is rapidly being transformed, and their access to marine territory and resources increasingly being constrained by competition from external sources. Whilst there may of course be a sense in which their stewardship of the marine environment would both camouflage and ultimately hasten their own demise, it also represents a pragmatic means of achieving two important objectives in the face of what might be seen as the inevitable and irrepressible forces of change.

Another pragmatic response, this time to the rapacious exploitation of the rainforests of Sarawak, is suggested by Mike Parnwell and David Taylor. Commercial logging has seriously depleted the stock not only of trees but also a wide variety of non-timber forest products which are of considerable importance to the lives and livelihoods of forest-dwelling societies in this Bornean state. Parnwell and Taylor describe the 'mining' of forest resources, and also the human impact. They also focus on how the societies concerned have responded to increasing resource scarcity, and draw lessons from this to suggest future development paths. One pragmatic means of dealing with the exigencies of change is to foster a 'capture-culture' transition which some Iban communities have spontaneously adopted. This involves the domestication of certain rainforest products and species, the pursuit of which in the wild is taking up more and more time and energy as they become increasingly scarce in their natural habitat. Several such products, in addition to improving standards of nutrition and health, could provide the basis for the development of non-farm activities such as handicrafts production for the burgeoning tourist market. As with the previous example, however, such a policy response may serve simply to camouflage the more fundamental processes which are being exerted on the forest ecosystem and forest-dwelling communities. But on the other hand, it represents a pragmatic means of coping with a seemingly inexorable and irreversible trend of rainforest destruction.

The chapter by Alan Dykes explores the role of environmental education and research as a means of promoting conservation principles and practice. Using the atypical example of the oil-rich Sultanate of Brunei Darussalam, Dykes shows how the absence of strong economic imperatives and pressures

18

to exploit its rich rainforest resources has allowed the Brunei government to manage these resources in a sustainable manner. The utilization of the Temburong Forest Reserve for educational purposes – a field centre was established there in 1990 – helped to counter pressures to flood part of the region for a hydro-power project. Equally importantly, the field centre has helped to raise awareness within Brunei's largely urban population, and especially among the younger generation, of the importance of environmental conservation. Awareness, concern and commitment may be seen as essential prerequisites to sustainable development in the longer term.

Finally, Gilbert Braganza considers the role of community-based management in underpinning sustainable development in the forest regions of the Philippines. This insightful account stresses the social, political and economic parameters of environmental exploitation and conservation. It emphasizes the importance of local-level approaches, as both an ideology and a strategy of sustainable development. The Philippines in the more open, tolerant and democratic post-Marcos era offers conditions which are conducive to such an approach. Indeed, one of the initiatives created by sustainable development in the Philippines has been the participation and involvement of all sectors of civil society in the development process. However, a number of contradictions and tensions has emerged as community-based initiatives have been put into practice. Local political structures have sometimes blocked the effective implementation of this planning principle (see also Rigg, 1991). Meanwhile, the government has occasionally lacked faith in its own rhetoric, and is seemingly reluctant to abandon completely the established top-down orthodoxy. Inherent weaknesses have also emerged in the government's own understanding of the sustainable development concept, and in the division of (competing) departmental and ministerial responsibilities. Nonetheless, the principle of place-based and bottom-up initiatives is now firmly established in the Philippines. In this way the poor, disadvantaged and marginalized are able to take the initiative in defining sustainable development.

We round off the discussion by considering the future prospects for sustainable development in South-East Asia. In order to present a balanced (or noncommittal) assessment, we consider three scenarios: 'worst-case', 'ideal' and 'middle path'. Only the last of these stands any realistic chance of achieving the underlying objectives of sustainable development. The consequences of *laissez-faire*, which underlies the worst-case scenario, are too grim to contemplate and, we believe, will lead (and, indeed, are leading) to the introduction of reactive, remedial and preventative measures. The best-case scenario, something akin to 'Ecotopia' (Pepper, 1984, p. 206), is seen as too idealistic to be realistic, given the deeply rooted socio-political barriers which exist in South-East Asia to a fundamental challenge to the *status quo*. Thus the compromise scenario, where movement towards the requirements of sustainable development occurs with a gradual but growing momentum, seems the most likely to prevail. With it rests a grain of optimism for a sustainable future.

We hope that this volume will make a valuable contribution to the sustainable development debate, not least by highlighting the importance of context and locally informed insight in facilitating our understanding of the complex and varied processes which are involved. These, we believe, are an important prerequisite for the effective operationalization of the sustainable development concept.

Part I
CONTEXT

2

FREEDOM TO PLANT

Indonesia and Thailand in a globalizing pulp and paper industry

Larry Lohmann

Success is measured by the freedom to plant fibre crops. . . . Our objective should be to create and move inside an ever-increasing friendly circle of public opinion.

(Fernandez Carro and Wilson, 1992)

Over the last decade some of the most important changes and conflicts involving the use of land and water in rural South-East Asia have stemmed from the regional and global expansion of the pulp and paper industry. Natural forests have been chipped, vast monocultures of eucalyptus and acacia established, and giant pulp mills built along major waterways, provoking rural strife and political debate throughout the region.

This chapter will sketch some of the pressures behind, and some of the dangers of, the expansion of the pulp and paper industry in South-East Asia. It will then describe some of the mechanisms by which the industry has enclosed land and water in two of the countries most affected, Indonesia and Thailand, and outline the various forms of opposition the industry is meeting. Finally, it will indicate some of the strategies by which the industry is attempting to manage this resistance.

AN EXPANDING INDUSTRY

The pulp and paper industry in both Indonesia and Thailand has been strongly influenced by patterns of regionalization and globalization of fibre production, consumption and trade. In the 1950s, most international wood fibre trade was cross-border between Canada and the United States (US) and among the European countries, and only a few countries which produced and consumed large quantities of paper were dependent on raw materials from distant continents. Since then, world trade in pulp has increased around five times, while the global wood fibre market has tripled (Dudley, 1992; Hagler, 1993). Today, high-consuming countries (see Table 2.1) pull in raw

Table 2.1 Paper Consumption, 1993

Country	Kilogrammes per capita
USA	313
Japan	225
Hong Kong	220
Singapore	218
Finland	215
Taiwan	205
Germany	190
UK	170
Australia	152
Italy	132
South Korea	128
Malaysia	62
Thailand	30
Russia	30
China	17
Philippines	12
Indonesia	10
Burma	5
India	3
North Korea	3
Papua New Guinea	2
Vietnam	1
Lao PDR	0.2

Source: Pulp and Paper International, July 1994

materials for paper from all over the world. The wood fibres in a sheet of paper in Western Europe or East Asia may well come from trees grown as far away as Brazil, Canada, Chile, Congo, Indonesia, New Zealand and South Africa. Prices for marketed wood pulp, moreover, are everywhere strongly influenced by Southern exporters.

One country which has pioneered the unharnessing of national consumption from national supply is Japan, which began importing large amounts of wood chips in the 1960s in specially built ships. Between 1965 and 1975, the proportion of Japan's pulpwood supplies which were imported jumped from 3 to 40 per cent. By 1990 Japan led the world in wood fibre imports, with 37.5 per cent of world trade. Historically, most of the country's fibre imports have been western North American softwood chips. Three recent shocks, however, have accelerated the Japanese industry's long-standing efforts to diversify its overseas supplies. First, in 1980, interest rates increased in the US, reducing new housing construction, saw mill production, and mill residue surpluses available for export, and doubling export chip prices overnight. Unable to exploit its own forest resources economically, the Japanese industry turned to other countries such as Chile, taking advantage of falling oil prices in the mid-1980s for cheap transport. Then, during 1987–88, a projected mill in Tasmania threatened to absorb Australian hardwood (eucalyptus) supplies which had been going to Japan, spurring the industry to turn to South-East Asia and the southern United States (in growing competition with other countries such as South Korea). A third shock in the early 1990s stemmed from decreased logging in western North American forests and an accompanying drastic reduction in waste chip supply, both due to the forest industry's depletion of old-growth forests without sufficient replanting and an accompanying rise of environmentalism affecting use of public lands. From 1988 onwards, the Japanese industry was working to assure itself of raw material flows not only from long-standing suppliers such as the Soviet Union, Indonesia, South Africa and New Zealand, but also from China, Vietnam, Argentina, Venezuela, the interior of northern Canada, Fiji, Papua New Guinea and other parts of Oceania (Schreuder and Anderson, 1988; Marchak, 1992; *Paper and Packaging Analyst*, 11.1993). As one industry figure commented, Japan had taken a 'new step to secure resources, that is, planting its own forests in other countries as long-term resource programs' as well as laying plans for the 'execution of pulp or paper production there' (*Japan Pulp and Paper*, n.d.). Competition among such a large number of countries for the Japanese market, of course, was seen to be reducing prices as well as providing guarantees of steady supplies.

As economically and politically available natural forests have been depleted and new hardwood pulp technologies developed, meanwhile, plantation fibre has grown in importance in relation to fibre from natural forests. This, too, has sharpened paper industrialists' interest in South-East Asia. Plantation species such as eucalyptus grow much more quickly in the tropics than in temperate zones, meaning both that their fibre is available earlier and that less land is required (*Know-How Wire*, 1.1989; Shell and WWF, 1993). Land is also cheaper in the South, particularly in big contiguous parcels – a decided advantage for huge chemical pulp mills which are most economically sited in the centre of large raw-material catchment areas. Small wonder, then, that the cost of wood suitable for manufacturing bleached hardwood kraft pulp in Indonesia may be half what it is in, say, Finland (*Know-How Wire*, 1.1993; *Pulp and Paper International*, 8.1993). Inexpensive labour is a secondary attraction.

One of the most important incentives for global investment in South-East Asian mills and plantations, however, is the widespread expectation that Pacific Rim consumption growth will be the fastest in the world over the next decade. Stimulated by tighter integration of the whole region into the world economy and lavish subsidies from public to private sectors, this growth will be driven largely by increases in industrial paper demand. Industry consultant Roger Wright calculates that by 1997, total demand in Asia and Africa will have outstripped that in North America, with Japan, the Asian 'Tigers', and China playing big roles (Wright, 1993).

Such growth projections are a powerful reason for integrating South-East Asian plantations with pulp mills, and pulp mills with paper mills, as soon as possible. Pulp has more value-added and is more efficient to ship than wood chips (which are half water) making it attractive to South-East Asian exporters, and if there is a huge local demand, paper produced in the region will have an advantage. Adding to pressures to build integrated projects are the blandishments of Northern machinery suppliers and mill engineering consultants seeking new export outlets, together with those of the 'aid' agencies with which they have a symbiotic relationship.

COMMON DANGERS

The course of pulp and paper development varies from country to country in South-East Asia. Several dangers created by the advance of the industry, however, are common to the region as a whole. One danger is associated with the industry's inordinate dependence on large, expensive machines. Except in China and a few other places, the industry has been locked into this dependence at least since the 1930s, when the prestige-obsessed North Atlantic newspaper companies of the day were vying over who could build the biggest mills. This competition resulted in each new machine being designed as a 'one-off'. Unit costs rose correspondingly: from 1930 to 1975 the cost per annual tonne of a newspaper machine increased at least forty-fold, while the price of newsprint increased less than twenty-fold. Yet once machine manufacturers had got into the rut of building big machines, it became difficult to fill orders for anything else, despite their high price, inferior cost-effectiveness and need for extraordinarily large supplies of fresh water and other resources. As paper expert A. W. Western wryly notes, building new paper machines

> became a luxury which could be afforded only by multinational giants or the governments of developing countries, advised by consultants that only scale to this degree could be economic! For the consultants it *was* economic; they were now essential for large mill design and co-ordination.
>
> (Western, 1979)

Combined with freely available technology, easy availability of debt finance, and little need for newcomers to buy into brand names, reliance on big machines has fed a chronic industry tendency to overbuild during boom times, resulting in slumping prices, the flattening and extending of supply/cost curves, inadequate returns on machinery investments, attempts to cut costs and stimulate demand, shakedown, closures and yet further concentration. In 1993, after the most recent bout of over-investment, pulp prices were half what they had been only four years previously in constant dollars, and 39 per cent of what they were in 1975, forcing producers such as Thailand's

Plate 2.1 Pulp and paper mill, Borneo: the mill has been established next to a commercial forestry scheme in Sipitang, Sabah

Phoenix Pulp and Paper to stop exporting and helping to drive Indonesian production down to 65 per cent of capacity (Wright, 1993; *Pulp and Paper International*, 2.1994). The increasingly international nature of the pulp and paper industry exaggerates uncertainty still further by giving exchange rate fluctuations the power to 'swamp all other factors in determining profitability' (Fletcher, 1988). Other factors difficult to control or predict, especially for Southern producers, include recessions and changes in European recycling laws and waste paper supply, the need to invest in machines capable of meeting Northern consumer demand for chlorine-free paper, unanticipated falls in yield from eucalyptus or acacia plantations after the second rotation, and unexpected increases in pulp capacity due to the whims of single manufacturers in Brazil, Russia, or anywhere else (Asian Development Bank, 1993).

It is ironic that such a cyclic, volatile industry – one which, as leading paper industrialist Hugh Fletcher (1988) of Fletcher Challenge, has remarked 'does not have a history of being logical' – is becoming so dependent on plantations, whose development requires an unusually long planning horizon of 10–15 years (Shell and WWF, 1993). Nowhere is this more true than South-East Asia, whose business élites and governments are staking a great deal on paper and pulp despite exorbitant machinery price tags – up to US$1 billion for a world-class, 2000-tonne-per-day export pulp mill (see

Plate 2.1) – and warnings in the early 1990s from experts that the Asian market was 'undeveloped', uncertain, and, in many grades and locations, over-supplied. One Western machinery supplier has openly proclaimed that many prospective Indonesian investors 'don't know what they plan to do with their pulp', while in Thailand, government targets for plantation acreage regularly overshoot market demand projections by 1,000 per cent (Sargent, 1990; Allen, 1992; Wright, 1993). South-East Asian producers can, of course, under-cut competitors in lean times, but this requires squeezing labour and land harder, and removal of the price umbrella provided by firms which charge more may result in a further collapse of prices.

A second set of dangers created by the pulp and paper boom in South-East Asia are ecological ones. In Indonesia, old-growth forests in West Papua and Kalimantan have been chipped for direct export, and plantations are also a direct cause of legal and illegal deforestation, devastating biodiversity, agricultural watersheds, soils and fisheries. In Thailand, although the industry has mainly coveted land which has already been deforested and thus cuts fewer trees as a prelude to plantations, displaced villagers have often been forced to dwindling frontiers, also causing deforestation. In many places in which eucalyptus has been commercially planted in South-East Asia, more-over, it has been alleged to have lowered water tables, monopolized nutrients needed by other crops and interfered with their root systems. It has also been accused of exacerbating erosion: not only is its leaf litter believed to release chemicals inhibiting the growth of other plants and their associated birds, insects and reptiles, but it is also harvested every five or six years, leaving the ground temporarily bare (Lohmann, 1991). The need to match raw materials to paper mill machinery is meanwhile creating pressures to make monoculture tree crops ever more uniform and high-yielding through restriction to particular strains, hybridization, clonal propagation and genetic engineering (Fernandez Carro and Wilson, 1992). Such pressures for uniformity carry with them a risk of epidemics and the need for heavy chemical applications. Pulp mills such as Thailand's Phoenix and Indonesia's Indorayon have polluted waterways with oxygen-hungry effluents and toxic chemicals, resulting in both contamination of drinking and bathing water and in large-scale fish kills which in some areas have devastated one of the rural villagers' most important protein sources and marketable products.

Many of these ecological effects are linked to the land and water rights conflicts which have made eucalyptus plantations and pulp mills a major source of rural strife in South-East Asia. For a closer look at these, it will be useful to examine the two contrasting examples of Indonesia and Thailand.

INDONESIA

Early in 1994, the countryside around what was formerly the little village of Kerinci, 10 kilometres south of Pakanbaru, Sumatra, was witness to an extraordinary scene. Working to a plan formulated by Helsinki-based consultants Jaakko Poyry, some 4,000 Indonesians laboured day and night, often in mud and rain, to finish the biggest single-line pulp mill in the world by October 1994. Under the eye of a Finnish project manager, an immense soda boiler built by Finland's Tampella firm was assembled on the 3 km² site, supplemented by complicated machinery from a score of other companies including Finland's Kone, Valmet, Ahlstrom, Sunds Defibrator, Sunds Rauma and Outukumpu; Sweden's Noss and ABB Flakt; Japan's Mitsubishi Heavy Industries and Nippon Sanso; Canada's Chemetics and Bailey; the US's Cranston and Solarturbines; Germany's Siemens; Britain's ICI; Brazil's Voith; Taiwan's Teco; and India's Ion Exchange (*Helsingen Sanomat*, 23.1.1994; *Pulp and Paper International*, 9.1994).

Built to be capable of converting 4,000,000 cubic metres of wood into 700,000 tonnes of pulp each year by the time it comes up to full capacity in 1997, the mill, known as Riau Andalan, will cost its owner, the Raja Garuda Mas company, US$750 million, equivalent to 1.25 million times Indonesia's per capita income. Two-thirds of its output is targeted for export. To satisfy the mill's appetite for wood, the equivalent of one log truck must pass through its gates every three minutes from 1,600 square kilometres of logged-over timber estates southwest of the mill and sources elsewhere. The mill's needs of approximately 4,750 cubic metres of water per hour will be supplied by the nearby Kampar River. According to company executives, the mill, which will require the establishment of a port and 45 kilometres of railway for the transportation of wood and wood products, will employ a mere 1,000 people – US$750,000 of capital investment per mill job (*Financial Times*, 14.12.1992; Wright, 1994; *Down to Earth*, 4.1994; *Pulp and Paper International*, 7.1994; 9.1994).

Riau Andalan, as one of a number of enormous pulp mills which have been springing up across Indonesia's 'hinterland' island of Sumatra, is a fitting symbol of a boom which has seen the country's pulp production rise from 167,000 tonnes in 1983 to 1.3 million tonnes in 1993, and paper production increase from 377,000 to 2.6 million tonnes (*Pulp and Paper International*, 7.1993, 8.1995; Wright, 1994). For one thing, the mill exemplifies a trend towards extreme concentration. As average Indonesian pulp plant capacity has leapt from around 5,000 tonnes per year in 1970 to 85,000 in 1991 and 145,000 in 1993, the industry has gravitated ever more clearly into the hands of Indonesia's biggest and most notorious business families, many of whom have built their fortunes through commercial logging or plywood or are business partners with members of President Suharto's family. Riau Andalan's parent conglomerate, for instance, is headed by Harvard-educated Sukanto Tanoto, the 'timber king' of Northern Sumatra. Pangestu Prayogo's Barito Pacific Timber Group, currently co-operating with Suharto's daughter Siti Hardiyanti Rukmana (Tutut) and Nippon Paper/Marubeni in the planned 450,000 tonne per year Tanjung Enim Lestari

29

mill in Musi, South Sumatra, is Indonesia's largest wood-based industries group and one of the world's largest plywood manufacturers, holding over 45,000 square kilometres of Indonesia's forest lands and a 10,000 square kilometre concession in Vietnam. Bob Hasan, Indonesia's top timber businessman, chair of the loggers' and wood-processing trade association, and long-time Suharto crony, has also moved into pulp and paper with the collaboration of Suharto's son Sigit Harjojudanto and US and Korean firms (*Paper Asia*, 8.1991; 4.2.1992; Soetikno, 1993; *Pulp and Paper International*, 2.1994).

Second, although Indonesia's pulp and paper industry has high hopes of building a huge domestic market among the country's 190 million people, new capacity for both pulp and paper of the type which Riau Andalan represents is focused largely on exports. Paper exports rose from 200,000 tonnes in 1990 to roughly 600,000 tonnes in 1993, and pulp exports, in 1993 around 100,000 tonnes, are expected to multiply nine-fold between then and 1996, after which Indonesia will be shipping 60 per cent of its total production abroad (Wright, 1994; *Pulp and Paper International*, 2.1994, 7.1994). Part of this market is being created by Indonesia's own Sinar Mas conglomerate, as it invests in paper plants in Bombay and in Ningpo and Cheng Chiang, China, which can take Indonesian pulp as raw material (*Pulp and Paper International*, 8.1993; 7.1994; 8.1994; 10.1994).

A third industry characteristic exemplified by Riau Andalan is heavy Northern involvement. Most of the hundreds of millions of dollars spent to build and plan the wood supplies for such new mills winds up in the hands of firms from the industrialized countries, with the lion's share going again and again to Scandinavian, Japanese, and North American suppliers and consultants and their consortia (*Down to Earth*, 1991; Allen, 1992; *Paper Asia*, 4.2.1992; *Pulp and Paper International*, 8.1993; 2.1994; 10.1994). Lubricating this process are bilateral 'aid' agencies (unsurprisingly, mainly from Scandinavia, Japan and North America), export-credit organizations, multilateral development banks and, in the case of Japan, the *sogo shosha* with their attractive finance offers (Westoby, 1987; Overseas Development Administration, 1992; United States Department of the Treasury, 1993). Acting as brokers and co-ordinators are forest industry consultancy companies such as Finland's Jaakko Poyry, who are skilled at using 'aid' money and corporate contracts to bring together Northern machinery suppliers and Indonesian élites with power over land, forests, labour and finance in a way which benefits both. Acting through a network of old colleagues, friends and like-minded technocrats in overworked donor agencies who are happy to be relieved of onerous planning and monitoring duties, such consultants are able to use public funds to build up a private forestry industry sector in Indonesia which will ensure them a continuing stream of future contracts at a time of rising unemployment in Europe for professional forestry industry personnel. For example, a 1984 contract with the World Bank and the Indonesian government to do a sector analysis of the country's paper and pulp industry helped Jaakko Poyry land over thirty subsequent contracts to plan or implement public and private sector projects to supply mills with pulpwood from natural forests or plantations. The company has also picked up dozens of contracts – some of them subsidized by Finnish taxpayers through Finnish Export Credit and the bilateral 'aid' agency FINNIDA – to plan or engineer pulp or paper mills for Indonesian corporate clients or do market surveys for Western machinery manufacturers (Jaakko Poyry, n.d.). Benefiting from Canadian government lobbying and handouts, Canadian forestry consulting firms such as H. A. Simons and Sandwell are competing to perform the same role. However important indebtedness, poor terms of trade, declining oil revenues, corruption, and the status ambitions of élites may be in driving Indonesia to rush to cash in its forests through plywood, lumber and pulp manufacture, it is through the concrete activities of such firms and their governmental helpers that Indonesia's forest lands are in practice integrated into the world economy.

Despite the immense wealth of Indonesia's pulp and paper tycoons, foreign investment and finance are often also necessary for the industry's immensely capital-intensive projects to get off the ground. The Taiwanese firms Yuen Foong Yu and Chung Hwa, for example, own a quarter of the shares in Indah Kiat, Sinar Mas's biggest pulp and paper subsidiary, and Barito Pacific and Sinar Mas have recently successfully floated international stock and bond issues (*Down to Earth*, 1991; *Pulp and Paper International*, 8.1993, 10.1994). Backed by the state Finnish Fund for Industrial Co-operation, meanwhile, the partly state-owned Finnish paper giant Enso Gutzeit is teaming up with state-owned forestry company Inhutani and Gudang Garam in a ten-year plan to develop an approximately 1,390-square-kilometre acacia pulpwood plantation in Sangau, western Kalimantan on a site riven by conflicting land claims (*Jakarta Post*, 22.7.94; *Pulp and Paper International*, 8.1994; Junus, 1994).

A fourth characteristic of the new Indonesian pulp mills is that they are almost all fed in their initial stages by natural (though often logged-over) forests. While plantations are often advertised as improving 'degraded' or 'unused' lands, the most obvious candidates for this treatment – anthropogenic, homogeneous *alang-alang* grasslands colonized by *Imperata cylindrica* – are prohibitively difficult and expensive to convert, due to low soil fertility, the grass's toughness, and their tendency to be scattered in patches over a large area (WALHI and YLBHI, 1992). It is far easier for pulp tycoons with privileged access to the Ministry of Forestry, which has jurisdiction over about two-thirds of the country's surface area, to take out what are essentially large, contiguous logging concessions at a rent of approximately US$0.30 per year, clear-cut them, and then replant them with pulpwood monocultures. Plantation entrepreneurs who do so are entitled to equity capital and no-interest loans from the Ministry of Forestry, together with revenues collected from logging companies and earmarked for rehabilitating logged-out concessions. In fact, so easy has it been for business to gain control over forest land by claiming to be interested in pulp and paper (by 1992 thirty-seven companies had submitted applications for 70,000 square kilometres of pulp estates, although only eleven bothered even to undertake feasibility studies on estate development) that the government recently had to stop granting the concessions (*Pulp and Paper International*, 9.1994). The government has also assisted corporate land grabs by suppressing attempts to enforce customary *adat* claims to many of the territories affected. Such claims, although they are recognized in theory, are in practice overridden by the demands of 'development'. As former Forestry Minister Hasjrul Harahap said in 1989, 'In Indonesia, the forest belongs to the state and not the people. . . . they have no right to compensation.'

Thus Riau Andalan will be harvesting fifty species of native tropical hardwood from its concessions over the next decade while waiting for *Acacia mangium* plantations established on cut-over sites to mature (*Pulp and Paper International*, 9.1994). Elsewhere in Sumatra, Sinar Mas's 790,000-tonne-per-year Perawang installations will consume 200 square kilometres of old-growth forest per year until the year 2000, much of it cut by transmigrant labour, before switching over to acacia; the company's 410,000 tonne-per-year Jambi mill further south will denude its 2,400-square-kilometre concession of logged-over forest at perhaps half that rate (*Pulp and Paper International*, 10.1994). Indorayon's 250,000 tonne-per-year pulp mill in North Sumatra, which started off using 860 square kilometres of old stands of indigenous *Pinus merkusii* planted during the Dutch colonial period, soon began clear-cutting mixed hardwood forests from a 1,500 square kilometre selective logging concession and planting eucalyptus there. Some 1,000 square kilometres of pine forests in Central Aceh, meanwhile, will feed the Kertas Kraft Aceh cement sack mill in Northern Sumatra (in which Bob Hasan holds interests) up to the turn of the century. Legal restrictions on clear-cutting or on logging on steep slopes or near watercourses, such as they are, carry little meaning in this context: the Department of Forestry itself admits that 86 per cent of timber concession holders violate government logging rules (*Down to Earth*, 1991; Zerner, 1992; *Jakarta Post*, 8.9.1993; 27.9.1993; 18.2.1994; *Kompas*, 27.9.1993; *Forum Keadilan*, 6.1.1994; World Bank, 1994). If plantations do not mature as expected – which is not unlikely given Indonesia's scant experience with them on the generally poor soil of the Outer Islands – then pressure on natural forests will increase yet further (WALHI and YLHBI, 1992; *Pulp and Paper International*, 8.1993, 10.1994).

Subsidized land, wood and technology, in sum, have helped make Indonesia into one of the cheapest pulp and paper producers in the world. But the subsidies do not stop there. Wage levels are also among the world's lowest, kept that way partly through state-supplied suppression of labour movements. Near-captive labour, meanwhile, is provided by the transmigration programme for projects such as Barito Pacific's Tanjung Enim Lestari pulp installation. At Sinar Mas's Perawang complex in central Sumatra, inadequately housed transmigrants required to do illegal logging have had to wait up to three months at a time for their wages, on one occasion being paid only after they took some of the company vehicles hostage (*Forum Keadilan*, 6.1.1994).

At a time of tightening controls in the Northern countries, lax enforcement of pollution laws (however strict some of them may be on paper) is also an attraction for pulp and paper investors. Sinar Mas's Indah Kiat – the subject of a lawsuit threat from Indonesian NGOs for its pollution of the Siak River – uses its wastewater treatment facility when there is an inspection, for example, but otherwise only sporadically (*Suara Pembaruan*, 31.8.1991; 4.9.1992; WALHI and YLBHI, 1992; *Jakarta Post*, 22.9.1992; 24.9.1992; 8.10.1992; *Kompas*, 17.11.1993). Little is done, in addition, to regulate dioxins, the pulp industry's most toxic products, whose regulation in the US is pushing the industry towards a likely multi-billion-dollar refit there.

Many ordinary Indonesians, unsurprisingly, experience the integration of local land and forests into the world pulp and paper economy as a *dis*integration of local livelihoods and relationships, and their government's gifts of low-cost forest land and riverine waste sinks to the industry as something more closely akin to theft. In Northern Sumatra, for example, Indorayon's clear-cuts, roads and plantations have displaced thousands of native Batak people in the Lake Toba area through usurping their traditional lands and degrading the environment which sustains them. Logging-related droughts have depressed rice harvests over wide areas, and Indorayon has also provoked local resentment by blocking access to common pastures essential for buffalo-raising and woodlands which villagers rely on for rattan or wood for carving; by planting eucalyptus on an ancestral graveyard; and by demeaning Batak villagers by forcibly overriding traditions of hereditary land transfers important to clan identity. Farmers from one village who agreed to grow eucalyptus for Indorayon now regret becoming involved since they are no longer allowed to pasture their animals on their land and fear Indorayon will set low prices for the wood they produce. The Indorayon pulp plant's pollution of the Asahan river, meanwhile, has resulted in degradation of fisheries and loss of village water supplies.

Protests have been met with a classic blend of threats, beatings, co-optation and delaying tactics. In April 1989, when two plantation employees tried to rape a young girl in the village of Sugapa, village women chased the men away and ripped up eucalyptus seedlings that Indorayon had planted. At a February 1990 trial at which the ten women involved were sentenced to six months' imprisonment (later reduced on appeal to probation), they vehemently defended their rights to the property: 'The land is the only source of income that the people have. If it is planted with eucalyptus, how are we going to eat? How are we going to feed and herd our cattle?'

Villagers have also been displaced by Sinar Mas's Perawang operation, where Sakai people were resettled from their ancestral lands on the company's Riau concessions. In South Sumatra, PT Musi Hutan Persand, one of Barito

Pacific's timber estate management companies, when it found that a concession it had been awarded overlapped with one given to another timber company in the area, took over fertile land belonging to 200 farmers in Muaraenim without warning, destroying durian trees, rubber plantations and wet-rice fields. Protesting farmers were arrested and their claims dismissed out of hand by the provincial governor, who claimed they were merely seeking financial gain. Minister for Forestry Djamaloeddin Soeryohadikoesoemo, for his part, claimed it was the responsibility of the company to settle the dispute – despite the fact that the concession was awarded only on the condition that fertile or privately owned land would not be used (*Tempo*, 18.12.1993; Brooks, personal communication, 1994).

In Kalimantan, meanwhile, in one of dozens of such examples, logging roads constructed parallel to a river on one plantation concession blocked tributaries, halving the river's flow and creating a malarial swamp. Filled with sediment from erosion from logging and road-building, the river became useless for transport to market or for fishing. No compensation was paid, meanwhile, for the logging of lands which had been owned and managed by the local community for 400 years. The company involved also put up a sign indicating that subsistence swidden farming on community land was now a crime, and ordered local residents and Javanese transmigrants to use hoes instead. Although only the transmigrants obeyed, catastrophic erosion was the result. Another disaster was precipitated when the company tried to 'develop' the villagers' traditional rattan cultivation methods by getting them to plant rattan in straight lines without regard for site conditions, with the result that 90 per cent of the crop died (Zerner, 1992).

Such cases expose a second way in which plantation projects in Indonesia degrade forests, namely, by displacing people and disrupting community-based forest stewardship systems. Such systems, as a suppressed World Bank report points out, are not only effective in sustaining forests in the Outer Islands, but in some cases have even enhanced biological diversity (ibid.). In resisting the actions of plantation and pulp companies – whether by pulling up stakes put down by survey teams, tearing down signs criminalizing traditional agricultural practices, cutting bridges built by encroaching companies, pelting company houses with rocks, refusing to co-operate with firms which do not consult with residents according to locally approved political processes, or taking firms to court – local villagers are also attempting to defend the only feature in this social landscape that could by any stretch of the imagination be termed 'sustainable development'.

THAILAND

Incentives for establishing commercial pulpwood plantations are as strong in Thailand as in Indonesia: a booming economy, good geographical location, lavish subsidies, and local and foreign élites eager to invest. But competition for available land and forest, a less repressive political climate, and, most of all, heavy resistance, have so far prevented a plantation boom on the scale of that in Indonesia. Continuing pressure from the paper and pulp industry and its allies, however, has led to a see-saw battle for high stakes across large areas of the country between plantation promoters and local villagers and environmentalists.

Much of the impetus for establishing new plantations stems from the Thai economy's growth rate – one of the highest in the world over the past decade – and the associated increase in paper and pulp production, which has more than matched it. Paper manufacture shot up from 294,000 tonnes in 1985 to 889,000 in 1990 and 1.28 million in 1993, with around 3 million tonnes expected in 1997. This boom has been fed largely by the rapidly expanding manufacturing sector, with its needs for packaging, which absorbs more than two-thirds of production (Puntasen, *et al.*, 1992; *Pulp and Paper International*, 10.1993 *Paper and Packaging Analyst*, 11.1994). Pulp production, meanwhile, doubled in the decade to 1993, to about 200,000 tonnes. Among the companies recently making large new investments in pulp and paper have been the Soon Hua Seng Group, one of Thailand's leading rice traders, which is planning to start up a 220,000 to 250,000 tonne-per-year paper-making plant in East Thailand in 1996 to be followed by a large pulp mill; and Siam Pulp and Paper, of which Siam Cement and the Crown Property Bureau hold over half the shares (*The Nation*, 14.4.1994; *Bangkok Post*, 13.7.1994). Over-investment has already pushed container board capacity to 50 per cent over demand, and net paper self-sufficiency is expected to climb to around 125 per cent in 1996, when the country will be producing twice as much short-fibre (eucalyptus) pulp as its paper-makers require (Udol, 1993; *Pulp and Paper International*, 10.1993; *Paper and Packaging Analyst*, 11.1994). Fresh long-fibre (conifer) pulp, however, cannot be produced in large quantities in the country and will continue to be imported.

As in Indonesia, foreign machinery suppliers are important beneficiaries of the boom, with Scandinavian, Japanese and North American suppliers again dominating the market. Foreign investment is also significant. Phoenix Pulp and Paper, currently Thailand's largest producer of pulp with 200,000 tonnes per year, is a venture of Ballardur, India's largest pulp and paper producer – which holds over 13 per cent of the shares – and the European Investment Development Corporation based in Luxembourg (see Plate 2.2). In good years, Phoenix exports 30 per cent of its bamboo pulp production to India and other destinations, and plans to ship 30–35 per cent of its eucalyptus pulp output to South Korea, the Philippines and Japan. Early Japanese and Taiwanese efforts to set up joint ventures to secure new raw fibre supplies, meanwhile, have recently been supplemented by increased investment in pulp capacity, and Shell and Kimberly-Clark have also launched ventures in Thailand.

Plate 2.2 Processing kenaf, Roi-et Province, North-East Thailand: kenaf is one of the materials used by the Phoenix Paper Mill

Anticipated future growth in both exports and domestic production, together with the need to keep machines running even in times of surplus, has compelled the industry to push for more and more of the country's land area to be planted to pulp trees. *Eucalyptus camaldulensis*, known since the early 1970s to be easily adaptable to Thailand's wide range of habitats, and increasingly attractive to paper manufacturers, has been the tree of choice since the appearance of cheap seed on the market in 1978. By 1992 around 800 square kilometres of eucalyptus had been planted by a wide variety of firms and private owners, mainly in the North-East and Central regions, and the industry has continually lobbied the government for further land, recently claiming it needed another 1,280 square kilometres under fast-growing trees by 2002. With its eye on increasing exports, the industry association has also requested soft loans with grace periods of six years, reductions in corporate income tax, waivers on sales tax, suspension of import duty on pulp- and paper-making chemicals and machinery, more government-underwritten training and infrastructure, and state collaboration in plantation ventures (Puntasen, *et al.*, 1992; *Bangkok Post*, 1.2.1993; *Paper and Packaging Analyst*, 11.1994).

Except when confronted by massive popular resistance, government officials have, by and large, been only too willing to co-operate. In the late

1980s, in fact, the government was calling for 43,000 square kilometres of commercial tree plantations in National Reserve Forests (or over 8 per cent of the country's land area), with communities and the government planting an additional 18,500. This eagerness caused some head-scratching even among corporate consultants, who, try as they might, could not locate a prospective national or international market for the products of more than 5 or 10 per cent of such a huge plantation area (Sargent, 1990; *Prachachart Thurakit*, 4–6.1.1989).

Such official enthusiasm for commercial eucalyptus plantations, however, is not as mysterious as it may look. The Royal Forest Department (RFD) has been overwhelmingly oriented towards commercial exploitation during its entire 100-year history. When logging was banned in 1989, it was only natural that it should turn its attention to promoting commercial plantations. This bias is no doubt reinforced by the close association many RFD bureaucrats enjoy with eucalyptus nursery owners or brokers and the equally close collaboration the state Forestry Industry Organization enjoys with the private pulp and paper sector (Puntasen, *et al.*, 1992). Both foreign and domestic eucalyptus-growing firms, in addition, are closely linked to political parties responsible for powerful ministries. Soon Hua Seng, for example, whose board chairman is Narong Mahanond, a former Director General of the Police Department, has helped finance the Democrat Party, which has often held the portfolio of the Ministry of Agriculture and Agricultural Co-operatives, which is responsible for the RFD. Members of Parliament from other parties also often have a finger in the plantation or pulp business. Promoting commercial plantations, moreover, allows the Thai government to portray itself as discharging its responsibility to 'reforest the country' in the wake of the catastrophic logging of the past 30 to 40 years. Finally, the RFD is eager to regain control over large areas of National Reserve Forest land, which, following decades of commercial logging and official promotion of upland export crops such as kenaf, maize, sugar cane and cassava, are now *de facto* in the possession of peasant 'encroachers' (Usher, 1990a). Just as the RFD once granted timber companies cheap logging concessions, it is, many believe, now eager to grant even cheaper concessions to eucalyptus firms in the hope that they will drive out the settlers that the logging concessions helped attract.

As in Indonesia, in addition, planting trees may not be the only thing tycoons have in mind when they demand access to more land for pulpwood plantations. While natural forests are not a leading source of pulp raw material in Thailand, businesses often hire or encourage villagers to clear forest so that it can be categorized as 'degraded land' legally eligible for planting with eucalyptus. The timber is then illegally sold for lumber. Many business figures who may not be particularly interested in wood fibre, moreover, by putting on a show of interest in pulp and paper, can amass land which can later be put to even more profitable uses such as golf courses or tourist resorts. RFD officials and political parties who stand to share the benefits are often only too happy to play along.

Influenced by Western models, Thailand's official institutions are structured in a way which cannot help but subsidize the plantation industry. The Royal Forest Department, for example, has legal jurisdiction over approximately 40 per cent of the country's surface area in the form of National Reserve Forests (NRFs), many of which are forested in name only. Until recently, transferable land documents could not legally be granted in NRFs, even in the 56,000 square kilometres which are estimated to be currently under cultivation, though many NRFs were gazetted in areas already occupied by villagers. This arrangement has enabled the government to label the more than 10 million people who live in the NRFs – some of whom have been present since before the NRFs were gazetted – as 'illegal encroachers' and to give away land they are occupying to plantation or other businesses at the derisory rate of US$2.50 per hectare per year, little more than 5 per cent of the (already low) typical going market rate (Puntasen, *et al.*, 1992). The Director-General of the RFD can now authorize the rental of up to 16 hectares of NRF per company, the Minister of Agriculture up to 3.2 square kilometres, and the Cabinet, still larger areas. In September 1994, in fact, the Minister of Agriculture was given authority to allow state agencies to use parts of National Reserve Forests without cabinet approval 'if their projects involve national and economic security'.

Other apparatuses have been set up for granting special privileges to pulpwood plantations which no other agricultural crop enjoys. In the late 1980s, the RFD set up a separate office devoted specifically to promoting commercial tree farms, complete with a 'public relations' budget of over US$24 million, and in 1994 asked for approximately US$130 million from the 1995 fiscal budget to subsidize private sector plantations and other 'reafforestation' projects, including one in which farmers in National Reserve Forests are given soft loans to plant fast-growing trees. The Board of Investment has meanwhile granted some firms tax holidays and exemptions from duty on imported machinery and raw materials (Usher, 1990b; *The Nation*, 9.4.1990; Sukpanich, 1990; Puntasen, *et al.*, 1992).

Providing additional subsidies are foreign governments. Particularly prominent, unsurprisingly, are those of Japan, Canada, and Finland, three of the countries which are likely to benefit the most either from sales of machinery and consultancies or (in the case of Japan) also from imports of Thai chips and pulp. As long ago as 1981, the Japan International Co-operation Agency set up a trial eucalyptus plantation in North-East Thailand to support research and training in the field, and the Japanese Overseas Economic Co-operation Fund has supported a venture called Thai-Japanese Reforestation, which was designed to supply raw materials to a consortium of Japanese paper-makers (Nectoux and Kuroda, 1989; Sukpanich, 1990). CIDA, the Canadian 'aid' agency, has helped finance the Canadian consultant H. A. Simons' work with Soon Hua Seng and the tree plantation studies of the Thai Development Research Institute, a prominent think tank. Britain's Commonwealth Development Corporation, which draws money from the British 'aid' programme, has also provided loans and debt finance to Soon Hua Seng.

The Finnish government, meanwhile, has pumped millions of dollars of its taxpayers' money not only into export credits for Finnish forestry and pulp and paper equipment, but also into a political lobbying exercise known as the Thai Forestry Sector Master Plan, which was conducted by the Jaakko Poyry consulting firm between 1990 and 1994. Largely aimed, in the words of its Finnish team leader, at bringing Thailand's 'institutional and social frame into shape', the Plan included an ambitious attempt to rewrite the country's entire forestry policy in a way which would allow Western techniques of industrial forestry to be applied more fully (Laitalainen, 1992). While it was not successful in achieving all of its aims, the Master Plan, like similar exercises in Indonesia, did help weave new personal and business networks among Thai and foreign private sector, academic and bureaucratic plantation proponents, as well as add to the technical base for plantation development in the country. The plan also provided a free education in Thai politics for Finnish industry figures which is already paying rich dividends. Finnish machinery exports to Thailand – much of it in the forestry industry sector – rose from FM83 million in 1990 to FM484 million three years later, and a special Thai–Finn pact for economic co-operation was signed in October 1993 (Finland National Board of Customs, 1990–93).

Also significant has been the strong support of multilateral agencies. The Asian Development Bank has funded plantation research and project development in both Thailand and Laos. The commercially oriented Tropical Forest Action Programme, an international co-ordination mechanism for increasing investment in forestry in the South organized by the World Bank, Food and Agriculture Organization (FAO) and United Nations Development Programme (UNDP), has meanwhile provided an umbrella for Poyry's Forestry Sector Master Plan.

Arrayed against such pro-plantation forces have been various popular movements, most notably among the more than ten million people living in cleared areas of National Reserve Forests. Since the mid-1980s, small farmers opposing the spread of eucalyptus plantations have petitioned district officials, Members of Parliament, and cabinet members; held rallies; spoken out at national-level seminars; blocked roads; and marched on government offices. Where other means have failed and they are organized well enough, they have ripped out eucalyptus seedlings, chopped down eucalyptus trees, stopped bulldozers, and burned nurseries and equipment.

Such actions are motivated mainly by experience showing that large-scale eucalyptus plantations displace or impoverish farmers, violating a strongly felt right to a stable subsistence. Even where villagers are allowed to stay on or near plantations, *Eucalyptus camaldulensis* allows little inter-cropping (villagers say it is *hen gae tua* ('selfish') in that it hogs nutrients), is useless for fodder, damages local soil and water regimes in ways to which villagers are sensitive, and supplies little firewood to the community. In addition, it provides none of the varied natural forest products that rural dwellers on the edge of the market economy need to eke out a subsistence, while usurping the community woodlands that do. Plantation labour requirements, moreover, are neither plentiful nor steady, and plantation work is unattractive to most farmers. The land speculation often associated with plantation expansion also undercuts farmers' security. Using fraud and legal chicanery and their political and bureaucratic connections, speculators hoping to resell land to plantation companies often threaten villagers with violence, murder, or eviction on 'encroachment' charges; attempt to co-opt village elders; try to get villagers into debt; or cut off their access to their own land by buying up surrounding plots.

Migrating away from plantation areas, moreover, is not an attractive option. In Thailand the frontier is nearly gone – forest cover has declined from perhaps 70 per cent of the country's land area earlier in the century to about 15 per cent today – and government bureaucracies are less likely now to acquiesce to illegal forest colonization than they were in the 1970s, when the country's unsettled upland areas served as welcome safety-valves relieving potentially explosive land conflicts elsewhere. Official resettlement programmes, meanwhile, faced with a dearth of suitable unused agricultural land, often try to relocate villagers on already occupied land, touching off bitter land disputes. Moving permanently to the cities, similarly, is seldom an alluring or even viable alternative, and casual agricultural labourers' jobs are notoriously insecure and ill-paid. Awareness of their limited choices, not surprisingly, only stokes villagers' moral outrage at the abuses of the cosy business–bureaucracy–politics patronage system used to promote plantations.

In their struggle, villagers have made shrewd use of a wide variety of rhetorical, political and cultural tools. In many areas, villagers have wrapped their actions in a religious mantle by tying yellow Buddhist monks' robes around large trees to 'ordain' and thus protect them from being cleared to make way for plantations. They have also adapted the *pha pa ba* ceremony traditionally used to mobilize assistance for Buddhist temple construction to promote the anti-plantation movement instead. Well aware of the need to seize the environmental high ground, village plantation opponents have also planted fruit, rubber and native forest trees to pre-empt or replace eucalyptus, or have explained to sympathetic journalists the methods by which they have preserved diverse local forest patches for generations as sustainable sources of water, mushrooms, vegetables, small game, honey, resins, fruit, firewood, fodder, herbal medicines, frogs and edible insects and insect eggs, or as sacred ancestral woods and funeral grounds. Where such traditions do not exist, villagers have often created community forest conservation organizations to counter the spread of eucalyptus. They have also been careful to stress positive demands for individual land rights, community rights to local forests, and the right to veto commercial plantation schemes targeted for their localities. Finally, villagers opposing pulpwood plantations have made cautious use of the support of many Thai environmentalists and other activists among the intelligentsia, who have called for recognition of local land and forest rights, sharp separation of plantation and forest conservation policies, and cancellation of privileges for industrial plantations. Environmentalists in particular have helped drive home the message that huge increases in commercial plantation acreage can only increase the pressure on natural forests, through displacement and disruption of locally developed forest conservation practices (Lohmann, 1991).

In the early 1990s, plantation proponents and opponents alike had to struggle for purchase on a slippery, constantly changing political field. In May 1990, then Prime Minister Chatichai Choonhavan prohibited all commercial 'reforestation' in National Reserve Forests following the arrest of 156 Soon Hua Seng employees for logging a plantation site in Eastern Thailand. (In accord with normal practice, Soon Hua Seng had been allowed unofficially to have access to the forest in order to 'degrade' it before negotiations for the concession were concluded, but political intrigue against the company's president led to exposure.) In February 1991, however, a military *coup d'état* paved the way for a massive US$2.76 billion programme to evict 5,000,000 residents of National Reserve Forests in order to free approximately 14,700 square kilometres for private-sector tree plantations. Within a year, perhaps 40,000 families were forcibly displaced in the North-East, with repression especially severe in areas targeted by the military as hotbeds of resistance (*Phuu Jatkaan Raai Sapdaa*, 16–23.9.1991; Ekachai, 1992). When a massive popular movement overthrew the military junta in May 1992, resistance to pulpwood plantations emerged again in force. Following major demonstrations by Northeastern villagers, including the blockade of the region's principal highway, and prolonged negotiation with farmer leaders, the government scrapped the military's eviction programme, suspended 'reforestation' with eucalyptus, and imposed a ceiling of 8 hectares on any type of commercial tree plantation. Yet in September 1993, under pressure from the pulp and paper industry and its allies, Thailand's economic ministers decided once again to reopen National Reserve Forests to commercial reforestation, touching off student protests. By July 1994, private corporations and state enterprises were being invited by the Ministry of Agriculture and Agricultural Co-operatives to participate in a scheme to cover at least 5,400 square kilometres of land with trees between 1994 and 1996, ostensibly to mark the fiftieth anniversary of the King's coronation.

Continuing protests, however, have made the country a watchword in the industry for conflict over raw materials. This has discouraged both foreign and domestic investment, particularly in giant export schemes such as those to be found in Brazil and Indonesia. Shell, for example, was forced to drop plans to plant some 125 square kilometres of eucalyptus in Eastern Thailand after violent conflicts and scandals forced delays in governmental approval for the project in 1990. Soon Hua Seng, India's Birla, and other investors have also been compelled to abandon ambitious plantation or pulp schemes out of fears of local opposition or subsequent rejection by the government. Phoenix's pulp mill, meanwhile, lost 141 production days during 1992–4 due to controversies over pollution releases which have damaged local fishing livelihoods, and other pulp investors, including Panjapol and Siam, have experienced problems with licensing authorities. Such difficulties have added to investors' reluctance to move from the Bangkok area, with its good infra-structure, consumers, and easy access to waste paper for raw material (*Paper and Packaging Analyst*, 11.1994).

Facing clashes with landless farmers or governmental vacillation in its efforts to take over NRF land outright, the industry has turned increasingly to promotion of contract farming (the Thai analogue of Indonesian nucleus estate schemes) as a second-choice strategy. Here the industry, instead of dispossessing peasants directly, contracts with them to grow eucalyptus on their own land, often distributing free seedlings and promising to purchase their harvest in five or six years. Although this approach is unwieldy, requiring pulp or chip firms to deal with thousands of smallholders scattered across a large area, it has several attractions for industry. First, it is unlikely to provoke the organized resistance to eucalyptus which has followed from attempts at outright eviction, and may even encourage farmers to clear new areas in forested uplands to plant the tree. Second, many participating small farmers are likely to have to foreclose sooner or later. Not only are they likely to become dependent on plantation or pulp firms for seedlings, materials and cash, but eucalyptus appears to be economically unviable for small farmers as well (Usher, 1990b). In the end, plantation or pulp firms may simply be able to buy up their land at a bargain price.

Another strategy for Thai pulp and paper investors facing resistance within the country is to move abroad. Working with the Asian Development Bank (ADB), Jaakko Poyry, and the Finnish government, Phoenix plans to feed its mill with a 96-square-kilometre plantation in neighbouring Laos. Some 70 per cent of the Asian Development Bank loan for the project will go to the private sector – which the Lao state is ill-equipped to monitor (see also the chapter by Rigg and Jerndal in this volume) – with only 30 per cent to small farmers. Finnish government subsidization of planning and political manipulation in Laos – for example, a US$5.8 million grant for technical assistance as part of the World Bank's so-called Forest Management and Conservation Project, which will lead to tens of millions of dollars in grants and loans being put into a sector where government officials' salaries are around US$30 per month – is aimed at opening up further such opportu-nities in the future (Traisawasdichai, 1994). Meanwhile, like Indonesia's Sinar Mas, Soon Hua Seng is moving into China, participating in joint ventures to plant over 2,200 square kilometres of eucalyptus in Canton and also setting up pulp and paper mills in the country. Led by the Ministry of Agriculture and Agricultural Co-operatives, the Thai government has also signed a co-operation pact on industrial forestry science and technology with China which is aimed, among other things, at supporting that country's policy of planting 65,000 square kilometres of fast-growing trees such as eucalyptus and poplar by the year 2000 with investment from Japan, New Zealand, Australia, Indonesia and Thailand (*Bangkok Post*, 23.6.1993).

CONCLUSION: MANAGING RESISTANCE

The enclosure of land and water associated with the expansion of the pulp and paper industry in South-East Asia is not simply a physical phenomenon, an invasion of space. Nor is it the result of 'free market mechanisms' or an inevitable and impersonal 'drive for economic development' which must now be made 'sustainable'. It is, rather, a complex social, cultural and political struggle featuring a highly varied set of actors with diverse motivations, who, working in loose conjunction with each other in a contingent and *ad hoc* fashion, make paper and pulp machinery sellable, debt finance possible, political interests meshable, and centralization of resource control achievable. In both Indonesia and Thailand, this process of enclosure has both presupposed and facilitated an ability to shift resources within an increasingly regionally and globally organized system. In particular, transfers of resources between expanding public and private sectors in South-East Asia, Japan and the West have been used to break down boundaries between discrete smaller social units in Indonesia and Thailand which are neither public nor private – namely, highly localized commons regimes. Attempts are then made to reshuffle and modify the elements of these regimes in order to create an all-encompassing new framework whose parts respond to certain standardized, system-wide signals in ways more advantageous to powerful actors (*The Ecologist*, 1993). As in other industries, subsidization and globalization have proved mutually enabling.

Dealing with resistance is crucial to the pulp and paper industry's struggle, and it has a spectrum of ways of doing so. Where opposition does not challenge the industry's most fundamental interests, it will attempt to contain it by internally redistributing its considerable resources in various ways, relieving tensions in one area through slack in another, depending on cost-effectiveness and other factors. It may, for example, do the following:

- Buy off resisters or attempt to demonstrate to them how their concerns can be met within the industrial system.

  This strategy amounts to inviting opponents to accept 'translations' of their objections into the language of the industry itself, in which these objections become more tractable. Contract farming schemes, which appear to meet the common objection that pulpwood plantations are usurping individual land-holdings, are one example. Other examples include bribes and promises that the suffering the expansion of plantations brings about will be seen in retrospect as a mere 'cost' attached to the far greater 'benefits of economic development'.

- Help see to it that resisters are crushed by force.

This option is especially useful against local communities whose livelihoods are entirely dependent on commons regimes, and who, because the logic of such regimes runs counter to that of economics, have proved unreceptive to the first strategy (*The Ecologist*, 1993; Lohmann, 1995). To be a candidate for intelligent repression, opposition must also be isolated, small-scale, poorly co-ordinated, and out of the public eye. For this option to work, too, government bureaucracies must decide it is in their own interest to foot the bill for military operations. Both Indonesia and Thailand offer recent examples.

- Insist on discussing the issues only in the idiom of neo-classical economics and 'global demand' rather than in the languages of ordinary farmers or of institutional politics.

This strategy, an analogue of physical suppression, attempts to deny opponents the opportunity of formulating their objections in the conceptual framework of their choice. It is not always adopted calculatingly, nor is it adopted only by industry figures. Thus some academic economists will not tolerate any analysis of how paper demand is created because it undermines the model – with which they begin their analysis – of *Homo economicus* as a being with in principle infinitely increasing abstract needs. Industry leaders such as Britain's David Clark are less squeamish, telling their peers that 'we shall have to fight for our future and create our own growth' and that 'total demand has to be stimulated' (Clark, 1994).

- Acquiesce, where necessary, to certain demands made by opponents.

This becomes a strategic choice where opponents cannot be bought off or persuaded to modify their demands and where suppression is counterproductive or impossible due to the scale, co-ordination, intensity, inaccessibility or public visibility of resistance. Japan's paper industry, for example, has had simply to accept environmentalist resistance in western North America as an 'economic' datum beyond a political solution and accordingly shift its search for raw materials elsewhere, including South-East Asia. By the same token, Western industry is slowly capitulating to strong and widespread opposition to chlorine-using industrial processes, treating it as an 'economic' reason for investing in new technology. (Indeed, in doing so, it may be able to outcompete rivals with fewer resources.) Nor does the call for more recycling greatly trouble an industry long accustomed to using waste paper as a raw material. Rather, it is fairly easily fed into overall supply/demand equations, while public relations officers welcome it as a new opportunity to claim 'green' status.

Some opposition, however, presents deeper threats. No paper corporation possesses the resources to adjust itself to falling demand for all its products, nor, faced with community-based opposition to plantations across very large areas of the South, to buy it off everywhere it arises, smash it wholesale, or shift its search for raw materials to another planet. Such challenges, impossible either to accommodate or to crush outright, are met most intelligently by the ancient strategy of divide and conquer. Abandoning attempts either to conciliate or to wipe out groups with which it has irreconcilable conflicts at the grassroots in South-East Asia, industry instead concentrates its attention on keeping those groups divided from potential allies in bureaucracies and in urban and Northern middle classes.

Thus pulp and paper interests in Indonesia have resorted to repression and abuses at home while hiring public relations firms such as the US's Burson Marsteller to present a softer picture to customers and legislators in the West, as well as to infiltrate, undermine and monitor Western environmental groups (*PR Watch*, 1993, 1994). Similarly, Jaakko Poyry consultants largely abandoned public relations efforts with farmers and environmentalists midway through the company's master planning exercise in Thailand, smearing and breaking promises to them, while channelling money to the bureaucracy and taking pains to ensure that the nature and extent of opposition to the master plan within Thailand did not become news in Finland (Lohmann *et al.* forthcoming).

Industry-retained public relations firms also attempt to marginalize as 'radical' or 'irresponsible' movements for reduction of paper consumption in the West. By promoting the use of an idiom which identifies economic growth with livelihood, paper consumption with literacy, and large corporations as merely another group of 'concerned citizens', such firms strive to create an atmosphere in which the resistance of grassroots groups in the South will appear cranky or, even better, inconceivable. Arjo Wiggins Appleton executives O. Fernandez Carro and Robert A. Wilson (1992) sum up such strategies well when they urge their colleagues not to target 'apparent opposition' if that means 'forgetting the vast mass in between: the public'; not to 'respond to the mobile agenda of others' but rather to 'write the agenda and diffuse negative issues'. According to them, politics

provides the packaging and the vehicle to achieve the industrial objectives. Success is measured by the freedom to plant fibre crops, recognizing the sum total of all the political forces (in the broadest sense). There are two elements to the political subsystem [of the total quality system of industrial forestry]: the message and the target. The message needs to be short, nontechnical, and fundamental: for example,

'Trees are good. We need more trees not less'. Our objective should be to create and move inside an ever-increasing friendly circle of public opinion.

It will already be evident from the examples of Indonesia and Thailand how disingenuous the 'message' is that pulpwood plantations can bring 'more trees' into being, or can help take pressure off natural forests. Several other such 'messages', however, have also proved to be useful to the pulp and paper industry in its divide-and-conquer strategy of recruiting support from consumers, officials and environmentalists, and have been used repeatedly in public debates, newspaper columns, and governmental negotiations both within South-East Asia and among its trading partners. As attempts to justify to uninformed audiences, particularly Northern ones, the environmental changes induced by the pulp and paper industry as 'sustainable development', they too are worth some critical attention:

- Indefinitely rising paper demand is either inevitable or desirable or both.

  This message, together with the idea that further exploitation of old growth is undesirable, is intended to lead to the conclusion that plantations must be expanded and that side effects, however regrettable they may be, are unavoidable. The search for 'alternatives' to the current paper and pulp production model, instead of being concerned with such social matters as demand creation or the desirability of junk mail or mini-packaging, thus becomes restricted to the search for 'alternative technical means of increasing production'.

- Demand for paper comes not from particular groups, classes, or societies, but rather from 'the globe' or 'the nation' as a whole.

  This notion helps obscure exactly who paper consumers are and thus helps erode the common-sense observation that a huge demand in, say, Japan or the US does not necessarily justify plantation expansion in South-East Asia. By suggesting there is a type of demand which, being 'global', over-rides merely 'local' claims to land and water, this idea also helps justify siting plantations at any location on the planet that industry deems appropriate and elevates officials and business people who advertise themselves as addressing 'global' concerns into a superior moral position. It also licenses cross-regional and cross-class subsidies for the industry, as well as large forced evictions.

- Pulpwood plantations are an economically productive use of unoccupied, degraded land which would otherwise rest unused.

  This 'message' is effective principally with environmentalists unaware of industry thinking and practice at the grassroots. As the Asian Development Bank and Shell International have both pointed out, industry is not particularly interested in degraded land; what is required for competitiveness, rather, is contiguous chunks of 'land suitable for superior biological growth rates for those species the market wants' as well as 'year-round water' and easy access to transportation – one example being Shell's own ill-fated project area on the well-watered, relatively fertile soils of Chanthaburi, Eastern Thailand (Shell and WWF, 1993). Even more attractive is land already forested. As an Indah Kiat executive has forthrightly stated, '[b]asically we are looking for forest which can be clear cut and replaced with eucalyptus and acacia' (Marchak, 1992). The message also cannot be used with groups who understand that what counts as 'degraded' or 'unused' depends entirely on who is talking – for example, that land identified in official Geographic Information Systems as 'marginal' or 'unoccupied' may in fact be used by local people as woodland commons, pasture, or fallow for rotational agriculture.

- Plantation expansion helps make underdeveloped countries 'self-sufficient' in paper.

This 'message' can be usefully employed with audiences unaware, for example, that Indonesia's new pulp capacity is aimed largely at export; that Thailand is already self-sufficient in paper in aggregate terms and in those sectors in which it is not, new Thai plantations will not help make it so; and that nominal self-sufficiency counts for little in the face of the liberal trade policies advocated by the industry itself, which will push pulp and paper imports into any country not producing them more cheaply.

- Plantations are up to ten times more productive than natural forests.

This 'message' implicitly defines 'productivity' as 'productivity of trees with market value as pulpwood over two or three growing cycles'. It is useful mainly with audiences likely to overlook the value both to local communities and outsiders of other trees and of animals, vegetables, mushrooms, fodder, fertilizers, water retention capacity, medicines, and so forth, and unaware that monoculture tree plantations typically interfere with agricultural production.

- Fast-growing tree plantations can help curb global warming.

This notion has already legitimized plantations in Malaysia and Central America supposedly to 'offset' carbon dioxide emissions from Northern industries. It must be used cautiously, however, with audiences aware of the link between plantations and overall deforestation and of the fact that following conversion into paper, most if not all of the carbon dioxide temporarily sequestered in plantations will be released to the atmosphere. It also cannot be used effectively with audiences who question why the South, or poor regions of the North, should provide ever-expanding sinks for infinitely-growing industrial carbon dioxide sources in industrialized regions (Barnett, 1992; Sargent and Bass, 1992).

- Promulgating plantation 'guidelines' will make plantations 'sustainable'.

This message appeals particularly to Northern academics, technocrats and environmentalists whose history, institutions, and jobs give them incentives for believing that if theory, objectives, or sound legal measures can be formulated 'correctly', even by outsiders, good practice, implementation or enforcement will follow fairly straightforwardly through the efforts of existing effective, disinterested, benevolent institutions. One British advertising executive recently accused of making misleading claims for Indonesian forestry in a television advertisement, for example, got considerable mileage out of retorting that what actually happens in practice in Indonesia is irrelevant. Evidence of rampant industry deforestation in Indonesia should not disqualify advertisers from saying that forest use there is sustainable, he maintained, any more than the evidence for the existence of drunk drivers in Britain should prevent anyone from saying that drink driving is not allowed there (Brooks, personal communication, 1994).

Such 'messages', used selectively, foster the globalization of the pulp and paper industry by helping block alliances between South-East Asian grassroots groups fighting monoculture pulpwood plantations and environmental groups elsewhere, particularly in the Northern countries. Yet the converse is also true. It is only the global reach of the contemporary pulp and paper industry – its ability to exploit the spatial and cultural distance between residents of rural areas in South-East Asia and intelligentsias elsewhere – that allow it to deploy its mystifications – 'Trees are good. We need more trees not less' – to drum up support for industrial tree plantations among largely urban and Northern power bases. This support is crucial, since a ballooning 'free market' in wood fibre, pulp and paper can be constructed and co-ordinated only if the subsidies given to consultants, foresters, aid agencies, and non-governmental organizations to promote plantations can be justified before a large and diffuse public.

To use such mystifications, however, is always to gamble that they will not be exposed. Whether that gamble pays off – whether the pulp and paper industry gains its 'freedom to plant' at the expense of affected people in South-East Asia – depends largely on the skill in intercultural conversation of plantation opponents. The prospects are far from hopeless. In a world thronged with naked emperors, paper industry figures claiming sustainable environmental benefits for large-scale monoculture pulpwood plantations are more notable than most for their sartorial minimalism.

3

ENVIRONMENTAL NGOS AND DIFFERENT POLITICAL CONTEXTS IN SOUTH-EAST ASIA

Malaysia, Indonesia and Vietnam

Bernard Eccleston and David Potter

Environmental problems caused by natural resource exploitation always have a political context: this point has been clearly made in Chapter 2. Any such context has at least three aspects. First, there are those general processes of steering and choosing through time by governments, business organizations, NGOs (non-governmental organizations) and other political organizations at local, national and international levels which produce policies or courses of action that create or otherwise affect environmental problems. Second, there are the conflicting interests and values related to such policies or actions by various organizations from which certain interests benefit more than others. Third, there are more enduring structures of power that shape the actions of political organizations, help to determine whose interests and values prevail, and frame the processes of steering and choosing at different levels.

For example, the governments of Indonesia and Malaysia have policies which use the tropical forests there as an important springboard of national development by encouraging large-scale timber exploitation with little apparent regard for sustainable forestry, the economic value of non-timber forest products, or the value of intact forests as a biodiversity bank. Such actions take place within a global economic structure, e.g. a dynamic world of global capitalism dominated by private ownership, production for profit, competitive struggle between large corporations, reliance of such organizations on wage labour, and so on. Dominant ideologies also form a structure of power. They help people to make sense of, and accept, the prevailing social and economic structures from which ruling groups and other powerful interests benefit.

Environmental policy-makers also work within the constraints and opportunities provided by political structures of power. An environmental NGO in Indonesia or Malaysia is confronted by an international system of nation–states and other international organizations; even if it formed an alliance with NGOs in other countries, there is no way immediately that the alliance could get rid of that system and replace it with, say, an environmentally friendly world government. Governments, corporations, NGOs, other organizations and groups must work within that international political framework and understand its particular configurations of power in order to advance their particular interests. Within Indonesia and Malaysia, people in government, business, and the NGO sector inherit a political structure and set of political processes which they cannot do much about, at least not immediately. No two political structures are the same, however, and this can make a difference to NGO advocacy work – a matter we return to in a moment.

The preceding sketch of the broad political context of an environmental problem suggests that environmental NGOs are up against powerful government and business organizations buttressed by global economic, ideological and political structures of power. In South-East Asia, it can be said, roughly speaking, that NGOs represent the environmental interest whereas most of their adversaries in government and business tend to have other priorities in mind when it comes to the forests. Since government and business in South-East Asia are far more powerful than NGOs, the general picture for those who care about sustainable forestry is not a very encouraging one. But this is too simplistic. Environmental NGOs are not always 'the good guys', nor are all government agencies and business corporations indifferent to environmental considerations. Also, there are many different sorts of NGOs and there can be serious disagreements between them. Moreover, some environmental NGOs can be influential in certain political contexts despite their limited political resources.

An environmental NGO is defined as an organization that is non-governmental and non-profit and engaged with an environmental problem or problems. Our particular interest is the different political structures within which NGOs have to operate. We deploy four criteria broadly to distinguish one political structure from another:

1 More or less accountability of rulers to the ruled through representative assemblies and governments, formed at regular intervals through competitive elections based on universal adult suffrage involving multiple political parties providing a reasonable choice for the voters.
2 More or less plurality of power centres within and outside the state, including a plethora or paucity of voluntary associations into which the state does not normally reach.
3 More or less guarantees in law of civil and political rights and freedoms, including freedoms of expression and association.
4 More or less popular participation by people throughout society in collective decisions which directly affect their lives together with the principle of equality informing political life generally.

These are the main features frequently used to distinguish more or less democracy in a political structure. The first three features are normally found in conventional liberal democratic definitions. The fourth is different, and closer to the way Asian socialists and communists use the word. Our definition is actually a tough one. No large democracy anywhere, any time, has met all four tests very well, certainly not in Asia. But the criteria do enable one to talk about NGO advocacy work within four political contextual factors related to more or less democracy. We deploy democracy criteria in distinguishing between different political contexts in South-East Asia because the literature broadly assumes that more democratic political contexts are better than less democratic ones for the environment and the work of environmental NGOs (e.g. Clark, 1991; Hardoy *et al.*, 1992; Fowler, 1993).

In this chapter we explore this assumption about NGOs and political contexts in terms of more or less democracy by comparing environmental NGOs and their work on forest issues in three very different political contexts: Malaysia – one of Asia's more settled (semi) democracies; Indonesia – basically a non-democratic authoritarian regime; Vietnam – a Communist Party mobilization regime, mostly non-democratic in a very different way. Our attention is focused mainly on the comparison between Malaysia and Indonesia. Vietnam is referred to only briefly as an unusual 'marker' for the other two.

ENVIRONMENTAL NGOS AND THE POLITICAL CONTEXT IN VIETNAM

Most of Vietnam was covered with tropical forests until the French cleared a lot of it in the South to make way for rubber, coffee and banana plantations. By 1943, 43 per cent of the country was still forested; by the end of the Vietnam War, probably about 29 per cent (Quy and Hu Ninh, 1990; Government of Vietnam: Ministry of Forestry, 1991). During the 1980s and until early 1995 (when this chapter was written), there was probably further deforestation, partly due to official resettlement programmes to relieve population pressure in delta regions, and partly due to timber industry extractions. Some planting of new forests also occurred. Logging was in the hands of about 400 state forest companies and 150 state wood processing enterprises. A range of progressive state policies was aimed at promoting sustainable forestry, including a ban imposed in 1992 on the export of logs and sawn timber, but enforcement mechanisms were weak, and regulations and bans were, not infrequently, ignored (see also the chapter by Lang in this volume). Smuggling of logs between Vietnam, Laos and Cambodia was also common.

The political context in which these activities took place was certainly not liberal democratic. Forest and other state policies in Vietnam were determined principally by relevant government élites and people in the higher reaches of the Communist Party not accountable to voters through competitive elections. There were local elections of some significance, but national leaders were not accountable in this way. Also, the Vietnamese political structure was not noted for much in the way of diversity of power centres within the state or a plurality of more or less autonomous voluntary association in society. As for individual civil and political rights and freedoms, during the 1980s and early 1990s Vietnam was regularly, for example, placed on Level 3 (on a scale of 1 to 5) by Amnesty International. Level 3 was defined roughly as a situation where political imprisonment, political murders or executions, and unlimited detention (with or without trial) was fairly common for people with political views whose activities offended the state.

By contrast, Vietnam was unusually democratic on the fourth criterion of popular participation by people throughout society in collective decisions that directly affected their lives. The extent of such popular participation was striking even for a Communist Party mobilization regime, in which such participation is to be expected. Kathleen Gough's remarkable study of Vietnam's political economy in the 1980s (Gough, 1990) showed, with numerous examples, how the Peasant Union, the Trades Union, the Youth Union and other mass organizations and their local branches, although linked closely to the Party, did debate policy proposals vigorously and frequently modified them. The Vietnam Women's Union (VWU), for example, was formed in 1930, and by 1994 90 per cent of their recommendations had been formulated into laws by the government; the VWU in 1994 operated at national, provincial, district and commune levels, and had a membership of about 10 million women (Mulla and Boothroyd, 1994, p. 37). VAC Ho Chi Minh Association is another mass organization, established in 1988, with a membership in 1994 of about 70,000; its main activities involved training farmers in ecological agriculture, promoting environmental awareness through the mass media, and implementing credit schemes (ibid., p. 38).

This unusual political context has left little room for domestic environmental NGOs. Only twenty organizations were identified as development-oriented Vietnamese NGOs by the authors of a recent study of the subject (ibid.). Thirteen were profiled in the study, of which only four had anything to do with ecology or environmental programmes. Only one of these, CNRES (Centre for Natural Resources Management and Environmental Studies), comprising about twelve people within the University of Hanoi, had engaged in any policy advocacy work since being established in 1985; the interviewee said CNRES had persuaded the government to join CITES (The Convention on Illegal Trade in Endangered Species). The other three environmental NGOs in Vietnam, all very small and located in Hanoi, were established only after *doi-moi* (the liberalizing reforms) began to take effect in the late 1980s; ECO-ECO (Institute of Ecological Economy) in 1990, CERED (Centre for Environmental Research, Education and Development) in 1991, and the Centre for Gender, Family and Environment in Development in 1993. There were over two hundred representatives of international NGOs in Vietnam by early 1995, but only a few had any interest in the environment.

Doi-moi in the mid-1990s involved a continuing

> transition from a traditional socialist economy, in which the regime attempted to impose central planning as the main means of controlling the economic development process, to a more market-oriented form of socialism in which economic levers (e.g. the price mechanism, taxation) are used to plan the economy by setting certain key parameters and allowing individual enterprises freedom to make their own investment, production and marketing decisions.
>
> (Beresford and Fraser, 1992, p. 3)

As the consequences of *doi-moi* become more firmly established, it is possible that the mass organizations may wither or take different forms. In such a changing political context, one possibility is that the number of domestic environmental NGOs may increase. As of 1995, however, there was virtually no environmental NGO sector in Vietnam engaged on advocacy work on forest policy.

The Vietnam case suggests that environment NGOs are least likely to exist or be influential in political contexts where mass organizations of citizens are mobilized by a political party to participate directly in making binding rules and policies related to their environment. NGOs as more or less autonomous voluntary associations appear to have their *raison d'être* in more liberal democratic political regimes, a matter we come to next.

ENVIRONMENTAL NGOS AND THE POLITICAL CONTEXT IN MALAYSIA

Since the mid-1980s Malaysia has been second only to Brazil as a focus of global attention over the rapid depletion of her tropical forests. Partly this reflects scientific concern about irreparable damage to some of the world's oldest and richest forests. In addition, NGOs in the North specially targeted Malaysia because her political system was perceived to be more open to influence than more authoritarian regimes like Indonesia. Questions remain, however, about the openness of the Malaysian polity and this leads some to label the system as 'quasi'- or 'semi'-democratic (Mahmood, 1990; Case, 1993). How, using the criteria outlined above to distinguish more from less democratization, can we assess the impact of the local political context on NGO environmental campaigns?

Though regular elections are held, doubts about whether they are truly competitive or effective channels of accountability mean such contests are hardly important moments at all for environmental NGOs. Indeed, interviews by one of the authors of this chapter with a range of NGOs in 1993–4 suggest that becoming involved in electoral activity is the 'kiss of death' for their campaigns. Such attitudes are an inevitable reaction in a polity where the United Malay National Organization (UMNO) and the ruling *Barisan Nasional* (BN) coalition circumscribe the activities of Opposition parties such that there is 'little opportunity . . . to effectively challenge let alone replace, the government' (Crouch, 1992, p. 27). With so little chance of an Opposition party winning Federal elections, there is little to be gained and much credibility to be lost in campaigning with parties that confront the BN's institutional monopoly of the national interest. The Environmental Protection Society of Malaysia (EPSM), for example, claim that they had positively to exclude Opposition parties from a place in the campaign to protect the Batu Caves in 1986 or otherwise risk their campaign being submerged in party political strife.

In any case Parliament is not regarded as an effective forum for public policy accountability. MPs' questions are tightly controlled, the activities of parliamentary committees are limited and usually operate in private and legislation is frequently passed in great haste. While debate 'has usually been sharp and sometimes acerbic . . . it has seldom made any discernible impact on government policy or performance' (Means, 1992, p. 293).

Rather than targeting elected officials NGOs see their links to bureaucrats as crucial opportunities to influence policy. Some like the Malayan Nature Society (MNS) explicitly emphasize their value to bureaucrats as additional scientific expertise and their role as colleagues in policy formulation networks. As some MNS members are also bureaucrats, much of this co-operation takes place through informal contact rather than through institutionalized channels. In other cases an NGO might collaborate more formally in implementing forest policies. WWF (Malaysia) for example, counts assistance with the preparation of state Forestry Plans as a classic instance of their ability to influence an environmental issue from the inside, though they do recognize dangers to their autonomy. In the absence of direct influence channels through elected representatives, bureaucrats are seen as an important supplementary route to the final decisions of politicians. Despite such activity NGOs are circumspect about their impact because they appear to be asserting the prior claims of environmental protection against the dominant ideology of natural resource exploitation. An even bigger obstacle in Sarawak is the interpenetration between local political élites and timber business interests.

Occasionally NGOs have been able to exploit political differences between the Federal and state governments. Even though local states control timber revenues, the Federal government can affect the rate of logging indirectly by restricting the level of log exports through the authority of the Federal Timber Board. Indirectly then, Federal decision-makers can affect the overall harvest of logs or even impose a temporary ban on the export of logs. During efforts to stop further logging of the Endau-Rompin National Park in 1977, MNS used Federal opposition to the policies of the Pahang state government in their campaign with some success.

Plate 3.1 Punan Man, Batang Rejang Catchment, Sarawak: the plight of the state's hunter-gatherers became a *cause célèbre* in the late 1980s

Similarly, in the mid-1980s, NGOs found allies in the Federal Ministry of Power to assist their efforts to halt the first proposals for the Bakun dam in Sarawak which would have taken over 695 km^2 of primary forest. (This 'success', though, has to be qualified. Construction of the dam began later in 1994, only this time NGOs have lost their Federal allies as Ministry of Power officials and politicians changed their stance (Eccleston, 1995)). Switching the focus of a campaign to take advantage of an alternative power centre was also a feature of the campaign to publicize the plight of the Penan forest peoples in Sarawak. To compensate for the intransigence of the state government, *Sahabat Alam Malaysia* (SAM) organized the appearance of Penan representatives in the Federal capital and arranged meetings with politicians and officials in order to exert indirect pressure on the Sarawak government. Federal pressure eventually led to an invitation for the International Timber Trade Organization (ITTO) to send an investigatory mission to Sarawak in order to assess the impact of local forest management. Even though the outcome of the mission left many NGOs pessimistic (Pearce, 1994), SAM was able to take advantage of the ITTO presence to gain more global attention for the Sarawak forestry campaigns. Whether this global dimension enhanced NGO influence in the long term is an open question as we argue later (see also Eccleston, forthcoming).

The wider question of using the media in environmental campaigns produces waves of pessimism from NGOs almost as though the media is an entirely unimportant power centre. Open knowledge of United Malays National Organization media ownership links (*Far Eastern Economic Review*, 19 August 1993, p. 16) does set limits to an independent and critical press but there has developed a definite press interest which is not confined to environment feature pages. Reports of illegal logging abound. Press releases from NGOs are printed even on highly sensitive issues and readers' letters have provided a very important medium of influence for public debate. In fact, this tactic of using a press letter-writing campaign by respected scientists was one further feature of the Endau-Rompin campaign (Aiken and Leigh, 1992, p. 123). It does have to be said that legislation to control press coverage remains on the statute book and though this acts as a spur to self-censorship by journalists (Mahmood, 1990, p. 38), there is evidence that NGOs do keep trying to use the press to mobilize public awareness.

There are, though, major problems with press coverage in the main centres of logging like Sarawak where coverage in the *Borneo Post* during the Penan campaign was very clearly defined and shaped by the state government. Though nearly 90 per cent of that newspaper's stories were based on government press releases compared to field reports, there was at least one alternative newspaper which reversed the source proportions (Ngidang, 1993, p. 95). Although the alternative *Borneo Bulletin* is published in Brunei, it does have a significant circulation in northern Sarawak which highlights the need to consider not just the Malaysian press but the global media. SAM mobilized a more global campaign by taking Penan representatives to the Federal capital in 1987 and EPSM has used similar tactics with the *Asian Wall Street Journal*. Despite various efforts to control the circulation of foreign publications, the government on the whole has not totally blocked access to alternative sources of comment which have been used by NGOs not just for domestic purposes but also to raise global awareness and thus hopefully exert indirect pressure on policy-makers.

Using the judiciary as an alternative power centre is not common in NGO campaigns. As elsewhere, targeting policy-makers via the courts does risk delaying campaigns, though there are times when efforts to establish basic rights have seen NGOs take the risk. In Sarawak, SAM did support the test cases brought on land rights partly, it seems, to maintain the public profile of their campaign but also to ensure that the crucially related issue of land rights was not lost in the wider campaign on forests (Hasegawa, 1993). However, as an alternative independent power centre to the Federal executive in particular, the judiciary suffered a particularly severe blow to its prestige in 1988. After a dispute with the Prime Minister, courts were restructured followed by 'the impeachment of the more outspoken proponents of judicial autonomy . . . appointing in their place those judges willing to assume a more passive role acceding to the current views of the executive' (Means, 1992, p. 302). In addition then to delaying their campaigns, NGOs have little cause for optimism that a judicial target would be any different from an executive one. What is more, taking the judicial route would be one certain way to revive the view of politicians that NGOs by their very nature are disruptive subversives seeking illegitimately to challenge the views of elected representatives.

As a competing power centre to safeguard civil and political rights generally, the judiciary, even before the erosion of its independence in 1988, 'never had a reputation for deep commitment to democratic values and human rights' (Crouch, 1992, p. 25). Senior judges coming from a conservative élite were as imbued with ideas prioritizing a deferential political culture within a tradition of respect for leaders as was the political élite itself.

In principle, rights and liberties are enshrined in the Constitution, unlike during the years of British colonial rule, but in practice the executive has accumulated a 'battery of authoritarian powers' (Crouch, 1993, p. 136) which heavily qualify citizens' rights. Most of these powers have been developed from emergency provisions inherited from the British and involve sedition laws, acts which regulate political societies and legislation to control the limits of public debate within an Official Secrets Act. The use of these authoritarian powers has been variable and most often associated with periods when inter-communal strife has been represented as a threat to political stability. But even though the environmental campaigns of NGOs have had no ethnic dimension, some of their leaders have not escaped arrest or harassment when all non-governmental opposition groups are labelled 'enemies of the state'.

One key underlying problem for all public interest groups in Malaysia is the overall view of democracy and individual rights developed under the UMNO leadership of Dr Mahathir. His views on the inapplicability of 'alien' Western notions of human rights have been well publicized in his opposition to global attempts to make aid, for instance, conditional on human rights guarantees. In particular he prioritizes the social good over individual rights: 'Democracy has come to mean individual rights. This is not what democracy is. Democracy is the will of the majority . . . and it is expressed through the vote' (speech at Chatham House, London, 21.7.1987; quoted in Means, 1992, p. 198). Thus without an explicitly political mandate through elections, public interest groups like NGOs are represented as arrogant intellectuals with narrow sectarian interests who have no legitimate right to influence elected leaders. It came as no surprise, then, to find various amendments to the Societies Act designed to force interest groups into the jurisdiction of the registrar of societies or otherwise lose their right to be a social association with 'political' objectives. 'Political' as defined in the 1981 Amendment, had an extraordinarily wide remit including attempts to 'influence in any manner the policies or activities . . . of the Government of Malaysia or of the Government of any state or of any local authority or of any department or agency of any such Government or authority' (Barraclough, 1984, p. 451). To maintain their existence most NGOs did register as political societies but then faced restrictions on the scope of their activities, the nationality of their membership and their access to foreign funds.

Even registration did not prevent further attacks on NGOs as anti-democratic subversives and saboteurs (Dr Mahathir, *New Straits Times*, 22.12.1986) in the lead up to what is called the crackdown of Operation Lallang in 1987. As well as their campaigns suffering in the general move towards greater press censorship, a number of NGO leaders were detained without trial despite having no connection with the ethnic clashes that were the immediate justification for the crackdown. There were other background factors behind this 1987 burst of authoritarian activity related to the economic recession, the intra-UMNO challenge to Dr Mahathir's leadership, and the uncertainty among élites about dealing with criticism from growing numbers of middle class, professional groups.

A revival of economic growth since the late 1980s and a more secure political position for Prime Minister Mahathir has lessened the tension around the NGO campaigning. However, the 1987 episode signifies that civil rights are not secure and that highly critical campaigns run the danger of renewed restrictions on their activities.

While overt repression of NGOs may have subsided at a Federal level in recent years, this trend is much less evident in Sarawak. Unwelcome global attention on the Penan issue remains a problem in NGO forest campaigns which are represented as undermining the local economy by encouraging Northern states to boycott imports of tropical timber. The prosecution of those who blockade the activities of logging companies continues, and the organizational presence of local affiliated NGOs is restricted. Where global campaigning involves the appearance of indigenous representatives at such important fora as the ITTO in Yokohama, individual rights to travel abroad can be revoked. The 13th session of ITTO in 1992 was, for instance, marked by NGO protests about the decision not to allow a Sarawak representative to leave Malaysia.

Forest campaigns in Sarawak highlight the constraints on NGO work where their precarious political rights meet an élite which jealously guards access to timber revenues against the contentious issue of customary land rights for indigenous forest people. It is precisely these problems of campaigning in such an area that show clearly why NGOs have to adapt to their uncertain status. Some, like SAM, maintain a critical local voice whereas others like MNS or WWF(M) have apparently made 'a move away from the traditional critical to that of a cooperative approach' (Woon and Lim, 1990, p. 14). Such competing strategies have to be seen in the light of recent trends to involve NGOs in a broader Federal dialogue on environmental issues; in return for less critical campaigns in Sarawak, some NGOs have become more involved in forest conservation planning.

Given the qualifications to Malaysia's supposedly more open political context referred to already, it is not surprising to find regular channels of participation for NGOs in political life hardly in evidence. While a dominant executive still eschews open policy-making institutions, there are some signs of greater political space for NGOs to at least get a hearing on environmental issues. Though the 'space' is controlled by political élites, a more visible dialogue has been seen, certainly since the 1992 UNCED conference in Rio. The formal seal of approval for more dialogue on environmental issues was formally confirmed by Dr Mahathir in September 1992 when at a post-Rio meeting with NGOs, he declared that NGOs were no longer enemies. In practice this meeting led to more open discussions on the previously denied problems of illegal logging and the implementation of existing logging regulations.

There are of course a whole host of factors involved in this apparent change of attitude by political leaders towards environmental NGOs, some of which are incidental to the ways in which the NGOs have campaigned. There is no doubt that Malaysia's booming economy and Dr Mahathir's more secure base within UMNO have softened attitudes to policy critics. Equally, Dr Mahathir's decision to take on a global environmental role on behalf of the South as a whole was to some extent his independent political choice. But in other ways the quality of the advocacy work of NGOs themselves has raised their legitimacy in the eyes of leading politicians.

First, there is growing awareness of the quality of NGO research and their professional and scientific credentials to such an extent that some are seen as useful partners for hard-pressed forestry officials. Then, while participating in the Rio debates, politicians found to their surprise that Malaysian NGOs were not at all afraid to criticize their Northern colleagues for refusing to relate environmental problems closely enough to development constraints. As this was also a main theme of the politicians themselves, it tended to dilute impressions of NGOs as being subservient to Northern interests which was previously a main factor behind charges of their being enemies of Malaysian interests.

One could also speculate on the impact of the global process which isolated the Malaysian government as a specially vulnerable target over the environmental impact of deforestation. This forced Federal officials and politicians to address some of the vociferous criticisms being aimed at them during which it might also have become clear that local NGOs could provide the expertise to reply to the foreign challenge. Publications by WWF International were, for example, used by the government to widen the forest debate to include not just tropical deforestation in the South, but also degradation of Northern forests.

Alternatively, the dialogue could also have been a strategic decision to try to divide local NGOs from global allies. But no matter what the speculation, the outcome has been less hostile relations between NGOs and their Federal government.

Overall then, the post-Rio scene for Malaysian NGOs has secured some improvement in their legitimacy as participants in the debate about environmental issues. There is still little evidence that this participation is becoming institutionalized or unconditional, especially where the state of Sarawak is concerned. Only recently, critical comments by NGOs on the revival of the Bakun Dam project in that state were at first ignored and then firmly rejected by the Energy Minister: '[E]nvironmentalists are not qualified to make comments on the economic feasibility of such projects' (*Far Eastern Economic Review*, 28.10.1993, p. 50). In other words, when it suits those in power, they feel free locally to separate economic from environmental issues which is the exact reverse of their global position in North/South environmental politics. There is also a danger of NGOs becoming too close an insider and being co-opted into government service so that the benefits of being able to participate in policy-making carry the cost of losing the right to be critical. NGOs are as aware as anyone of such dangers and appreciate that political élites remain firmly in control of the consultation agenda.

ENVIRONMENTAL NGOS AND THE POLITICAL CONTEXT IN INDONESIA

Malaysia's giant neighbour, Indonesia, had in 1995 the largest stand of tropical 'moist' forest in the world after Brazil and Zaire. Particularly since the mid-1960s and the arrival of President Suharto, the government has encouraged the exploitation of the forest for wood and wood products and the rate of deforestation has rocketed. Forest cover declined from an estimated 120 million hectares (63 per cent of Indonesia's land area) in 1973 to an estimated 95–104.75 million hectares (49–55 per cent of land area) only 20 years later; the estimated deforestation rate in 1992 of 623,000–1.3 million hectares per year was the highest in South-East Asia (Indonesian government and NGO figures as summarized in Belcher and Gennino, 1993, p. 15).

Most of the forest area was designated either as state forest lands, which included the state-managed plantations on Java (Peluso, 1992), or as logging concessions (on twenty-year leases initially) 'consigned on a non-competitive basis to individuals closely related to the military government and its senior officials, and to business organizations controlled by the military directly' (Rush, 1991, p. 36). Certain conditions were laid down in the leases aimed at ensuring forest survival for continuing commercial exploitation. The rights of indigenous forest people were ignored. By the mid-1990s most of the original 560 concessions were held by no more than fifty conglomerates which managed both upstream logging and downstream wood processing. They were powerful indeed and, as the *Asian Wall Street Journal* (2–3.2.1990) remarked, the concessionaires 'don't hesitate to bring pressure to bear in Jakarta if local foresters are too insistent on investigating breaches of concession agreements' or collecting unpaid taxes. Illegal, untaxed and unreported logging by cheap labour on a large scale, with government standing by, has meant windfall profits for the concessionaires.

An authoritarian government which either controlled its production forests directly with armed guards (Peluso, 1993b, p. 60–5) or was in league with ruthless forest concessionaires was a powerful juggernaut. It was not account-able from time to time to voters, as in Malaysia, nor required to bother about criticism from mass organizations, as in Vietnam. The electoral system was organized to ensure that GOLKAR won. There were opposition parties – the *Partai Persatuan Pembargunan* and *Partai Demokrasi Indonesia* – but they were strictly controlled. Political activity was banned between elections and the rural population was not allowed to belong to a political party. The lack of pluralism inside and outside the state in Indonesia was remarkable, unmatched in South-East Asia. There was a virtual 'fusion of state, local and foreign capital' (Chan and Clark, 1992, p. 44), with generals, politi-cians and bureaucrats pursuing power, patronage and revenues through their control of state monopolies, concessions, licenses and subsidies (Robison, 1993). In 1980, twenty-four of the thirty-four local companies engaged in the timber business involved military interests (Bresnan, 1993, p. 212). Other 'centres of power outside the government were eliminated or appeased' (Dauvergne, 1994, p. 505). Within the state, the fusion of executive–legisla-tive–judicial power was striking, together with military penetration of the civilian apparatus. In 1990, probably more than 60 per cent of the senior officials in central government ministries were military people; in 1987, twenty-one of the twenty-seven governors were 'either active or "retired" generals or colonels' (ibid., p. 504). These connections were reflected in the Indonesian joke: 'under colonialism we had a governor-general; now we're independent, we have general-governors' (Bresnan, 1993, p. 110). The civil service grew rapidly under Suharto, and most people in it depended on the regime for their survival. Even professors were civil servants. All civil servants were expected to support GOLKAR, and many were members; for foresters, membership of GOLKAR was a prerequisite to promotion. As for political rights and freedoms, Indonesia was regularly rated Level 3 by Amnesty International in the 1980s and early 1990s (East Timor was no doubt above that, Jakarta below). Given such an authoritarian political context within which forest policies were made, who could stop the juggernaut?

There actually are environmental NGOs in Indonesia, and they are not completely without influence. Belcher and Gennino (1993) profiled forty-four indigenous ones working more or less on forest protection issues: eleven in Jakarta and Bogor, four in the rest of Java, eight in Kalimantan, seven in Sumatra, five in Irian Jaya, and nine in other areas. About 300 NGOs, including some of these forty-four, were linked together in WALHI (Indonesian Forum for the Environment). WALHI conducted research, engaged in policy advocacy work related to the environment, and campaigned on forest and other environmental issues. Its basic stance was to promote decentralized management of Indonesia's natural resources and to strengthen local NGOs. The other main national forum or networking secretariat for NGOs on forest issues was SKEPHI (NGO Network for Forest Conservation in Indonesia). It linked student and other groups and organizations working to stop forest destruction. It supported the rights of local communities to manage the forests upon which they depended for their livelihoods. One of the authors of this chapter interviewed WALHI, SKEPHI and other NGO people in Jakarta and North Sumatra at provincial and local levels in January-February 1994, and the remainder of this section is based mostly on interviews and other material collected during this visit.

WALHI and SKEPHI and other Indonesian NGOs transparently represented a position on forest and forest use at odds with that of the state juggernaut. So why didn't the state simply shut the NGOs down? It did close or harass certain environmental NGOs from time to time (in January 1994, for example, SKEPHI's offices were raided and all their computers were taken away). But it allowed the NGO sector to carry on, for at least three reasons. First, forestry people within the state regarded environmental NGOs as perhaps a nuisance from time to time but essentially of little consequence; in any event security people stayed well informed about what the NGOs were doing and could crack down if necessary. One measure of such NGO insignificance was that government reports on forests rarely even mentioned NGOs. Nor did their foreign friends; a major environmental sector review of Indonesia by the World Bank virtually ignored NGOs, remarking in passing that NGOs have 'relatively little voice in policy decision making' (SKEPHI, 1992, p. 173) in comparison to private sector entrepreneurs. Second, in an authoritarian political structure, environmental NGOs were seen by state leaders as useful pressure valves through which energetic middle-class people could let off steam. Third, a number of Indonesian NGOs were part of global NGO networks, coalitions or alliances. Their colleagues in Northern NGOs, operating in more democratic political contexts and more highly regarded there, were sometimes influential in relation to Northern governments, the World Bank and other Northern organizations whose policies profoundly affected the timber trade and international capital markets, and therefore affected the vital interests of the Indonesian state and the concessionaires. In such an international context, it would have been politically unwise for the Indonesian government to wipe out the Indonesian end of these global NGO networks.

Despite the authoritarian political context, NGOs have had some limited success in influencing forest policies. Probably the best WALHI example is its efforts over a period of years, together with local NGOs in North Sumatra – WIM (*Wahana Informasi Masyarakat*) at Medan and KSPPM (a local NGO) – to change the environmentally damaging activities of the concessionaire PT Inti Indorayon Utama, or Indorayon, in the forests around Lake Toba (for details, see WALHI, 1992; Tanjung, 1992; Purnomo, 1994: see also the chapter by Lohmann in this volume). At one point (5 November 1993), for example, a chlorine tank at Indorayon's pulp and paper plant at Poresea village near Lake Toba exploded; subsequent local struggles by affected villagers, local NGO activity and international NGO protest and letter-writing during December 1993 and January 1994, together with some pressure from one or two government departments that had been lobbied by WALHI in Jakarta, led the government to order a full audit of the company's activity, including its management of the forests that supplied the factory. In an unprecedented move, Indorayon officials met local NGO representatives on 1st February, admitted their environmental record was poor, and promised to improve their forest practices and to communicate more openly with the NGOs and the local community. During the Spring of 1994 company policy noticeably shifted. It helped also that Indorayon was currently negotiating with CS First Boston Corporation for a loan of US$110 million who were in turn being lobbied by international NGOs demanding that First Boston withdraw from the negotiations because of Indorayon's appalling environmental record.

One of the best SKEPHI stories concerns its continuing advocacy work in relation to the rapacious logging by PT ANS (*Alam Nusa Segar*) in the outstanding rainforest on Yamdena Island in the Molluccas (recent issues of SKEPHI's journal *Sethiakawan* have covered this). In 1992 SKEPHI sent action alerts to their foreign friends in the WRM (World Rainforest Movement) and beyond, urging them to write letters. International NGO pressure was matched by, and grounded in, pressure from below. Hundreds of angry people from villages on Yamdena Island attacked several logging camps; buildings and equipment were burnt. The police fired on the people. They also jailed and tortured some of the leaders. ICTI (Tanimbar Intellectuals Association, a group of people from the islands in which Yamdena is located who work in Jakarta) also lobbied in Jakarta; IMIM (Maluku Students Association) lobbied at provincial headquarters. SKEPHI helped to bring Yamdena people to Jakarta to attend sessions of the House of Representatives when Yamdena was being discussed. The matter was covered in the press. The outcome of this local, national and international effort was a policy decision announced by the Ministry of Forests in February 1993 to order PT ANS to cease logging on Yamdena for at least six months, to allow a detailed study to be carried out to assess the impact of logging on the island, its economy and ecology. NGO people were ignored during this enquiry and local people were not consulted. In October 1993 the government announced that the study had been completed and PT ANS would be allowed to resume logging, subject to certain new conditions (e.g. the rate of cut). So a slight shift in policy had occurred.

The political context is profoundly important to explaining these and other limited NGO 'successes'. Prior to the 1990s, NGOs tried to work in what they referred to as 'a climate of terror'. This had manifold consequences. For example, it produced what one NGO respondent referred to as 'an ambiguous organizational image' which weakened their advocacy position. Policy advocacy work *vis à vis* government policy meant being critical of government policy, but in that political context of 'terror' NGOs also had to be supportive of government; advocacy objectives became less than clear because they had to be expressed in politically acceptable language; advocacy meant taking sides, but the 'climate' weakened any firm position, leading to internal contradictions, e.g. NGOs operated within élitist advocacy concepts like 'working with progressive bureaucrats and political leaders' while mouthing populist positions like 'strengthen democracy', 'empower the grassroots', create 'participatory forest management structures', etc. Also, NGOs believed that effective advocacy campaigns required a sharply focused series of actions leading to a definite 'win or no win' result. But in the political 'climate' it was less than clear what success meant, and this weakened their work. In addition, communication skills and working with the media were seen as important in advocacy work, but in the political context the role of the media was to support the interests of the ruling élite and the dominant ideology of *pancasila*. Not only was media coverage of environmental issues slanted, NGOs also had limited access to the media.

By the beginning of the 1990s, however, a little 'political space' had opened up for NGOs, a little movement towards a plurality of conflicting and changing power centres providing points of leverage for NGO lobbying. It wasn't much, but there had recently been an opening, such that the government, the military, the courts and corporations were no longer completely at one. The NGOs still usually 'lost' against this formidable array, but they were beginning to achieve in some cases at least temporary shifts in policy due mainly to this shifting political context.

Finally, NGOs in Indonesia frequently 'went international' in their environmental advocacy work. They were part of international processes that moved beyond the authoritarian political context and put some pressure on it. For SKEPHI, such international network support was probably essential to its very existence, let alone any success it achieved. WALHI was a little different. They were the Indonesia representative of Friends of the Earth International, and activated that and other international networks from time to time. Sometimes, however, they were influenced by their analysis of the political situation locally not to 'go international'. For example, in the Indorayon story, they perceived that government departments were to some extent in conflict, which was advantageous to WALHI's interests. To have brought international pressure to bear, they said, might have meant driving these conflicting government factions together again in the face of an external 'threat'. Generally speaking, NGOs in Indonesia have been conscious of working at both the grassroots and international levels, and have seen such links as providing a potentially powerful combination for environmental advocacy, given the particular characteristics of Indonesia's political context.

CONCLUSION

This chapter has demonstrated a range of political obstacles faced by NGOs in campaigning to reduce the rate of deforestation. In general, NGOs face severe problems in confronting the dominant values of national élites who give priority to the short-term benefits of forest exploitation over environmental costs. With the timber trade in particular, the dominating power of consumer states in the North, especially Japan, continues to value timber at well below its social cost which in turn encourages faster rates of logging. Where logging concessions are part of local patronage systems as in Sarawak and where the military are an additional economic and political interest as in Indonesia, forestry campaigns by NGOs have to confront very powerful forces.

More regular elections in Malaysia appear to offer more political space for NGOs than in Vietnam or Indonesia, but the limits both to the openness of party competition and the power of parliament means that we would argue for differences of degrees rather than kind. Though the mobilization of political participation through *pancasila* in Indonesia or through mass organizations in Vietnam appears quite different, there are similarities with the attempts of the *Barisan Nasional* coalition in Malaysia to monopolize political activity. NGOs need guarantees of a truly independent civil society to legitimize their existence but in all three countries such guarantees are not forthcoming.

Despite these formidable barriers, NGO campaigns in Indonesia and Malaysia reveal a range of coping or even survival strategies to exploit the appearance of the slimmest of political spaces. Acute political awareness to exploit differences between Federal and state governments in Malaysia or between environmental and other bureaucrats in Indonesia, is essential. Closer involvement with the state does, however, bring with it the need to compromise, which in turn may reveal contradictions between environmental and local empowerment objectives. At another level, any shift in forest policy that emerges from closer co-operation with progressive bureaucrats requires stringent implementation if it is to be effective. Restrictions on the export of logs, for example, have been attempted in all three countries but enforcement problems at the local level in forests distant from national power centres has undermined the overall objective of regulating exploitation.

The problems of policy implementation are a very clear reminder that the political context within which NGOs campaign should incorporate not just the national but the local and global context as well. One of the key outcomes of our research is to show the limitations of focusing only on the nation–state as a unit of political analysis. For Malaysian NGOs the local political context in Sarawak places different constraints on their forestry campaigns when compared to their work at the Federal level. Similarly, campaigning in places such as Irian Jaya or East Timor means that Indonesia NGOs have to accept different constraints than when dealing with policy-makers in Jakarta. Campaigns in either country that seek a louder voice for indigenous forest peoples have to operate in a political context that denies the local claims of such peoples to land rights or forest livelihoods.

Forging local–national collaboration between NGOs is one strategy used to bridge the gap between differing political contexts within a nation–state as the work of SKEPHI in Indonesia and SAM in Malaysia has highlighted. Nevertheless, limits to the power of policy-makers in Hanoi, Jakarta or Kuala Lumpur still mean that NGOs have to operate in different ways in different parts of the same country.

The global political context intrudes into national NGO campaigning in various ways not least because of the connections made between tropical deforestation and the international problems of climate change or biodiversity loss. This has led to a varied pattern of North–South collaboration amongst NGOs to exert joint leverage on external political actors who finance forestry projects or oversee the tropical timber trade, and of those in the media who can spread awareness of local environmental degradation. Leverage of this kind is designed to exert indirect pressure on governments in Malaysia or Indonesia and compensate for the limitations in national political systems.

As we have shown, however, NGO campaigning on the global stage does not always provide such compensation and in some cases it may be counterproductive (Eccleston, forthcoming) The internationalization of the Penan campaign has made the work of local NGOs in Sarawak much more difficult and similar possibilities encourage WALHI to be circumspect about how far they involve international networks in their Indonesia campaigns.

Another element of uncertainty concerns just how much leverage can be exerted through external institutions. In a country more reliant on foreign investment, Indonesian NGOs can at least find accessible targets through Northern NGOs working around the World Bank or the US Congress. In contrast, more development projects that affect Malaysian forests are internally financed or are the product of impenetrable bilateral funding arrangements especially by Japan. The gross limitations on political space for NGOs in Japan make them much less significant as partners even at the headquarters of ITTO in Yokohama where the Malaysian state is, in any case, a dominant influence.

The work of environmental NGOs is undoubtedly affected by the constraints imposed by their national political context for which sometimes global campaigning may compensate. But the outcomes of their campaigns are perhaps more importantly dependent on the local context within which policies are implemented. All three spatial dimensions of the political context should be considered in relating NGO advocacy work to their influence in resolving environmental problems.

4

JAPAN AND SOUTH-EAST ASIA'S ENVIRONMENT

Owen Cameron

The aim of this chapter is to discuss the Japanese influence in the dramatic environmental transformation of South-East Asia, and the parallels with and lessons from Japan's domestic experience.

Throughout South-East Asia development and population growth have resulted in extensive non-sustainable exploitation of the region's coastlines, forest, rivers and wetlands. The scale and rate of destruction have varied among the ten nations that make up the region, but in general has increased as development has proceeded and as the impact of international markets has been felt more keenly.

The natural environment has undergone a rapid transformation. Species extinction and habitat destruction, combined with decreased yields of fish, crabs and clams, indicate the stresses on the natural environment. The gradual loss of commons is a constant theme, and vast tracts of tropical rainforest have been cleared. The introduction of high technology resource extraction has disrupted traditional resource-use patterns.

These patterns of environmental change also have social and political dimensions. As economic activity in South-East Asia shifts from an agricultural to a manufacturing base, accompanying social changes such as urbanization and increased consumption of energy and resources have had major environmental impacts. Environmental management strategies have been heavily influenced by the political framework in which they are expressed, and this has influenced the extent of sustainable use of the region's natural resources and equitable distribution of the benefits resulting from their development.

Japan has been an important influence in environmental change in South-East Asia through the impacts of consumption, overseas economic activity, and international environmental policy. For sustainable development in South-East Asia to be a reality it is vital that these impacts, and the lessons of Japan's domestic environmental experience for the South-East Asian region, are much more fully discussed and understood.

THE POLITICAL ECONOMY OF JAPANESE ENVIRONMENTAL MANAGEMENT

Japan is a major economic force, contributing approximately 15 per cent of the world's economy (Lauber, 1994). Japan's consumption of resources and status as the world's foremost creditor and aid donor nation mean that, along with the United States, Japan has a profound impact on the world's environment. Japan has traditionally had a major presence in South-East Asia (Armour, 1985), although the Japanese impact has varied from country to country depending on a number of factors including resource abundance, political connections and relative stage of development. Economic development within Japan has been accompanied by a dramatic drop in self-sufficiency (Ui, 1992) and an increasing reliance on imported resources, many of which have come from South-East Asia (Kirby, 1980).

Japan's influence on South-East Asia has increased as the Japanese economy has developed, and as Japan has progressively become more internationally orientated (Hook and Weiner, 1991). Since the 1970s the Pacific Rim economies have become progressively more integrated, with Japan leading the way (Rudner, 1995). Currently, two-thirds of all Japanese trade with developing countries is with Asia, and much of this is with South-East Asia. Negotiations towards forming economic trade blocks in the Asian region, in which Japan's economic influence has a pivotal role, have serious implications for the environment.

Japan is also a vital player in the geopolitics of the region and is often, albeit cautiously (Murdo, 1994), discussed as a potential political leader to match its economic position (Narasimha, 1990; O'Tuathail, 1993). However fears of economic imperialism by Japan remain prominent in some South-East Asia circles (Bartu, 1993), and extensive debate and criticism have surrounded the charge that 'Japan. . .is deeply implicated in the environmental destruction of Asia' (Japan Federation of Bar Associations, 1992, p. 31). Most recently, Japan's international environmental activities have come under intense scrutiny, with the country declared an 'eco-outlaw' (Begley, 1989) in the late 1980s. Demonstrations against Japan's environmental impacts shocked Japanese officials who subsequently moved to reform environmental policy, and were successful to the extent that the early 1990s have seen discussion surrounding Japan's potential as a regional and global environmental superpower (Pitman, 1992).

Japanese environmental management strategies have reflected Japan's 'resource-poor' status, and been shaped by specific domestic constraints. The environmental history of Japan has been heavily influenced by several geo-physical factors, including Japan's geography, the high incidence of natural disasters in Japan, and the influence of concentrations of population and industry. All these factors have influenced attitudes towards nature and the environment among the general public and policy-makers. Environmental change has been a function of social, political and economic change. However, environmental problems have generally been addressed in a reactive fashion, emphasizing technological solutions whilst neglecting social and political dimensions. The Japanese political establishment has consistently neglected environmental considerations during development, resulting in extensive, long-term, and often irreversible and unnecessary social and economic costs. This neglect has been encouraged by the system of political funding (whereby politicians require large cash donations in order to get elected), and extensive close links between politicians and development interests. The formation of a political democracy after the Second World War was vital in allowing public opinion over pollution to be expressed and acted on, especially at the local level.

Environmental degradation, pollution and conflicts over resource-use have been an intrinsic part of Japan's industrialization (Ui, 1992). Japan's environmental problems have shifted qualitatively as the Japanese economy has developed and diversified. Pollution has varied as the industrial base of the Japanese economy has changed, and as social and political structures have affected the degree and types of pollution acceptable to the Japanese public (Cameron, 1995). Environmental issues have increasingly been influenced by international political and economic factors, and recently increasing environmental concern has been reflected in a changing Japanese stance both domestically and with respect to international policy (ibid.).

JAPAN AS AN AGENT FOR ENVIRONMENTAL CHANGE IN SOUTH-EAST ASIA

Given Japan's record of extensive domestic environmental degradation and the country's growing international and regional impact, this chapter now examines the role of Japan in environmental change in South-East Asia and suggests three dimensions to this role: first, as a consumer of, and market for, the region's natural resources; second, in terms of Japanese economic activity and investment in the region; and finally, the impact of Japan's international environmental policy on South-East Asia.

Japan as a consumer of natural resources

Japanese consumption of resources from South-East Asia has had a great impact on both the economic development and environment of the region. At issue here is one central fact – namely that whereas South-East Asia is rich in natural resources, Japan is one of the world's most resource-poor and least self-sufficient countries (Ackerman, 1953). Hence resource acquisition has been a vital part of Japan's national economic planning and foreign policy throughout the history of the country's industrialization (Ozaki and Arnold, 1985), and South-East Asia has been a natural focal point in this strategy (Olsen, 1970; Shiraishi and Shiraishi, 1993).

To appreciate this situation, Japan's role as a consumer society needs to be emphasized. Japan has a population of 124 million, approximately 2.19 per cent of the world's total, yet in a fashion similar to the United States, and developed countries in general, consumes a disproportionate amount of the world's natural resources. Whilst remembering Japan's consumption patterns are similar to those of other developed nations, it is important to acknowledge that geophysical realities and resource limitations have constrained domestic agricultural production, increasing the reliance of Japan's modern society on food, energy and raw material imports. However, most of Japan's transition to mass importer and consumer has been a direct consequence of industrialization, development, and accompanying degradation of the domestic environment. The attraction of primary industry as a profession has decreased dramatically as traditional social and economic structures have changed within Japan to produce an affluent consumer high-technology society. This has led to increased imports and consumption of natural resources from South-East Asia.

Currently Japan produces only 30 per cent of required substantial foods and 49 per cent of the country's calorific intake (NGO Forum Japan, 1992), the lowest of any advanced nation despite using a significant proportion of global chemical fertilizer output. To meet domestic requirements Japan imports as much grain as all Africa needs. Similarly, the emergence of an affluent consumer high-technology society has led to dramatic increases in energy consumption. Japan is the world's second-largest consumer of fossil fuels, virtually all of which are imported (Sun, 1988). Approximately 80 per cent of Japan's energy is supplied by foreign sources. To place the impact of Japan's 124 million human consumers in perspective, Brazil (1992) notes that the imported food requirements of Japan's 5.3 million pet cats alone are seriously affecting the environment and farming in South-East Asia. Japan's overseas economic activity has also led to the repetition throughout South-East Asia of patterns of social and economic change which contributed to environmental degradation and decreased self-sufficiency within Japan itself.

These issues are well illustrated by Japan's consumption of marine and wildlife products from South-East Asia. Japan is the world's largest consumer of marine products, which traditionally supply much dietary protein. Japan with 2.2 per cent of the world's population catches 15 per cent of the world's total fish catch. Japan also is the biggest importer of fishery products, with a total of 2.54 million tonnes in 1990 (NGO Forum Japan, 1992). Imports have risen for each of the seven years to 1992, and given the fall in Japan's deep sea catches over the last ten years it is likely this position will continue. Whilst the implications of the drastic decline in global fish stocks are serious for Japan's discerning marine product consumers, they are potentially disastrous for many of the world's regions, including South-East Asia, where marine protein is vital for survival and economic revenue to pay off foreign currency debts. In Asia as a whole over a billion people rely on fish as their main source of protein, and this consumption has global implications (Marsh, 1992).

Coastal environments in South-East Asia, traditionally of great nutritional importance, are being rapidly devastated. Mangrove forests are being cleared, often for fuel or new land, and South-East Asia's vast wetlands are under strain from similar anthropogenic disturbances, most notably in Malaysia (Rush, 1991). Rapid coastal degradation has accompanied dramatic drops in yields from devastated fish stocks, with catches increasingly being composed of juvenile fish. Coral reefs have declined due to the impacts of pollution, tourism, fishing and other commercial activities.

Much of the region's coastline is now taken up by business geared for export, such as intensive fish and shrimp farms, and this transformation has been partly driven by demand for marine products for the Japanese market. Some 90 per cent of Japan's shrimp and prawn consumption is imported from South-East Asia, and like most marine imports to Japan the bulk of this is consumed in restaurants.

South-East Asia's coastline has also been affected by social and techno-logical transformations facilitated by Japanese activity. The use of intensive fishing techniques by highly capitalized fishing companies has threatened the sustainable traditional use of resources by small-scale fishers, and Japanese tourism and resort development have introduced environmentally damaging technological activities throughout the region.

Japan is also a leading importer of wildlife products, and is the second largest consumer of wildlife products after the USA. Many of these wildlife products come from South-East Asia, with brightly coloured birds and fish imported as food, ornaments and pets. International anger has erupted over Japan's reluctance to ban the trade in the endangered hawksbill turtle. An average of 18 million tons of hawksbill turtle shell were imported annually to make traditional ornamental hair combs (Begley, 1989), and the trade was only banned in June 1991 after the threat of sanctions from the United States (Lauber, 1994). Recently Japan's role in the Asian trade in bear gall bladders and body parts has come under extensive criticism (de Wok and Pearce, 1995). Endangered Sun and Black bear cubs have been seized in Thailand, and bear cubs are being illegally smuggled from Burma to satisfy Korean culinary preferences (*Geographical Magazine*, 1995).

When examining the influence of Japanese consumption patterns on South-East Asia, it is important not just to consider resource constraints and political and administrative structures within Japan. The role of socio-political and economic factors in the producer country, and the impact of the interna-tional political economy on market structures in both countries, must also be taken into account.

A case in point is that of the tropical timber trade. Japan has a long history of innovative pre-industrial forest management strategies (Totman,

1989). Despite this tradition, and the fact that Japan is still 67 per cent forested today, the second highest level of any industrialized nation, the domestic self-sufficiency rate in timber had fallen to 50 per cent in 1969, and by 1988 had reached 29.2 per cent (NGO Forum Japan, 1992). For the last 20 years Japan has been the world's top importer of tropical timber (Bartu, 1993).

Within Japan the emergence of the international trading system, labour and economic changes accompanying development after the Second World War, and characteristics of the Japanese forestry ownership and management system, have conspired to leave the domestic forestry sector in a slump whilst domestic demand is met by imports. Thinnings from plantations that have taken 20 years to grow are left to rot in Japan, whilst tropical forests, notably in the Amazon, Siberia, and South-East Asia, are disappearing completely under logging for importation to Japan. This trade has resulted in the degradation and loss of economic value from Japan's domestic forests, whilst at the same time decimating South-East Asia's invaluable stocks of genetic diversity (Wilson, 1992).

To meet Japanese demand, vast areas of rainforest have been logged first in the Philippines and Thailand, and subsequently in Indonesia and Malaysia (Hurst, 1990). Extraction has been highly inefficient – on average around 50 per cent of forest has been destroyed to extract 10 per cent of timber (McDowell, 1989). The Philippines used to have 16.5 million hectares of forest, now only 5 million is left, and of this only 1.2 million is virgin forest. Up to 1973 the Japanese timber industry annually turned 20 million cubic metres of Philippine mahogany into cheap Japanese plywood. Thailand had an annual rate of deforestation of 2.6 per cent between 1965 and 1985 (*New Internationalist*, 1995), and it is estimated that there is only around 15 per cent of primary forest left. Malaysia's forested area has declined from around three-quarters of national land in 1958 to under half today (McDowell, 1989).

Producer nations have often received minimal economic benefits from this extensive deforestation of their natural resource base. Japanese trading corporations have focused on short-term returns, thereby placing the onus on the producer countries to ensure that it receive adequate recompense for timber resources, something they have not always been able to do. Kuroda and Nectoux (1990) estimate that between 1970 and 1982 total taxes from the Philippines forestry sector represented about only 14.8 per cent of the total reported export value of timber products. Deforestation has also had profound social and environmental impacts – soil erosion, floods and landslides have all increased, and indigenous people who depend on forest products have had their livelihoods threatened. In Cambodia natural irrigation systems have been altered by deforestation, leading to the failure of the rice harvest and the threat of starvation for many communities (*Geographical Magazine*, 1995), and logging in Sarawak has been accompanied by extensive soil erosion. As a result deforestation has become a major political issue in many countries (Hirsch and Lohmann, 1989).

Corruption is also widespread in the timber industry, and the forestry sectors of both Japan and South-East Asia as a whole have suffered in the economic long term because of the non-sustainable use of resources by self-enriching corporate and political élites. Bribery, illegal exports and price fixing have all resulted in an extensive loss of revenue for producer countries, yet many government officials, politicians and private enterprises have made fortunes from the timber trade to Japan. In the Philippines cynics have claimed that the ban on log exports introduced by the Marcos regime in the 1970s was designed to allow total control over kickbacks and concessions in the timber trade (Bartu, 1993). Japanese companies have been accused of fuelling this corruption by paying bribes and kickbacks to officials in South-East Asia (ibid.), although this is hardly a surprising revelation given the history of corruption in Japan's domestic political system.

Further, non-sustainable logging has spread throughout the region, and practices initiated by Japanese companies are now being followed by other nations whose domestic forests have been devastated. Chinese-Malay businessmen have become involved in logging around South-East Asia, following Japanese practices in focusing on raw logs, discriminating against processed forest products (Holliman, 1987), and moving operations once a country introduces regulations to exert some control over its timber exports. Thai firms are operating in Laos, and the Thai and Burmese governments reached an agreement to log Burma's forests in 1989. However due to rapacious logging by Thai firms the logging concessions were cancelled in late 1993. Thailand is now a net importer of timber, with the Philippines close to being one, and Peninsular Malaysia is relying on imports to maintain milling production levels. Only Indonesia has large untapped reserves, and these are being rapidly depleted, so much so that 'a complete exhaustion of economically viable wood stocks in the region has been predicted for the year 2000' (McDowell, 1989, p. 311). Continued high levels of raw log production are thus not a reflection of improved efficiency or management of forest resources, but represent the fact that an increasingly wide area has been subjected to non-sustainable short-term extractive logging.

International market pricing systems have facilitated non-sustainable exploitation of the resources of the least politically and economically advanced countries at a fraction of their real value, and Japan's extensive utilization of imports reflects these short-term price advantages. Thus whilst acknowledging that Japanese consumption is at least indirectly responsible for much deforestation, it is necessary to consider the problems of the timber trade in light of the political economy of the region as a whole. The demand for imported timber has recently been fuelled by a drive to increase domestic consumption within Japan, initiated in response to international trade tensions focusing on Japan's balance of payment deficits. Further, since the 1970s the economies of the Pacific Rim countries have become more interdependent, facilitating expanded intra-regional trade.

The political economy of resource use within the producer country must also be taken into account, at both the national and local levels. Deforestation in South-East Asia has been a function of both the timber trade and population pressures (as marginalized communities search for land and fuel). The distribution of any economic benefits from logging is thus a key area. Moreover, the structure of local politics among the resource producing areas has played a crucial role in facilitating non-sustainable rates of timber extraction (King, 1993c; Bartu, 1993; Cleary and Shaw, 1994), and the relationships and priorities among different levels of government in the producer country are also important. For example, in Malaysia, the state of Sarawak obtains the majority of its income through forest development, whilst the federal government controls oil/gas reserves (King, 1993c; Tsuruoka, 1994). Changing this relative balance and giving the Sarawak state government more income from energy resources might remove some of the short-term economic pressures encouraging non-sustainable silvicultural practices.

Any solution to unsustainable logging must involve changing Japan's discriminatory policies towards developing countries' processed wood products (Holliman, 1987), the market systems which allow companies to log timber at a fraction of its real worth in South-East Asia whilst Japanese timber rots, and the socio-political structures in producer countries which have reinforced trends towards non-sustainable use of timber resources. Vietnam, where forests were devastated as a result of the war, illustrates that practical steps can have profound results. Every schoolchild in Vietnam planted one tree each year, and the country's forests recovered (*Network First*, 1995), even if they are now under threat from the same short-term market forces that have devastated forests throughout South-East Asia. Conservationists in Japan have proposed a timber protection act (*Japan Times*, 1995), and Japanese importers of tropical timber could follow the lead of West Germany and agree to take logs only from sustainable concessions.

Thus, as the case studies of marine and wildlife products and the timber trade have illustrated, the increasing impact of Japanese consumption on South-East Asia's environment is tied to a number of factors. Most of Japan's domestic laws, social institutions and academic fields reflect values based on the exploitation of resources (NGO Forum Japan, 1992). These values have combined with a lack of appreciation of the importance of biodiversity, comparatively low levels of self-sufficiency worsened by domestic environmental degradation, and the desire for development in producer countries, to result in non-sustainable exploitation of South-East Asia's resources. Yet the example of the tropical timber trade illustrates how, in considering the Japanese influence on South-East Asia, it is necessary to place consumption in a broader context, considering social and political structures in South-East Asia as well as the role of international markets.

It is also important to place Japanese consumption in a global context. The environmental impact of Japanese consumption of resources is generally more prominent than that of other developed nations because of Japan's domestic resource constraints, and perhaps also because of historical legacies left over from Japan's rapid and extensive use of resources during the periods of rapid industrialization. However, Japanese per capita consumption of resources is generally less than other developed nations, most notably in comparison to the USA, and the Japanese example only really serves to highlight the structural macro-economic and socio-political problems inherent in consumption issues between developed and developing countries. Despite changes in attitude as a result of the United Nations Conference on the Environment and Development (UNCED) and increasing evidence of the economic costs of bad environmental management, the central problems of over-consumption and structural constraints on sustainable use of marine and wildlife products, timber, and other resources remain. Until these are addressed, the future for South-East Asia's forests and abundant resources looks far from sustainable, and the deleterious impacts of Japanese consumption on environmental change in the region will continue.

Japanese overseas economic activity/foreign direct investment (FDI)

In addition to consumption, Japan's influence on environmental change in South-East Asia also extends to economic investment in the region. Japanese economic activity has had a dramatic impact globally, and has played a vital role in South-East Asia's development, especially in the case of the seven members of the Association of South-East Asian Nations (ASEAN). Japanese overseas investment has focused on South-East Asia in particular, where it has taken many forms but in general concentrated on resources, factories and distribution. The economic relationship between the South-East Asian nations and Japan has changed as the respective economies have grown and developed, generally becoming more interdependent since the 1970s. However, this relationship has also had profound environmental implications, as Japan's economic activity casts an 'ecological shadow' (Maull, 1992) across South-East Asia.

Japanese–South-East Asian relations are underpinned by geographical, historical and cultural ties, but it has been since the 1960s that resource scarcity, market and international considerations have led to rapid integration and environmental degradation. Japan relied on the emerging nations for raw materials, sites for new investment, and markets for increasing exports. Responding to the New International Division of Labour, processing industries relocated to South-East Asia. Overseas investment in the region grew 3.6 times between 1974 and 1981 (Kitazawa, 1990), and this period saw a first wave of Japanese investment and economic activity concentrated on resource development and labour-intensive manufacturing. 'Development-import' schemes took advantage of cheap labour, land, energy, and resources in the host country (Park, 1979). To facilitate an industrial structure suited to these schemes much Japanese Overseas Development Assistance (ODA) concentrated on developing energy supply and infrastructure in the South-East Asian nations. However, the environmental and socio-economic impacts of the Japanese presence led to anti-Japanese riots in Malaysia, Indonesia, Thailand and the Philippines in 1974.

In response, Tokyo's 'New Policy' pledged increased aid to, and exports from, the region, and since then Japan has consistently been the largest provider of economic assistance to South-East Asia as a whole. Following the shifts in Tokyo's attitude, South-East Asian leaders became interested in copying the Japanese model for development. Singapore ran a 'Learn From Japan' campaign from 1978, and the Malaysian Prime Minister encouraged his nation to 'Look East' in 1982. The 1980s also saw increasing trade tensions resulting from Japan's failure to increase its imports of light industrial products from the rapidly developing economies in South-East Asia (Kitazawa, 1990).

Following the Plaza Accord of 1985 (an agreement between the United States and Japan whereby the yen was allowed to appreciate in order to address Japan's balance of trade deficit with the US), a second wave of investment went to the region, notably in manufacturing of consumer and electronic products, and tourism and resort development. Between financial years 1986–1990 Japanese foreign direct investment in ASEAN nearly doubled (Grimm, 1992). Whilst not always having as obvious an environmental impact as the primary extraction and processing industries, the shift throughout much of South-East Asia to a manufacturing export-led economic base has been accompanied by social changes which in turn have profound implications for the environment.

Although most FDI in the region has originated from Japan, since 1986 the Newly Industrializing Economies (NIEs) have started to invest along lines similar to Japan in China and the ASEAN developing nations (Rudner, 1995), and environmental problems resulting from this investment are spreading throughout the region. Japan's economic activity and investment have stimulated industrialization in other countries in the region, notably in the NIEs, which in turn affect other nations in a 'flying geese' (Craib, 1994) formation of development. The Vernon–Akamatsu 'product-cycle' model of economic development predicts industrialization in these nations is tied to the pace of industrial development in Japan, and that they follow Japan through a cycle of industrialization as Japan relocates production in which domestic facilities are no longer internationally competitive (McDowell, 1989). As South-East Asian nations move through the cycle they will encounter environmental problems similar to those faced by Japan. However, there is no guarantee that an entire cycle will be completed and the economic benefits received in full. 'Develop and clean-up' strategies may not be viable where a nation's economy is heavily reliant on natural resources of natural processes (as in the case of Malaysia). Alternatively, the domestic politics in a country may mean that only so much pollution will be tolerated by the public in the name of economic development. Moreover, the increasing influence of the international political economy and international environmental movement acts as a confounding variable, meaning the model may have limitations in its ability to explain the links between Japan's industrial development and environmental degradation in South-East Asia.

The preceding discussion illustrates that Japanese foreign economic activity and investment have reflected both the impact of domestic resource constraints on consumption and industrial activity within Japan; and the rationalization of labour and production during structural changes in the Japanese economy in response to the international political economy and domestic socio-political change within Japan. Japan and the South-East Asian nations have had specific economic objectives which were not always complementary. Criticism from South-East Asian nations that Japanese economic activity in the region has encouraged dependency, taken a short-term perspective, and focused on resource acquisition, can be viewed as an inevitable consequence of these tensions and Japan's comparative economic strength. Japanese FDI can actively dissuade, or at least remain indifferent to the development of industrial sectors in the producer country. Japan has used it in South-East Asia to facilitate resource control and acquisition in key sectors (Wu, 1977), and even to act as a disincentive to cartel formation among producing nations (Kitazawa, 1990).

Two main waves of overseas investment have been highlighted, each following structural readjustment in the Japanese economy and each representing the relocation of specific economic activity to South-East Asia. As well as facilitating a growing interdependence between the regions' developing economies and Japan, each wave has had profound direct and indirect implications for the region's environment. Direct implications stem from the nature of the industrial activity itself. Indirect implications reflect the short- and long-term effect of changes in the social structure of countries, and the resultant environmental impacts of these changes.

The rapid growth of the Japanese economy during the 1960s had been due to the expansion of heavily polluting energy and raw-material intensive industries. As a result of public outrage and court rulings on pollution, the early 1970s saw a rapid tightening of pollution controls in Japan, with victim redress and the installation of pollution control equipment placing an economic burden on some of the highly polluting industries. The 1973 oil crisis led to an increase in the cost of raw material processing and imported energy which, in conjunction with the changing socio-political conditions due to Japan's pollution experience, speeded up structural adjustment in the Japanese economy and reinforced the already prominent trend of primary processing industry moving overseas. Local subsidiaries and joint ventures were set up, notably in South-East Asia, and supported with government ODA finance and technical assistance.

This first wave of Japanese corporate activity did bring benefits, and was of great importance in the emergence of South-East Asia's NIEs, but in many cases it was accompanied by extensive environmental destruction and economic tensions which did little to further sustainable development in South-East Asia. As would be expected from product cycle theory, environmental destruction in many cases has been very similar to that experienced within Japan during comparable stages of industrialization. Many industrial practices and stagnant sectors no longer tolerated or efficient within Japan were exported to South-East Asia with little consideration of environmental and social impacts. Lessons from Japan's pollution history were ignored as pollution was exported for short-term economic gain.

An illustrative example of the environmental problems with this investment is that of Mitsubishi's involvement in the Asian Rare Earth ore-refining company in Malaysia. The political climate against pollution in Japan necessitated installing expensive new equipment to process dirty ores (Gross, 1989). In response, Mitsubishi moved its ore-processing operations to Malaysia in 1972. Mitsubishi Kasei Corporation, a Japanese-Malaysian joint venture ore-refining affiliate of Mitsubishi Chemicals Ltd. was set up near Ipoh City in Perak State (the northwestern part of the Malaysian peninsula). No environmental impact assessment (EIA) was conducted, and local people were not consulted (interview with Kojima Nobuo, Japan Federation of Bar Associations Pollution Committee Member and lawyer involved in ARE case, Tokyo, 15.4.93). It began operating in 1982 and discarded radioactive thorium waste from the production process, contaminating the local vicinity. This was followed by the appearance among the local community of leukaemia victims, an increase in the rates of miscarriages and infant deaths to three times the national average, and migration of local residents because of anxiety over radioactive contamination. Workers at the company plant have registered radiation levels in excess of limits recommended by the International Committee for Radiation Protection. In desperation village residents resorted in 1985 to lawsuits to get the company closed down. Mitsubishi reduced their share in the company after the story broke internationally around 1991. A team of Japanese lawyers went to visit the village, and subsequently took local victims back to Japan to raise publicity and meet corporate and government officials (Kojima interview, 15.4.93). Whilst expressing 'regret' at what had happened, the Japanese Foreign Ministry said it could not interfere in another country's domestic affairs. Mitsubishi denied any liability as it was now no longer a major shareholder.

This case is particularly interesting as it provides not only an example of Japan's economic investment leading to severe degradation, but also a clear example of how to exert pressure on Japan, and how reluctant the Japanese authorities have been to pass on the lessons of their own polluted past and even ensure similar incidents do not occur in other countries. Furthermore, it is by no means an isolated example. Slag outflow from copper mines, in whose operation Japanese trading companies are involved, has caused serious downstream damage in a number of South-East Asian countries. The patterns of pollution are almost identical to those in Japan's early phase of industrialization, when air and water pollution from mining activity led to serious environmental damage and social protest. Mercury poisoning in Djakarta Bay has been reported to be very similar to the early stages of what happened in the 1950s and 1960s at Minimata in Japan's Kumamoto Prefecture, during the expansion of Japan's chemical industry. Asahi Glass has also been criticized for overseas pollution export (Pearson, 1987), and textile factories which discharge harmful acetate dyes have been moved to Thailand (McDowell, 1989). Kawasaki steel transferred heavily polluting sintering operations to Mindanao in the Philippines (NGO Forum Japan, 1992), but neglected to install basic pollution prevention procedures and technology used in Japan, resulting in extensive pollution and damage to human health.

This first wave of overseas economic activity and FDI has thus not always aided the sustainable development of the recipient nation. However, once again Japanese activity needs to be viewed in a wider context. Japanese companies were of course by no means alone in taking advantage of weak environmental regulations in developing nations – subsequently, other NIEs have followed Japanese investment patterns in the region, with similar environmental consequences. It is also necessary to acknowledge that in many cases developing countries failed to obtain the full benefits from FDI due to their vulnerability in the international political economy, and corruption and patronage within their national governments and polity (as well as the investing Japanese firms). The ability to co-ordinate policy was hampered by conflicts between international market structures and national development priorities, and national versus regional economic preferences.

Following the Plaza Accord of 1985 a second wave of Japanese investment and economic activity moved to South-East Asia, again with profound environmental implications. Much Japanese manufacturing relocated to South-East Asia, a move which has reinforced the trends towards increased urbanization, energy and resource consumption and development of infrastructure in South-East Asia. Again, ODA has been used to facilitate this shift. Another prominent trend since 1985 has been increasing Japanese investment in tourist and resort development throughout the region. The potential impacts of this investment type are of particular relevance given current debates on the sustainability of tourism as a form of economic development (Cater, 1995). It is important to consider the dimensions of distribution of economic benefits from tourism, and whether there has been adequate financial payment for the long-term value of lost environmental commons (Hitchcock et al., 1993). Uneven frameworks for resolution of environmental conflicts mean that small landowners and farmers/fishers often receive small compensation for loss of land or natural resources, and, indeed, are often not aware of the true commercial value of such resources, as they have relatively little information and political power in comparison to organized and well-connected corporations promoting development.

As was shown in the case of pollution export, our understanding of Japanese overseas tourist investment is greatly aided by reference to domestic events within Japan. The environmental impact of Japanese tourism has increased sharply following the yen appreciation in 1985 under the Plaza agreement, which strengthened the spending power of Japanese consumers. In June 1987 the Comprehensive Leisure Area Facilities Law (Resort Law) was passed in Japan, which promoted resort development by introducing the deregulation of development restrictions.

The general patterns of resort development within Japan are well illustrated by golf resort construction, which was one of the most prominent growth areas under the Resort Law (Yamada, 1990). Golf is the most rapidly growing property sector in the world, and its epicentre is Japan. The game was originally developed in Scotland, and in the tropics uses vast amounts of water, chemicals and often non-native greenery to create an entire artificial ecosystem (Kingsnorth, 1995). As of January 1995 at least 2,016 golf courses were operating in Japan with 800 currently under construction (ibid.) and many more planned. In comparison Montana, which has a similar land area to Japan, has only 79 golf courses. The clearing of land for golf in general, and in state forests and forest reserves in particular, has aroused much public concern, as has the safety threat from chemicals and the drain on water resources implicit in course management. For example, in Hatoyama golf courses have replaced 20 per cent of forests, degraded habitats to the extent that one endangered species of hawk has moved from the area, and polluted rice paddies (Seymour, 1993).

Golf is big business in Japan and has become a national obsession. The total revenue of Japanese golf clubs in 1988 was over US$6.6 billion (Einarsen and Rodgers, 1993). Average membership fees are US$100,000, rising up to 3 million dollars for the top clubs, and green fees are on average around US$270. The dominant system of selling club memberships by demanding large deposits in advance has also encouraged much speculative development, as well as misuse of funds and corruption. In Tokyo dealers have specialized in trading in golf course memberships as a kind of speculative commodity, and the daily turnover in officially traded golf club memberships and debentures in golf courses has reportedly reached a volume that is far greater than the combined daily volume of the share markets of the Philippines, Malaysia and Indonesia put together (Bartu, 1993).

Over 15 million people play golf in Japan, but there are only 1,600 clubs. Licences to build new golf courses in Japan are proving hard to obtain. Most Japanese never have a chance to play golf (other than on a driving range), and the demand for places has pushed the price of club memberships up and created a new kind of tourist – the Japanese golf traveller. In conjunction with the continued strength of the yen, this has meant that, for domestic Japanese consumers and Japanese companies, the attraction of resort development and tourism around South-East Asia has been great.

Tourism in general is a boom industry in South-East Asia. Malaysia's Ministry of Tourism aims to attract 18 million visitors a year throughout the 1990s, and Sarawak's tourist sector is growing at a rate of 18 per cent per annum (Tsuruoka, 1994). The Vietnamese tourist industry has attempted to lure visitors away from beaches in Thailand and Indonesia, and Burma is also expecting a boom in tourist visitors (Wheat, 1995). Golf resort construction is one manifestation of this phenomenon, and the environmental costs associated with it. 'Cheap land, weak regulations and feeble local opposition' (Traisawasdichai, 1995, p. 16) have stimulated a wave of Japanese-funded golf course developments in South-East Asia. Most courses are either built by Japanese companies, with Japanese money or with Japanese golfers in mind.

Golf in Asia is generally associated with privileged status, being viewed as a 'power sport' (ibid.). The boom in golf courses has caused club memberships in South-East Asia to spiral so much that, as in Japan, memberships are currently being traded as a speculative commodity (Bartu, 1993). It is debatable as to how much of the valuable foreign currency earned from such resorts remains in the country. In Indonesia half the existing golf courses are owned by President Suharto and his family, and golf is the preferred sport of the Thai officer corps (Traisawasdichai, 1995). The game's association with the *nouveaux riche* of the industrializing nations of Thailand, Malaysia, Indonesia and the Philippines has meant that investment and development are no longer dominated by Japanese funds, as was originally the case. Burma, Laos and China are 'new converts' (Sexton, 1992, p. 109) to golf tourism development.

The first golf club resort opened in Malaysia in 1991, and the negative environmental effects are already so clear that non-governmental organizations (NGOs) and the public are protesting (ibid.). Over 160 courses have been built in Malaysia, often at the expense of tracts of rainforest, and each course uses enough water to meet the needs of 100 Malaysian farming families (Kingsnorth, 1995). In Indonesia traditional farming wetlands and nature reserves have been replaced by ninety-one golf courses (Traisawasdichai, 1995), which each year consume more water, pesticides and fungicides than all the country's farms combined (Kingsnorth, 1995). The Philippines has eighty courses, and golf courses have also started to appear in Vietnam, the Hanoi Golf Club having opened with membership fees of US$15,000 to US$50,000. The customers are predominantly corporations, and there are currently only four Vietnamese members (*Network First*, 1995). In Sarawak a course and hotel are under construction in a primary rainforest to cater for ecotourists' amenity desires (Kingsnorth, 1995).

Thailand illustrates the problems associated with golf development particularly vividly. Pleumarom (1992) notes courses have influenced land speculation, changed land ownership patterns and resulted in extensive dispossession of local farmers' lands. Golf resorts are very land-intensive, élitist, and have acted as a serious drain on water resources and polluted the environment. Pesticide poisoning has killed local wildlife, and the intensive use of chemicals has been blamed for illness and death among caddies (Traisawasdichai, 1995). Thailand has just experienced two consecutive years of record droughts and water shortages. The country's 200 golf courses are a drain on water resources vital for rice crops, yet in 1994 the government prevented farmers from growing a second rice crop whilst allowing golf courses to continue pumping water from depleted reservoirs. Golf managers also go beyond legal means, with water-theft tactics on the increase.

The impact of golf courses has been such that 1994 was declared the year of 'citizens against golf courses' by concerned NGOs. Japanese groups with more experience of the problems have been lending their experience and skills to their counterparts throughout South-East Asia (Morita, 1993). A Global Anti-Golf Movement (GAGM) was formed in 1993, linking anti-golf networks in sixteen countries. Whilst global in action the movement is biggest in South-East Asia and Japan, where it claims to have stopped more than 300 courses being built since its formation (Kingsnorth, 1995). As with other environmental issues, local concern over tourist and golf development has been increasingly expressed as political action. However, in the context of increasing desires for leisure pursuits, the increasing affluence of many sections of developing Asian nations, and the general population pressures in South-East Asia and Asia as a whole, it is unlikely that pressure for resort development will decline in the immediate future.

Overseas investment in tourist development is a function of a combination of factors in the local, national and international political economies. Structural economic change in the Japanese economy and the international economy have also encouraged overseas resort development, a move facilitated by opposition within Japan and the desire for development throughout South-East Asia. As with the timber trade and marine/wildlife resource use, local élites are benefiting disproportionately from golf resort construction, as common and limited natural resources are degraded and sold in a non-sustainable fashion. Other countries are following the lead of Japanese finance, building courses which have in general exacted a similar environmental toll throughout South-East Asia to that experienced within Japan and Singapore. There are also parallels with the socio-political dimension of overseas pollution export, the boom in course construction in South-East Asia being a reflection in part of the success of strong local opposition within Japan to golf resort construction.

Again, as was the case when considering Japan's consumption, analysis of Japan's overseas economic activity and its impact on South-East Asia shows the need to consider domestic developments within Japan, the influence of the international political economy, and the political framework of resource use within South-East Asian countries at the national and local level. Whilst Japanese economic activity has undoubtedly been invaluable in contributing to the region's development and creating economic growth, it has also been designed above all to meet Japan's needs. This has led to conflicts of interest where Japan's economic clout and the expression of vested interests in South-East Asian countries have combined to neglect environmental considerations and encourage non-sustainable resource use. Investment has also facilitated social and infrastructure change to meet industrial requirements, which in turn bring new environmental problems.

The example of golf resort construction and the pattern of manufacturing investment in the region since 1986 shows how, in a similar fashion to the timber trade, Japanese economic strategies are being followed by many developing South-East Asian nations. Moreover, industrial activities challenged in Japan because of social and environmental impacts have been consistently relocated to South-East Asian countries, a pattern that some NIEs in the region are also following.

However, it is important not to see Japan's influence as 'unique'. The problems associated with Japan's economic activities are manifestations of wider problems associated with environmental issues. Competitive factors in the international political economy, and the extent to which Japan has integrated production vertically in the region, combine to make it very unlikely that Japan will reduce its investment in ASEAN and South-East Asia as a whole. Given the impact of Japanese companies on their domestic environment, further change is required both within Japan and in South-East Asia.

Japan's international environmental policy

The final aspect of Japanese influence on environmental change in South-East Asia to consider is that of Japan's international environmental policy. Japan's economic strength and increasing political influence make it a vital player in the development of international environmental protection (Lauber, 1994), so much so that Maurice Strong has stated that the keys to the success of the new United Nations Commission on Sustainable Development are Japan and the United States (Murdo, 1993).

When considering Japan's international environmental policy it is important to remember that Japan has the interests of a developed nation, albeit with serious resource constraints. It is at this fundamental level that Japan offends many environmentalists, who see the country's development as representing an 'unbridled pursuit of economic growth' (Cope, 1990, p. 15), no matter what the social and environmental costs.

Lauber (1994) reviewed Japan's pronouncements and legislative stance with respect to the international environment from the 1972 Stockholm Conference on the Human Environment up to the 1992 UNCED Conference. She identified political rhetoric claiming a will 'to actively resolve global environmental problems', yet alongside this found 'convincing evidence that Japan's commitment to its own economic development runs more deeply than its commitment to protecting the international environment' (ibid., pp. 38–9). Her conclusion is supported by an overview of Japan's stance on a number of issues pertaining to South-East Asia. Japan is a major market for endangered species, many of which come from South-East Asia, and has generally been reluctant to embrace biodiversity concerns. This reflects a lack of domestic experience in managing biodiversity, and the dominant ideology among government circles of 'continuing to exploit species for economic gain' (ibid., p. 46). Japan has traditionally been weak with regard to accepting and complying with international conservation treaties, belatedly signing both the RAMSAR and CITES conventions and, after initial hesitation, recently ratifying the UNCED Biodiversity Convention and the World Heritage Convention.

Japan also initially showed reluctance to act on both chloro-fluorocarbons (CFCs) and greenhouse gases. The subsequent realization of Japan's geographic vulnerability to the effects of global warming, combined with technological innovation in environmental management technology as a result of domestic air pollution legislation, have made Japan one of the global leaders in responses to atmospheric pollution (Cope, 1990; Lauber, 1994). Recent problems with acid rain from China have reinforced domestic determination to address the issue, as has Japan's desire to be seen as an environmental leader. The issues surrounding greenhouse gas emissions also illustrate another point, namely, that it is slightly misleading to judge Japan's international environmental policies solely on the basis of domestic economic activity. As we have seen, much economic activity, particularly of a highly polluting nature, has been exported overseas to South-East Asia where environment quality standards are often much lower than in Japan. Japan's contribution to global environmental degradation is thus expressed beyond national boundaries, and comparative statements focusing on relative impacts with other developed countries are often misleading. International environmental policies should acknowledge Japan's overseas environmental impact and the crucial issue that, like other developed nations, Japan's emissions are high in comparison to developing countries. Given the increasing energy demands in the Asian region as a whole, and South-East Asia in particular, Japan's future stance on greenhouse gases is of vital regional importance in terms of setting an example for other nations.

Japan's influence is also vital in setting an example on other international environmental issues of importance to South-East Asia, such as fishing stocks, the proliferation of nuclear energy and weapons, dumping of waste at sea, and burden-sharing in relation to development debt (where Japan's role is of particular importance given its dominance as a creditor nation (Nishikawa, 1988)).

However, in considering the South-East Asian region, the area of Japan's international environmental policy which has had the greatest impact is that of Overseas Development Assistance (ODA). Recent years have seen extensive criticism of Japan's environmental impacts, creating an 'image problem' for Japan as the nation attempts to take a more prominent global role and be seen to meet its international responsibilities, especially with respect to environmental problems. In response to this criticism the government has taken steps to counter Japan's image as an environmental predator, a shift also illustrated by changes within Japan's ODA programme.

Japan's ODA has traditionally had various objectives, many of which have been highlighted as environmentally damaging. Criticism has centred on the structure (Orr, 1990) and motivation of aid (Ensign, 1992), and more specifically on commercialism (Forrest, 1991), a lack of burden-sharing (Kubota, 1985; Islam, 1991), and the transference of Japan's development model (Stallings and Sakurai, 1993; Craib, 1994) with its accompanying social and environmental costs. In response, Japan's ODA has increasingly been used as a vehicle to address criticisms focusing on Japan's international environmental impacts (Rix, 1993), especially in South-East Asia.

To understand the problems associated with Japanese ODA, especially when considering the impact on South-East Asia, one needs to take a long-term perspective of the historical structure and aims of the aid administration, which in turn has been a reflection of Japan's domestic economic priorities and resource constraints. Japanese aid has focused to a great extent on growing economies in South-East Asia (Rudner, 1989), reflecting their strategic and economic importance to Tokyo. ODA has been closely linked to trade, FDI and resource acquisition (Tanaka, 1986). The character of these linkages has changed as the structure of the Japanese economy has altered and the international political economy has developed. Aid was originally tied to the purchase of Japanese exports, and has frequently been given to a recipient country to ensure the acquisition and control of resources or to create an industrial structure which is amenable to Japanese overseas economic requirements. ODA must thus be considered in light of the extension of Japanese economic activity beyond national boundaries. Overseas export of heavily polluting processing industries facilitated by aid, and ODA strategies focusing on maximizing short-term resource acquisition and control, have represented an extension of environmentally damaging industrial activity beyond Japan. Bartu (1993, p. 134) sees Japan as having used its ODA 'as a sort of "seed money" for the industrialization of Asia'. Notable examples in South-East Asia include the Leyte industrial development project and the Asahan hydro-electric copper smelting project, both of which met Japan's economic needs, but brought highly questionable benefits to the Philippines and Malaysia

respectively (see Kitazawa, 1990). The establishment of eucalyptus plantations in Thailand, and the introduction of intensive fishing technologies, are
further examples where Japanese ODA has been used to secure national
economic interests.

Evidence from around South-East Asia suggests that in many cases Japanese
aid has had a notably deleterious effect on the environment – pollution
export, non-appropriate technology transfer, and a lack of consideration of
social and ecological impacts being recurrent themes. Japanese aid has often
favoured commercialism and protecting Japan's economic interests, as opposed
to encouraging sustainable development in the recipient country. In the 1980s
the Sarawak-Limbang logging road (Malaysia) and Kedung Ombo dam
(Indonesia) projects attracted great criticism because of their ecological and
social impacts. Other schemes have been criticized for a lack of consideration of local society, culture and commerce; an insistence on using Japanese
goods and materials during projects; and for encouraging the transference of
Japanese values and procedures. Examples include the construction of the
Ayutthaya historical study centre and the Bangkok culture centre in Thailand,
and the Philippine general hospital project. Such problems are of course by
no means limited to Japan, as illustrated by the UK's project for the Pergau
dam in Malaysia. Similarly US aid has often been ideologically disbursed to
various dictatorships, and much is linked to military debt forgiveness.
Moreover, the proportion of tied US aid is higher than that of Japan, and
US aid also has many commercial dimensions.

As noted, the reasons for this focus on Japan's self-interest relate to the
historical structure of Japanese aid. However its ideological motivation is also
relevant. Much Japanese aid is based on Japan's own developmental experience, with a resultant emphasis on infrastructure. Many of the problems
caused by Japanese aid are the same as those faced by Japan's domestic environmental administration, and have resulted in environmental degradation
similar to that experienced during Japan's development.

When the impacts of Japanese aid in South-East Asia are criticized, the
role of the recipient country must also be considered. Problems exist with
the request system for aid and corruption in the recipient country (as has
also been highlighted in the cases of FDI and consumption). Paternalistic
patronage-based political structures in Japan and South-East Asian nations
leave much room for business interests to transform projects into pork-barrel
schemes which do little to further sustainable development in the recipient
country.

ODA policies have been changing recently, and following UNCED the
Japanese government promised to increase overall aid and the proportion of
environmental aid to total aid. There has been a new emphasis on and increased
funding for population issues, and the proportion of grants to loans has
increased (although Japan continues to have one of the lowest grant-to-ODA
ratios). The traditionally frosty attitude of the national government to NGOs

has thawed slightly, and a small amount of support money has been put forward for NGO aid. In April 1993 the Japan International Co-operation Agency (JICA) elevated to division level status their office responsible for environment, women in development, and other global issues (Murdo, 1994). Most recently the new Environment Bill passed in the Japanese Diet in 1994 may also have further implications for Japanese ODA, as it defines environmental protection as Japan's contribution to the global community, and places ODA at the heart of this commitment (Sakurai, 1993). However whilst environmental issues are being addressed more effectively and sincerely in the aid administration, it would seem that Japanese officials are likely to continue to advocate Japan's developmental model without full reference to its social and economic costs. There have also been recent reports of public 'aid fatigue' in the Japanese press, and whilst there are some measures to encourage NGOs in the new Environment Bill, it does not allow for tax deductible donations to such groups (Sakurai, 1993), and has been extensively criticized for lacking 'opportunities for public participation in drafting environmental measures' (Murdo, 1993, p. 7). Criticism has also come from recipient countries throughout South-East Asia that aid still does not meet their needs, especially with reference to technology transfer, and is focused too heavily on Japan's economic requirements. Further changes are required to meet the continuing criticism of Japanese ODA.

The preceding overview of the impacts of Japan's international environmental policy highlights several key points which apply in the South-East Asian context. Four main recurrent themes can be identified in the impacts of Japan's environmental policy in South-East Asia – commercialism; resource acquisition and control; pollution export (directly and indirectly through technology transfer); and the fact that Japan's approach to environmental policy is changing in response to pressures from international bodies and governments and domestic groups in Japan and South-East Asia.

Criticisms of Japanese international environmental policy tend to be focused on charges relating to an economics-first and self-interest approach, with an accompanying lack of concern for biodiversity issues, and the fostering of dependency and not self-reliance in affected countries. However, this is true to a certain extent of all developed nations. Many of the problems associated with Japan's impact on South-East Asia are in fact more structural problems associated with political and social dimensions of resource use, development strategies, and distribution of the risks and benefits associated with environmental management. Comparing Japan to the USA it can be seen that each country has specific policy areas where it has predominantly acted in its own self-interest (Murdo, 1993). For Japan this is reflected in the stance on biodiversity, whereas for the US in the protection of excessive patterns of energy consumption, waste and consumption – the 'American way of life'.

Japan's international environmental policy, and criticisms of its impact, are strongly linked to domestic policies and economic requirements. The prominence of criticism of Japan is in part a reflection of domestic constraints acting on all dimensions of Japanese environmental policy. Japan's reliance on imported resources, emphasis on technological solutions, and lack of consideration of biodiversity are themes in domestic environmental policy which have in general permeated international environmental policy. Such continuity is hardly surprising as many of Japan's overseas environmental policies are in reality an extension of the ideology behind and economic requirements of Japan's own development model.

However, as has been illustrated, many elements of Japan's environmental administration and international environmental policies have been slowly evolving as domestic and international pressures lead to reactive reform. Apart from biodiversity issues and areas where Japanese resource requirements are threatened, Japan has often been at the forefront of change and some issues such as recycling are being discussed much more seriously in Japan than in the USA. Further international and grassroots pressure will be required to continue this process of reforming policy and attitudes throughout South-East Asia and in Japan.

FUTURE DIRECTIONS IN JAPAN–SOUTH-EAST ASIAN ENVIRONMENTAL RELATIONS

As this chapter shows, Japanese consumption patterns, economic activity and international environmental policies have played an integral role in the degradation of the environment in South-East Asia, and these impacts are in turn a reflection of Japan's own domestic practices. Yet both within Japan itself and throughout South-East Asia changes are occurring which augur significant shifts in these relationships.

First, in Japan itself there has been much activity with respect to environmental issues recently, with changing stances among government, scientific, bureaucratic and business circles, as well as a revival in the fortunes of the Japanese environmental movement and increased public awareness.

At the national level a new Basic Environment Law was passed in 1993, and at the local level ordinances have been passed in response to Agenda 21. There is also evidence that the perception of environmental issues is changing among some government officials, to an extent that would have been unthinkable even a few years ago in Japan. However, this is not to say that all has changed. Details of an energy tax and an environmental impact assessment law in the original draft of the Basic Environment Law were watered down in the face of opposition from business interests and the Ministry of International Trade and Industry. Japan is still dominated by money politics and it is unlikely there will be any radical change in favouring development and economic concerns over environmental ones. In 1992 the Environment Agency suffered setbacks at the hands of other ministries (*Japan Times*, 1992a), illustrating that whilst appearances and dialogue over environmental issues may have greatly changed recently, the *realpolitik* position is much as it ever was. In interviews officials stressed in mitigation that their environmental policy is rapidly evolving, and that Japan has focused since the war on rebuilding and development.

NGOs in Japan have been very active recently. During 1991–1993 there has been a rapid emergence of events and networks in Japan (Suzuki, 1993), many in response to the UNCED process, and the social status of NGOs has risen significantly. A Society of Environmental Science in Japan was formed in 1988, and Japan's technical expertise will have a crucial role in developing pollution prevention and environmental monitoring technology both for South-East Asia, and globally (Cope, 1990). However, research into the social and political dimensions of environmental problems is not as well represented (Ui, 1992), and under-emphasized in the Japanese education system.

The Japanese corporate sector has been slower to respond to environmental issues than its equivalent in the West (Cope, 1990). Keidanren, the Japanese federation of industrial organizations, released its global environmental guidelines in April 1990, a document aptly summed up by the Japan Federation of Bar Associations (1992, p. 25) as 'waffling' on corporate responsibilities. The importance of changing attitudes among Japan's business community has been recognized in the Basic Environment Law (which called for Japanese corporations overseas to take the environment of their host countries into consideration) and by international conservation organizations. The United Nations Environment Programme (UNEP) recently held a conference in conjunction with Keidanren to raise the consciousness of Japanese businessmen with regard to the importance of conservation and the role of business. Schemes set up post-UNCED to allow Japanese industry to contribute towards funds set up for environmental protection and to promote sustainable development have reportedly met with a disappointing response (Sakurai, 1993), and it is likely that of all the sections of Japanese society with the potential to influence environmental policy, Japanese industry will be the last to reform. However, the continuing emergence of the eco-business sector in Japan (*Japan Times*, 1994) does offer hope.

Many current problems relate to Japan's consumer society, and it is not sufficient to explain these solely in terms of business or government. Real change requires a shift in public attitudes to consumption, which does not seem likely to be forthcoming (despite complaining about the impacts of traffic pollution and consumption on their lives, many Japanese consumers are reluctant to consume or drive less). Such consumption-related problems are perhaps the ones least likely to change, although the activity of Japanese NGOs and consumer co-operatives indicates some willingness to act.

Further questioning of Japan's role in international society, and more specifically in the Asian region, will be vital in shaping Japan's future environmental impact. In this context the current restructuring of the Japanese political system (in response to voting imbalances between urban and rural regions, and public concern and anger over political corruption and a lack of adequate representation) has great environmental implications. Similarly, the increasing prominence of international environmental problems and agreements should reinforce the domestic trend towards considering global environmental issues. The expression of Japanese consumption and industrial patterns will also continue to be influenced by international trade factors, and the increasing interdependence of the Pacific Rim economies. However, the key to change in Japan's environmental impacts will involve increased consumer and voter awareness, as well as changing attitudes and policies in bureaucratic and political circles.

Concurrently, environmental protest groups have sprung up throughout South-East Asia since the 1970s (see also the chapter by Eccleston and Potter in this volume). Groups range from large national organizations to regional groups, with strategies ranging from litigation to direct action. Most are relatively young, or older organizations that have recently taken up environmental issues (see McDowell, 1989; Rush, 1991; Howard,1993). The influence of the Western environmental movement, and increasing international debate over environmental matters, have been put forward as influencing factors (Rush, 1991). However, the Japanese experience would suggest that the impact of progressive degradation of domestic environments, in conjunction with international environmental conferences from Stockholm to most notably the UNCED, have also played a large part. South-East Asian environmental movements are more than transplants from the West. They have been shaped by the local culture and experiences, reflect traditions not represented in the mainstream western movement, and are of growing importance in relation to global environmental problems and national politics. Thailand's water shortage and the logging bans introduced in Indonesia and Thailand are examples of the increasing role of NGOs in the region. Green groups have even gone to the lengths of threatening to sue President Suharto (*Guardian*, 1994), something not imaginable in earlier climates.

Since the 1970s South-East Asian national governments have responded to the increasing environmental crisis by passing legislation incorporating environmental goals into national plans, and creating environmental ministries and departments. Unfortunately this focus on legislation has in general failed due to a lack of relevant, appropriate and enforceable laws, as well as the fact that vested interests have often ensured loopholes have been inserted in environmental regulations. However, as the social and economic costs of non-sustainable resource use become more apparent and an increasingly affluent populace increasingly addresses post-materialist concerns, the attitude of national governments has been changing. The period since the 1980s has also seen the development of co-ordinated ASEAN/APEC (Asian Pacific Economic Caucus) environmental management forums, with co-operation dating from the 1972 Stockholm Conference (McDowell, 1989). Thailand's water crisis (Rigg, 1995) illustrates how debate has changed radically recently, with more discussion on issues of ownership and management of public goods and the increasing influence of the environmental movement in the country. Indonesia has been putting forward eco-standards (*Japan Times*, 1994), and Malaysia has claimed that polluting industry is no longer welcome. Countries are becoming more vocal in their criticism of trade and aid practices, as their export economies try to break into the Japanese markets. Quotas and export bans have been introduced on raw resources to try and encourage local industry, and prevent value being added beyond national boundaries.

However, the objective of sustainable development is still being thwarted by vested interests and corruption, and conflicts between environmental protesters and some authoritarian national governments pose an increasing potential threat for the future. Moreover, the natural resources of less developed South-East Asian countries such as Burma and Laos are under threat from the same structures and agents that have so devastated much of the environment in Thailand and elsewhere. As has been illustrated in the cases of deforestation, relocation of industrial activity, golf resort development, and ODA, Japanese policies and practices are being copied by other developing nations throughout South-East Asia. Furthermore, whilst citizens are generally more aware of environmental issues as they have been exposed to the costs of environmental degradation, there still is a great desire for continued development and increased consumption.

Thus changes in the extent of Japan's environmental impact in South-East Asia are related to the progression of an industrial consumer society in both Japan and South-East Asia, as well as the influence of the international political economy. These socio-political processes in Japan, South-East Asia, and globally, have considerable implications for the region's environment.

Japan experienced qualitative and quantitative changes in type of pollution, and the same can be expected in South-East Asia. As the region's economies continue to develop, and the local environment is further degraded, environmental problems associated with industrial consumer societies will become more prominent, while those associated with underdevelopment continue to cause serious degradation, most notably in rural regions. Energy demand poses a serious threat to the region's stability, with nuclear power proliferating (*Daily Yomiuri*, 26.2.1994). Japanese energy consumption has been expanding recently, and rapidly increasing energy demand in South-East Asia and China is predicted to continue growing. The extensive and continued degradation of Japan's domestic environment is a pattern that observers are worried is being repeated in South-East Asia. Ui (1992, p. 175) notes that Japan's loss of self-sufficiency is 'an important factor that should be taken into consideration by other nations of Asia as they head inexorably towards industrialization'. If the agricultural sectors of developing countries suffer a decline similar to Japan's and other nations in the region, this could result in food supply problems. The Worldwatch Institute has highlighted the importance of China to this equation, noting that a similar drop in

91

agricultural productivity there as has been experienced throughout the more developed nations in South-East Asia would have serious repercussions for the region (Worldwatch Institute, 1995). This potential problem has been highlighted by the recent series of poor harvests in China, Japan and the Korean peninsula (Lloyd-Parry, 1995).

Further development is likely to be accompanied by further rapid social change, and the continued breakdown of traditional social structures. As urbanization progresses, more problems associated with a lack of space, amenity facilities, and increasing population density are likely to occur. Japan's experiences of environmental management in crowded areas should be invaluable here. As income gaps widen and problems of regional variations in development continue, it is likely tensions over resource use policies will continue between urban and rural regions, and national and local economies. These tensions are likely to have the worst environmental impacts in countries lacking local democracy, given that Japan's environmental history shows clearly the importance of local accountability and democracy in ensuring adequate protection of the local environment in the face of national development aims. Finally, it is likely there will be shifting environmental impacts associated with increasing internationalization in a political and economic sense, increasing awareness of global environmental problems, and increasing integration of South-East Asian economies into the global economy. However, it is likely that the under-emphasis by Western nations of the importance of the contribution of the region towards creating and solving global environmental problems is likely to continue.

CONCLUSION

Japan has been an agent inducing environmental change in South-East Asia as a consumer, through overseas economic activity and FDI, and through international environmental policy. In all three cases Japanese impacts reflect domestic resource constraints, socio-political constraints resulting from Japan's own pollution history, and a prominent dimension in policy towards protecting Japan's economic interests, as well as acting as an example of structural problems between developing and developed nations.

It is likely that increasing urbanization, industrialization, energy demand, and population density will result in qualitative and quantitative shifts in existing and emerging environmental problems facing South-East Asia. The domestic experiences of Japan will provide vital lessons towards addressing these problems.

Environmental problems have social and political dimensions as well as technical ones. However, the impression given by Japanese authorities is that environmental problems in Japan have been 'solved' by technological solutions. This is misleading, as technological solutions are the easiest to find, and in reality environmental problems have shifted as economies and society have changed. In discussing and disseminating the lessons of its experiences with pollution to further the aim of sustainable development in South-East Asia, in addition to highlighting the important role of pollution control and abatement technologies, Japan also needs to emphasize that preventative steps make economic sense, and that democracy and the expansion of human rights in the region facilitate expression of grassroots experiences with pollution. These lessons should be increasingly reflected in Japan's ODA and international environmental policies, and is an issue that governments and individuals in South-East Asia should carefully heed.

Continued domestic change will be required within both Japan and South-East Asia, especially among consumer (and voter) attitudes. Further economic progress, environmental education, population control, and open and accountable societies are a prerequisite for the reality of sustainable development throughout the region. NGOs have a vital role to play is seeing these aims made real, and given the reliance of many of South-East Asia's economies on natural processes and resources, proper environmental management is essential to the region's prosperity. In this context, the lessons for environmental management from Japan are not all downbeat. Japan shows that within limits of political constraint, when public opinion is angry enough or the economic need for change is clear enough, that it is possible to have rapid change and introduce effective legislation. Such change is urgently needed in South-East Asia today.

Part II
PROCESS

5

THE SEARCH FOR SUSTAINABLE LIVELIHOODS IN INDONESIAN TRANSMIGRATION SETTLEMENTS

Colin L. Sage

INTRODUCTION

Rising concern over the contribution of deforestation to global warming has increased public attention on the scale of environmental destruction resulting from shifting cultivation and agricultural land settlement in tropical moist forest regions from Amazonia to Indonesia (Houghton, 1990; 1991; Williams, 1990; Sanchez *et al.*, 1993). In addition to the environmental consequences of rainforest colonization, the experience of land settlement schemes in the humid tropics has demonstrated the difficulty of maintaining productivity and ensuring sustainable livelihoods (Helmsing, 1982; Hulme, 1987; Oberai, 1988). Many of the principal consequences of land settlement are well known and can be briefly summarized into two general areas of concern for sustainability: the use of environmental resources and the building of livelihood security.

Environmental resources

The clearing of vegetation from a tropical moist forest biome leads to the breakdown of nutrient recycling systems, and sites experience a rapid loss of soil fertility through leaching and surface run-off. Without a protective canopy, soils are exposed to heavy rainfall, winds and intense solar radiation which can result in impermeable laterite crusts. Elsewhere, invasive weeds and grasses quickly take hold inhibiting the recovery of soil fertility, and can be difficult to manage. One such species, *Imperata cylindrica*, is estimated to cover one-fifth of the Philippines and one-tenth of the former forested area of Indonesia's outer islands (Blaikie and Brookfield, 1987).

97

Livelihoods

The arrival of farm families as settlers/colonists from other regions with quite different agro-ecological conditions often results, in the absence of adequate training and preparation, in farming practices which are inappropriate to local conditions. For example, the promotion of high external input monoculture, or Green Revolution-style agriculture, has usually resulted in widespread failure whether economically, ecologically, or both as the costs of controlling pest infestations spiral. It has been found that settlers do not usually possess an appropriate repertoire of techniques and experiences on which to fall back in times of stress. Many abandon their original plots and return home to their place of origin; others move further into the forest in order to 'mine' new sources of soil fertility; others make increasing use of off-farm employment for income generation to supplement declining agricultural returns; while for a few, moving into a more extensive land use strategy such as cattle ranching represents a short-term option in the absence of pasture improvement.

This brief review suggests that the experience of land settlement schemes has been disappointing at best, with some critics maintaining that certain programmes have resulted in widespread destruction of environments and cultures (*The Ecologist*, 1986; Hecht and Cockburn, 1990). The expectation that schemes would make significant contributions to national economic development has generally not been realized in practice, with the notable exception of the FELDA (Federal Land Development Authority) schemes in Malaysia, due largely to an absence of linkages to the wider economy (Oberai, 1988). Indeed, the cost of schemes has proved so expensive and rates of return so low that many donors and governments have had to re-evaluate their level of support for government-sponsored resettlement programmes. For example, during Indonesia's third five-year plan (Repelita III, 1979–84) expenditure per settled family reached US$12,000 (Oberai, 1988), although according to the World Bank the overall cost of resettling 366,000 families was US$2.3 billion, amounting to an average cost of US$6,300 per family (World Bank, 1988). Such costs have usually been justified on the grounds of resulting in an improvement in living standards for migrants, generally measured by cash income levels. Yet, as this chapter will demonstrate, this is no indication that sustainable livelihoods are being established.

The concept of sustainable livelihoods has proved an especially useful one with the capacity to integrate a range of important variables at the household level. These include the meeting of basic needs, access to productive resources and other food- and income-generating opportunities, and recognition of the varied repertoires of activities within the household. Chambers and Conway have defined sustainable livelihoods as follows

> A livelihood comprises the capabilities, assets (stores, resources, claims and access) and activities required for a means of living; a livelihood is sustainable which can cope with and recover from stress and shocks,

maintain or enhance its capabilities and assets, and provide sustainable livelihood opportunities for the next generation; and which contributes net benefits to other livelihoods at the local and global levels and in the short and long term.

(Chambers and Conway, 1992, pp. 7–8)

From the above, it is clear that the notion of sustainable livelihoods is a flexible, actor-oriented framework of principles reflecting the specificity of diverse and complex contexts. It is possible to distinguish livelihoods which are largely pre-determined or ascribed from birth, such as within the traditional caste system of India, from those which people improvise with varying degrees of desperation: 'what they do being largely determined by the social, economic and ecological environment in which they find themselves' (ibid., 1992). Consequently, although a livelihood is, at its simplest, a means of securing a living, its elements and structure are complex and deserving of analysis.

From this perspective, therefore, rural people are presented as agents, actively exploring options and formulating strategies in accordance with their tangible and intangible assets. Chambers and Conway define tangible assets as stores – food stocks, cash savings, jewellery etc. – and resources, which cover land, trees, livestock (which are also stores of value). A household's intangible assets, on the other hand, include claims (which are demands and appeals for material, moral or other practical support), and access, which is the opportunity in practice to use a resource, store or service to obtain food, income, employment and so on. It is by deploying these tangible and intangible assets, together with their skills, knowledge and labour, that enable rural people to contrive a livelihood. Rural people, then, are agents but not (echoing Marx) under conditions of their own choosing: they are engaged in the generation and application of knowledge within specific cultural, economic, socio-political and agro-ecological contexts which are themselves the products of local and non-local processes. Moreover, these processes have had a socially differentiating influence with the result that there is a multiplicity of capabilities, of identities and aspirations, and consequently of livelihood strategies (Bebbington, 1994).

However, it is not sufficient simply to recognize the complexity and diversity of livelihood strategies developed by rural people. Within the new professionalism of agricultural and rural development research, most popularly associated with the work of Robert Chambers and the notion of 'Farmer First' (Chambers, 1983; Chambers, Pacey and Thrupp, 1989; Chambers, 1993; Scoones and Thompson, 1994), is a commitment to a more interactive process of social learning and participatory enquiry. Beginning from a position of respect for rural people's knowledge and their capacity for agricultural experimentation, the challenge for researchers is to engage in collaborative programmes of action research that offer new practical options that might strengthen local livelihoods. This 'soft systems' approach (Ison, 1990) stands in sharp contrast to Transfer of Technology (TOT)-style thinking in which a research scientist or extension agent is the purveyor of 'new' knowledge to farmers. Besides, developing diversified and complementary farming systems appropriate to the circumstances of small farm households and to the agricultural potential of the local resource base will require a good deal of collaboration, involving multidisciplinary teams of researchers, extension workers, official agencies and grassroots organizations, as well as the full participation of local people themselves who ultimately are the final arbiters of whatever technical solutions might be developed.

Consequently, it was within this perspective that the research described by this chapter was developed. Forming part of a long-term, multidisciplinary project involving UK and Indonesian university researchers and an international agricultural research centre, the wider project is concerned with the development of sustainable farming systems and improved soil management in areas of recent agricultural settlement. As a first step in this process, the research described in this chapter sought to assess the nature and scale of the environmental constraints facing transmigrant households, and then to examine the strategies and activities developed by farmers to overcome or mitigate these constraints, or to compensate for their effects on household income. The concern was for the possibilities of strengthening livelihood security and this, of course, requires paying close attention to the individual roles, needs and aspirations of men and women within the household.

THE INDONESIAN TRANSMIGRATION PROGRAMME

Indonesia's transmigration programme is acknowledged to be the largest land resettlement scheme in the world (*The Ecologist*, 1986; Leinbach, 1989). Since its inception under the Dutch in 1905, over 4 million people have been moved from Java, Bali and Madura to the outer islands of Sumatra, Kalimantan, Sulawesi and Irian Jaya. The term transmigration refers to four different groups of people and it is helpful briefly to distinguish between them:

1 Sponsored general transmigrants who receive extensive support from government agencies in the form of transport, land, housing, food and other services for an initial period.
2 Registered spontaneous transmigrants who receive land and perhaps some other limited support.
3 Unregistered spontaneous transmigrants who receive no support whatsoever other than access to a plot of uncleared land.
4 Local transmigrants who are individuals relocated within a province from areas which have been designated protected forests or watersheds (Leinbach, 1989).

In most cases the general pattern is for transmigrants to receive a house plot of approximately 0.25 hectares together with land up to 1.75 hectares beyond the village for cultivation, although official allocations have varied between different areas and schemes. In practice, however, many migrant families have either not received a full allocation or have been unable effectively to utilize land to which they have received titular rights, because of conflicts and disputes with indigenous populations. This is discussed later in the chapter.

Although planned settlement schemes became a feature of Dutch colonial policy, the number of people moved was limited until the 1930s and even then was insignificant compared to the rate of natural population increase in Java (in 1941, for example, the number of people moved was less than one-twentieth of the increase of Java's population in that year (Hardjono, 1977)), although population redistribution was and remains the most important factor in transmigration. From 1950 to 1970 some 420,000 people were moved, to be followed in the first five-year plan (Repelita I, 1969–74) by 240,000, in Repelita II (1974–79) by 465,000 and by the end of Repelita III (1979–84) a further 1.49 million people. Although relieving population pressures in Java has been the overriding concern – nearly 65 per cent of Indonesia's 190 million people live in Java – the transmigration programme has evolved over each successive five-year plan. In Repelita I, for example, the emphasis was upon raising food production and promoting the goal of self-sufficiency in rice. By Repelita II, however, more attention was being given to regional development, with the objective of achieving national integration encouraging a broadening of transmigrant receiving areas. Much greater financial investments were being made in the transmigration programme by this time, with the state diverting part of its rapidly increasing oil revenues together with support from international agencies, most notably the World Bank.

Despite difficulties in meeting past resettlement targets, Repelita III had an unprecedented and most ambitious target of resettling 2.5 million people. Demographic preoccupations remained central, as illustrated by the *transmigrasi bedol desa* programme in which whole village populations were moved as a unit, from critical lands or areas earmarked for development in Java to resettlement sites in the outer islands. The *translok*, or local transmigration, programme was also developed to protect sensitive forest and watershed areas, and involved the resettlement (enforced) of former spontaneous migrants and local people to new areas. In Lampung Province over 42,000 spontaneous migrant families and a further 22,000 indigenous families were resettled from designated production forests and critical watershed areas under this programme (Zainab Bakir and Humaidi, 1991).

The types of land resources destined for transmigration have inevitably widened given the limited extent of suitable land. For example, an Overseas Development Administration study produced in 1985 concluded that only about 75,000 hectares in Central Kalimantan were suitable and available for three transmigration sites, yet government targets projected the clearance of almost seventeen times as much forest for settlement (Secrett, 1986). Consequently, land with medium to high agricultural potential and suitable for irrigation represents a tiny fraction of that which is settled which, in the main, comprises rain-fed uplands with limited soil fertility which accounts for 80 per cent of transmigrant families (World Bank, 1988). Efforts have also been made to bring wetlands and tidal sites into cultivation requiring considerable investment in drainage and water management systems, not always to good effect. Secrett (1986) notes that 80,000 transmigrants were moved into the Air Sugihan swamp forests of South Sumatra between 1980 and 1982, but the poor soil conditions led to crop failures, the settlement sites lacked potable water and the migrants suffered a cholera epidemic. The sites are now officially classified as failures by the Department of Transmigration. Yet, 'in complete contrast, the local indigenous people had sustainable systems of agriculture, exploiting the peaty soils and forests of the swamps without damaging them. Their model example was ignored by resettlement authorities' (Secrett, 1986, p. 82). Up to the end of Repelita III in 1984 the Indonesian government had moved 563,000 families of which almost two-thirds were resettled in Sumatra. This number excludes, of course, spontaneous migrants whose numbers are not accurately known, although almost 170,000 families were estimated to have moved with little or no government assistance during the period of Repelita III (1979–84). Under Repelita IV (1984–89) ambitious targets were still set – a total of 750,000 families – but the intended geographical focus was encouraged to move eastwards, towards Irian Jaya, as national security considerations have increased in importance. Nevertheless, Sumatra still remains an attractive destination for many, accounting for 61 per cent of the 477,000 families moved under Repelita IV up to June 1987, with two in every three families moving entirely unassisted (Leinbach, 1989). Repelita IV also saw the expansion of more diversified farm models, including the expansion of the Nucleus Estate and Smallholder programme in which migrants are tied into a contract farming and wage labour system with a central plantation enterprise (Sajogyo, 1993).

THE RESEARCH LOCATION

[Since the] time when land was rather loosely [sic] occupied – the density being less than 5 inhabitants per km^2 in 1900 – a systematic colonization of the forest has been underway since the turn of the century. Due to the combined effect of a voluntary populating policy and of spontaneous and massive pioneer fronts, the [Lampung] Province has been subject to a land mutation which is quite spectacular for the tropical world. Within fifty years, an area of 33,000 km^2 has been radically transformed and developed, while the population increased more than tenfold, rising from 376,000 inhabitants in 1930 to 5,250,000 in 1986.

(Pain *et al.*, 1989)

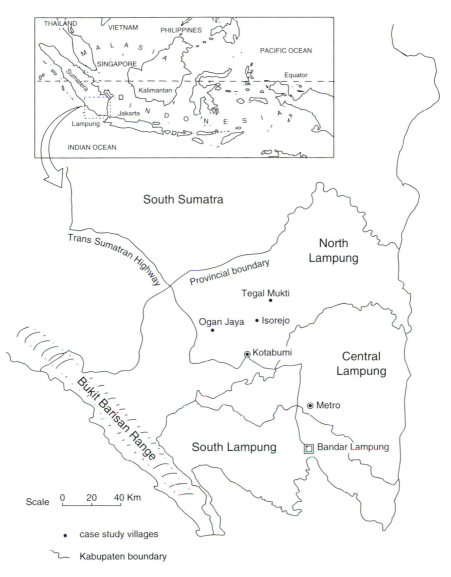

Figure 5.1 Location of case study villages, Lampung Province
Source: Adapted from Pain *et al.*, 1989

The proximity of Lampung Province (situated at the southernmost part of Sumatra: see Figure 5.1) to Java, a short ferry journey away today, makes it somewhat surprising that population densities remained so unequal for so long. In the mid-nineteenth century, for example, Lampung mustered barely 80,000 inhabitants, a density of just 3 per square kilometre, while population density on Java as a whole was 70 people/km^2. At the end of the century average densities were still less than 5 inhabitants/km^2 in Lampung while they exceeded 200 per km^2 on Java. As one might expect, the economy of Lampung was largely a simple, predatory one involving the export of forest products such as rattan, resins and animal products including elephant tusks, rhinoceros horns and swallows' nests (Pain *et al.*, 1989: this work is a study of the history of settlement of Lampung Province and, consequently, has provided the source of much information in this section, unless otherwise attributed). Nevertheless, the native population did engage in agricultural production – pepper and coffee becoming increasingly important export commodities – cultivating land along the coastal plains and along the main river valleys. There were, however, vast areas devoid of human habitation: the swampy plains, the estuaries, the mountain range and much of the interfluves.

By the turn of the century Lampung's population had grown to 160,000 people, but now began to swell more quickly with the introduction of the *kolonisasi* programme under the Dutch administration. In 1940 it had reached around 750,000, and then more than doubled to 1.67 million by 1961 and 2.78 million ten years later. The average annual rate of population growth during this period was around 5.8 per cent, of which the natural rate of increase was between 2–2.3 per cent, so that in-migration accounted for around 3 per cent or so. Yet this was by no means dominated by government-supported transmigration. During 1971–80 the total inflow of migrants to Lampung was 1.95 million, of which only 100,000 were sponsored transmigrants. As one official report put it, the 'wave' of spontaneous transmigration to the receiving provinces in general was, in fact, a 'flood' into Lampung, 'leading to saturation point so that the provincial government wishes to see the region become a net sending province' (Transmigration Advisory Group (TAG), 1989, pp. 3–4). Indeed, the central government terminated assisted transmigration to Lampung in 1980, although because of the scale of spontaneous movement, the average annual rate of population growth in the province still exceeded 5 per cent – the highest in Indonesia, including the Jakarta region (TAG, 1989). Perhaps unsurprisingly, by the end of 1988 the first transmigrants originating in Lampung were sent to Riau.

Analysis of population data at a sub-provincial (*kabupaten*) level reveals the uneven pace and spatial pattern of growth during the last twenty to thirty years. Between 1961 and 1980 Central Lampung grew by a factor of 3.25, to 1.69 million, and South Lampung by a factor of 2.5, to almost 1.6 million. Yet North Lampung, 'because of its poor soil and exclusion from the transmigration programme until 1970, only had a population of 880,000 in 1980' (Pain *et al.*, 1989, p. 133). This was to change under the influence of the 'local trans-migration' programme (*transmigrasi lokal* or *translok*) involving the resettlement of families from designated areas in South and Central Lampung. Growing recognition of the extent of soil erosion in critical watersheds and of increasing land hunger in the south of the province lay behind the resettlement of families in the north. Indeed, land reclamation work in South Lampung was hindered by far greater population densities than officially recorded, as well as widespread absentee landlordism, something specifically prohibited in the 1960 Basic Agrarian Law (Tjondronegoro, 1991).

By 1990 North Lampung had grown at an average annual rate of 6.5 per cent to a population of 1.35 million. Of this increase, some 65,500 families were moved under *translok* between 1979/80 and 1989/90 (representing around two-thirds of the total increase during the decade), while between 1986/87 and 1992/93 it is estimated that a little under 40,000 families settled in the region as spontaneous migrants. It is also expected that spontaneous and local transmigration will continue given that population densities are still low relative to South and Central Lampung.

However, these general figures reveal little about the spatial distribution of the new arrivals. For this it is necessary to examine data at the level of the district (*kecamatan*), which is neither comprehensive nor consistent but which shows that several districts in the North and North East of the *kabupaten* have grown extremely quickly. For example, Menggala more than doubled its population between 1986 and 1991 to 181,000 people, yet its density remains below the average for North Lampung (77 as against 95 people per km^2). Mesuji Lampung, the largest district by area, remains sparsely populated at 27 people per km^2 and a natural destination for people in search of land. In the district of Pakuon Ratu, the population has grown from less than 12,000 people in 1980/81 to almost 78,000 in 1991/2 at a density of 60 people/km^2. Finally, Tulang Bawang Tengah increased its population by almost 30 per cent over the period 1986–91 to become the third most densely populated *kecamatan* with 194 inhabitants per km^2 (Government of Indonesia, 1993; Elmhirst, 1995).

These figures demonstrate a very rapid rate of inward migration, of land settlement and, consequently, of environmental transformation. Yet the anticipation of securing land on which to produce food crops for family subsistence has, in the majority of cases, certainly been followed by disappointment at the productive potential of the soils. In general the rain-fed uplands of North Lampung with their highly weathered, low fertility, acid soils offer limited prospects for annual crops without irrigation. Indeed, an earlier study of the physiographic potential of Lampung warned that areas that could not be irrigated should not be settled, unless perennials such as rubber were made the principal crop (Verstappen, 1956). This has clearly not been the case and so it is hardly surprising to find that the districts which have received much of the local transmigration are amongst the poorest in the region (according to data on poverty rankings obtained from the regional development agency, BAPPEDA, although these provided no clear indication of the criteria that were used to identify poverty).

Despite the recent flood of migrants into the area, North Lampung nevertheless displays a diversity of circumstances reflecting location and quality of the resource base, age of settlement, type of farming system, the knowledge and adaptability of farmers, their level of capital and so on. In order to understand the nature of the environmental constraints facing transmigrant households and the strategies they develop to overcome them, it was decided to work in three different localities which would offer a degree of complementarity and a diversity of social and physical circumstances. The locations in which Rapid Rural Appraisals were conducted were: Isorejo, in District Sungkai Selatan; Ogan Jaya, in District Sungkai Utara; and Tegal Mukti, in District Pakuan Ratu. Both Ogan Jaya and Tegal Mukti are ranked amongst the poorest villages in North Lampung (though note the caveat in the previous paragraph).

Isorejo was established in 1976 with government-sponsored transmigrants from East and Central Java. It is the most prosperous of the three communities and benefits from some irrigation – except during the driest months – which supports one crop of paddy rice. Problems with pests and the low prices obtained for food crops (principally maize, cassava, peanuts) have encouraged many farmers to begin planting sugar cane, and this is likely to increase in importance given the relative proximity of the village to the Bunga Mayang sugar factory.

Ogan Jaya is a considerably more heterogeneous community comprising Ogan people from South Sumatra, Balinese, and migrants from East, Central and West Java, all of whom arrived and settled without government support (i.e. spontaneously). There have been three main waves of immigration to Ogan Jaya. In the early 1950s migrants from West Java (Sundanese) first settled in the area, securing and clearing land and planting food (*palawija*) crops. From around 1965 onwards Ogan farmers then began to arrive from South Sumatra, buying land from nearby Lampung villages and establishing their pepper and coffee plantations. Finally, Javanese and Balinese farmers arrived in 1977 having moved from Central Lampung where access to land was becoming increasingly difficult and where most had been dependent upon casual wage labour. Thus, to some extent, Ogan Jaya comprises several separate communities of different age and ethnic composition, practising different farming systems. Gradually, however, these distinctions are beginning to break down as farmers adopt new practices learnt from their neighbours, and as the second generation begins to link the different groups through inter-marriage. There is, nevertheless, still a very strong sense of ethnic identity in the locality and this is manifest in the settlement pattern.

The third locality is Tegal Mukti, a village established under the local transmigration (*translok*) programme in 1983 and which was formally recognized as a legal entity in 1987. This is still, consequently, a young community which has *de jure* rights to around 900 hectares of relatively poor undulating land on the North-eastern edge of the Bunga Mayang estate, but 300 hectares of which are in dispute with native Lampung farmers. Previously, most of the community of Tegal Mukti had lived in the area of Way Jepara in Central Lampung, where they had land under rubber, cloves and food crops, before its designation as a protected area. The economy of Tegal Mukti is closely tied to the activities of the sugar cane plantation, both for contract production of cane and as a source of labour for the company (for whom, it is estimated, half the village works for some part of the year).

KEY ISSUES CONFRONTING TRANSMIGRANTS IN NORTH LAMPUNG

Arising from a series of discussions with individuals, households and groups of men and women within the three communities, a range of factors that severely constrain the development and consolidation of sustainable livelihoods in the transmigration sites of North Lampung was identified. These can be grouped into five main categories:

1 Price of farm products and access to credit.
2 Pests, disease and weed problems.
3 Health.
4 Problems associated with the ownership and management of land.
5 Transportation.

Each of these is examined in turn below.

Price of farm products and access to credit

During 1992–93, the farmers of North Lampung experienced a dramatic fall in the prices paid for their agricultural commodities, most especially cassava, much of which is sold to a feed processing company in the area. One of the principal reasons behind the decline in price, which had been quite buoyant until 1991, was that increasing numbers of farmers had switched to cassava from soyabean and maize which had become increasingly affected by pest and disease problems. Not only is cassava relatively resilient to pest and other kinds of infestation, but it has the capacity to withstand soil moisture deficits, and is usually expected to take advantage of residual soil fertility so that few inputs are applied. Yet it is not entirely a 'free' crop: fields of unharvested cassava dominated by *Imperata* grass are widespread in the area of *translok* settlements which occupy some of the poorest soils, and testify to the costs of squeezing soil fertility without adequate inputs or appropriate management. This landscape of *Imperata* out-competing food crops has been used as evidence to argue that cassava is becoming a potent factor in land degradation (Thorbecke and van der Pluijm, 1993).

Farmers working with estate crops experience similar frustrations, indeed, more so given the fluctuations and unpredictability of market prices and the lack of flexibility which perennials present. For example, farmers in Ogan Jaya enjoyed prices of £2 per kg. for pepper in 1985, yet in 1993 these had fallen to just £0.35 per kg, while coffee prices had, for the first time, dipped below that of sugar on a kilogramme to kilogramme comparison, a relative benchmark used by farmers to assess their terms of trade.

Farmers have also experienced the increasing cost of agricultural inputs as a result of the phasing-out of subsidies on chemical fertilizers. Previously, the rice intensification programmes had encouraged increasing levels of fertilizer applications so that average use of urea (to provide nitrogen) had risen to over 350 kg per hectare by 1985, and is still rising. Since fertilizer was so heavily subsidized and, by the late 1980s, represented 'only about 8 per cent to, at most, 15 per cent, of the total costs of production in areas of intensive cultivation' (Fox, 1991, p. 78), farmers often used more fertilizer, particularly urea, than was recommended. The subsidy on fertilizers has now been substantially reduced.

However, the perception of rising input costs is accentuated by declining crop prices: one Balinese farmer in Tegal Mukti said that in 1986 one 20 kg bag of fertilizer could be bought with 10 kg of soyabeans; now the same bag of fertilizer requires more than twice that amount of soyabeans. Consequently, farmers stint on the use of fertilizer and tend to concentrate applications on wet rice cultivation where *sawah* land has been developed (figures from 1983 show that 80 per cent of fertilizer was used on wetland rice, and a further 12 per cent applied to maize, leaving very little available for other *palawija* crops which are expected to depend upon residual fertility: Thorbecke and van der Pluijm, 1993). (*Sawah* land refers to levelled lowland fields which have contour bunds to collect irrigation or rainwater and to control inundation. It may be fully irrigated, semi-irrigated or wholly rain-fed, and while it is usually sown to rice as a main crop, may support another *palawija* food crop (such as maize) during the dry season.) Alternatively, farmers can make use of informal credit arrangements with sub-distributors of fertilizer or borrow from money lenders under the *ijon* system of crop mortgaging on highly unfavourable terms. Such arrangements, with all their attendant risks of deepening impoverishment and even loss of land, also illustrate the failure of the KUD (village unit co-operatives) system. Notwithstanding their name, KUD operate at the district level, making access to their services difficult. In Tegal Mukti farmers have to travel 25 km to the KUD office to register a request for inputs. But bureaucratic and other delays mean that farmers often do not receive the inputs promptly, sometimes not until after the harvest for which it was intended! This reinforces Thorbecke and van der Pluijm's observations (1993, p. 269):

KUD management is often in the hands of traders, small entrepreneurs and better-off farmers. The social distance between these managers, who are based in the *kecematen* town, and the farmers in the surrounding villages, can be fairly great. This explains the relatively low participation rates reported, and also the often heard complaint that many KUDs do not function well, do not serve the needs of the people and are characterized by mismanagement and corruption.

Pest, disease and weeds

It is apparent that pest and disease problems are increasing in the areas of agricultural settlement in North Lampung, although the type and intensity of infestation vary between crops and communities. No more than several years ago the biggest problems were caused by elephants, wild pigs and monkeys. With the continued expansion of the agricultural frontier, the elephants and monkeys have disappeared, although pigs remain a real problem, bedding in fields after sunset and trampling plants. However, the single most common problem is rats. They first became a significant nuisance in 1985 and during 1993 farmers in Isorejo declared that up to 95 per cent of the rice harvest in certain fields was lost to rodent infestation. (In 1987 a particularly tragic event occurred here when rice, previously mixed with *klerat*, a rodenticide and used as bait, was retrieved for consumption during a time of food shortage. One family died and twenty others became seriously ill by ingesting this poisoned food.) The problem may have worsened in the area since the expansion of land under sugar cane. Yet this has not triggered a collective response by farmers: while some lay poison, others believe it is just a temporary phenomenon and the rats will disappear eventually. Fortunately, it is a problem that is recognized by the Ministry of Transmigration, according to the Director of the North Lampung office, which has started to investigate the use of owls for the predatory control of rat populations.

Besides rats, rice suffers from many insect pests (plant hoppers, stem borers, stink bugs, ants) and diseases (Fujisaka *et al.*, 1991). Other food (*palawija*) crops also face increasing problems of disease. In Tegal Mukti soyabean was one of the first crops to be planted following the clearing of fields, and it produced well for the first three years before 'leaf curl' disease (*keriting*) and pest soon led to yields of around 10 per cent of those previously harvested. Today households in Tegal Mukti have to buy the soya they need to make *tahu* and *tempe*, key elements of their preferred diet. A similar problem has beset groundnuts, which in some areas produces only 4–5 cans per hectare, barely enough to provide seed for subsequent planting. In Ogan Jaya farmers spoke of the utter failure of food crops over two successive harvest periods resulting in real shortages and a breakdown in levels of self-provisioning. It is apparent that without the cushion of perennials, households in Ogan Jaya would have experienced acute deprivation.

However, perennials, too, suffer from pest and disease problems, but especially fungal infections affecting roots. There can be a loss of 20–30 per cent of pepper plants per year as a result of root infections, despite the widespread use of chemicals such as Cobox. Before replacing an infected or dying plant with new stock, farmers usually remove a small area of top soil around the host tree in order to try and reduce the spread of infection into the new plant. Usually kapok or durian are the species which are planted to provide support for the pepper vines, although they also produce economically useful seed and fibre material and fruit, respectively. These, together with the coffee and rubber trees that are also interspersed in the farmers' 'plantations', indicate the level of diversification of perennials which some farmers in Ogan Jaya enjoy.

Imperata cylindrica (Indonesian: *alang alang*) is also a major problem for agricultural production in North Lampung. Where it has been allowed to develop around estate crops – principally coffee and rubber – due to neglect or other priorities for labour, there is a marked decline in plant health and, naturally, of productivity. As *Imperata* is not shade tolerant it is usually relatively well controlled around mature estate crops. Within a field of cassava, however, it is a different matter. While herbicides such as Roundup are recommended to control *Imperata*, the cost is often beyond the reach of many farmers. Moreover, given the current low farm-gate prices, there is only very marginal benefit to be derived from hand weeding. This is why fire remains the most popular method for dealing with *Imperata*, with the attendant dangers that this sometimes presents, described below, and also why there is an urgent need to find a more biologically appropriate method of inhibiting the spread of this grass.

Health

It is appropriate to begin a brief discussion of human health in North Lampung with a consideration of nutritional well-being. According to Thorbecke and van der Pluijm (1993), protein-energy malnutrition was estimated by a 1984 survey to afflict 30 per cent of pre-school children and 9 per cent of pregnant women, although anaemia and other vitamin deficiencies were very much more widespread. Unfortunately it was not possible to identify a more recent nutritional survey relating to North Lampung, although the impression gained during field-work is of generally poor nutritional status.

This is supported by the extent of dependence on non-rice staples (*palawija*) which have been relatively neglected by the preoccupation with rice self-sufficiency. Lack of credit for agricultural inputs and the absence of technical support have meant low yields of *palawija* crops, which then fail to cover the nutritional requirements of the household. Thorbecke and van der Pluijm (1993) advance the argument that dependence on non-rice staples is a fairly reliable indicator of poverty, and that Lampung, amongst other provinces, is dependent upon non-rice staples because it cannot produce enough rice locally. They cite data from 1981 to show that annual per capita consumption of cassava in Lampung is the second highest in Indonesia. This is supported by discussions with informants who spoke of the need to eat *tiwul* (grated cassava mixed with a little rice) for some part of the year, despite its associations as a 'food of the poor', in order to eke out their limited supplies of rice.

The overall level of human health and well-being in North Lampung is also seriously compromised by the incidence of a cluster of diseases and infections. The most important is malaria which is prevalent in an area offering ideal conditions for the breeding of a mosquito vector. The impression was that the majority of households in the case study localities had at least one infected member, resulting in a reduced labour capacity for part of the year. Yet efforts to reduce the incidence of malaria were rather patchy: there appeared to be no assistance provided to villages to drain swampy areas nor concerted spraying campaigns, and there was little medical care beyond visits to paramedics.

The second most important form of ill-health is respiratory infections, and third is water-borne and water-related infections such as gastro-enteritis, diarrhoea, and skin diseases which particularly occur during the dry season when people use river water for cooking, washing and drinking. Given the poor sanitation and water-supply facilities which prevail in all parts of rural Indonesia, this group of problems is certainly not unique to Northern Lampung, but demonstrates the urgent need for improved infrastructure and health education.

A major health issue that arose during discussions with women, both collectively and individually, was the issue of contraception. There is a tremendous emphasis placed upon family planning in the country, and this has done much to reduce the overall fertility rate. Yet, typically, it is structured in a top-down fashion involving a range of organizations from provincial to village level, in which men take a leading role in setting targets and monitoring achievements, but which ultimately relies upon women adopting measures to control their fertility. As Smyth observes, there is nothing to indicate that the Indonesian Family Planning Programme counts among its main objectives that of providing women with the means autonomously to regulate their fertility, or of safeguarding and promoting their reproductive health which is particularly surprising in a country where maternal mortality rates are still extremely high (Smyth, 1991).

The Integrated Services Posts (POSYANDU) mobile clinic which visits communities monthly largely concentrates upon monitoring mother and child health, and making available different forms of contraception. There are four main methods which are promoted, excluding condoms which do not seem to be widely used: the pill, which costs approximately £3 for a month's supply; injections of Depo Provera that last for three months and cost £1.30; hormonal implants that allow a slow release of oestrogen over a much longer period and which cost £1.70; and IUDs, which are free. As Smyth (1991, pp. 797–8) notes:

Injectables and implants have come under a great deal of criticism for the health risks they present to women users. Although medical and other reports have so far been inconclusive, the fears raised in connection to cancer, infertility and suppression of the immune system, have not been answered.

The ownership and management of land

There is widespread uncertainty throughout North Lampung about the status of land ownership. It was found that land disputes between transmigrants and native Lampung farmers have existed since the 1970s and are growing in intensity and violence. In Ogan Jaya several lives were lost in an armed confrontation in 1987. In Tegal Mukti an estimated 30 per cent of land is in dispute, affecting around one-half of all households. Land which has been allocated to transmigrants continues to be claimed by native farmers, so that often we were told that after clearing and preparing a field ready for sowing, transmigrant farmers would return to find cassava already planted by local Lampung farmers. Transmigrants expressed fear and frustration about the circumstances in which they found themselves. Many had finally agreed to buy from a native farmer land that had originally been allocated to them by the government. Yet even after money had changed hands there were still difficulties with other members of the seller's family. Some transmigrants said that they had bought the same piece of land three or four times, thus illustrating the different notions of land ownership representing individual as opposed to common property rights.

Elmhirst, in work in progress, has sought to examine the implications of transmigration and plantation development on native Lampung communities. Through field-work in a Way Kanan village within the zone of *translok* settlement, Elmhirst has uncovered a high degree of internal conflict over customary land rights, which has divided an old 'aristocracy' of village chiefs and the new power brokers: 'business men and those able to capitalize on formal government links' (Elmhirst, 1995, p. 12). Naturally, this has implications for land use, given the sale or expropriation of land that was previously used by the Way Kanan for shifting cultivation, leaving them to intensify agricultural production in the remaining fertile river-margin land. It also meant that transmigrants were allocated, and possibly also purchased, land that was considerably less fertile and often in a somewhat degraded state (Tjondronegoro, 1991).

It also has consequences for social differentiation, a process that is also well under way in the transmigration communities. Besides those transmigrants who had simply purchased their own, government-allocated plot from the native Lampung in order to regularize their position, there were also other transmigrants able to make further acquisitions of land. This enabled some settlers, such as the Ogan, to accumulate substantial areas of land, some in excess of 10 hectares. Meanwhile, in Isorejo, around a quarter of households have still to receive their full allocation of land (and some are surviving on just 0.25 hectares), yet others have up to 10 hectares acquired through judicious purchase. Such purchases represent a significant accumulation of capital, either from agricultural production or derived from spells of migrant labour away from their village.

Yet, notwithstanding the real insecurity caused by conflicts over land, most transmigrants still saw more hope in their present position than either returning to their place of origin or moving elsewhere. A Javanese transmigrant in Ogan Jaya, when asked why he remains despite the problems, provided two arguments: first, it has already been very painful, so why move now when good fortune might appear? Second, he has never had any good news from those who abandoned Ogan Jaya and returned to Java, although he does not know whether they are in a better or worse position than when they left. Consequently, it seems that people stay and try to make the best of it in the hope that disputes over land can be resolved, and because they have no entitlements to join a new transmigration programme elsewhere.

Another issue which arose during discussions with some farmers related to the transfer of ownership of (undisputed) land. While it is prohibited for transmigrants to sell any of the land which they have been allocated for a period of 10 years, it became apparent that some had had to resort to this in order to raise money for a variety of reasons, e.g. urgent medical treatment for a family member, the discharge of cultural duties, or to purchase livestock or other productive assets. This suggests that a process of differentiation takes place rather quickly, in which poorer households relinquish control of their land bit by bit, while others are able to acquire extra land and extend their agricultural operations.

This, of course, raises important questions regarding the prospects for the children of poor households to remain on the land as farmers rather than simply as wage labourers. Today the population of Isorejo is approximately double the size of 16 years ago, yet during this time an irrigation system has been installed, which delivers water for nine months of the year, and this has contributed to raising the productivity of land. This is reflected in rising land values. But it is apparent that most parents do not foresee their children staying in the village, and, indeed, many have aspirations for their sons (and perhaps their daughters) in pursuing careers in the major urban centres: Elmhirst (1995) reports from her work in the Way Kanan village that nearly all the unmarried women from the village are working in garment factories in Tangerang (an industrial town in the Jakarta region) and sending remittances home every three to six months.

The deterioration in the productivity of land was ranked as one of the most serious problems confronting farmers in all three localities. Indications of land degradation were experienced through declining yields, the increasing prevalence of *Imperata*, soil loss on slopes and an increasing reliance on cassava. The impoverished nature of the soils, particularly in Tegal Mukti, were a stark contrast for most farmers compared with the productivity of land which they had previously cultivated in Central Lampung and in Java. Yet not only were they now struggling to create livelihoods from poorer land, but many farmers have been deeply troubled by the difficulties presented by the prevailing rainfall regime. Not only does a seasonal water deficit during the months of July to October place a constraint on crop production, but there appear to be considerable fluctuations in rainfall patterns at the local level such that periods of drought were frequently mentioned by farmers. Moreover, as mentioned above, the dry season leads to a fall in the water table and the drying up of wells, which forces people to make use of more distant river water for their domestic needs with resulting consequences for health.

Finally, fire represents another serious land management problem. During the dry season farmers burn off the *Imperata* without the necessary care and attention resulting in fires spreading to contiguous fields and forests. Sugar cane is especially vulnerable during this time, and fire represents one of the most serious hazards confronting contract (*plasma*) producers. In Tegal Mukti efforts have been made to create isolation corridors around the fields, although some farmers suggested that fires might be set deliberately by native Lampung people as part of the ongoing conflict over land.

Transportation

In the rainy season transportation is a serious problem in the transmigration areas with many villages entirely cut off from vehicular access. Given the dependence upon truck owner–operators as a means of moving products and people, this results in prices which are unfavourable to the farmer. First, because truck operators charge more for delivering goods to communities over slippery and flooded roads; second, because the delays in getting produce to market can depress the prices paid to farmers. The nature of the terrain, involving crossings of seasonally flooded rivers, creates considerable construction and maintenance logistics for bridges and other structures involving high cost for local government and local labour. For this reason the issue of transport remains a vexed one for the communities of North Lampung in common with other regions undergoing an expansion of the agricultural frontier.

CONSEQUENCES AND RESPONSES

The obstacles and constraints which have been discussed above give rise to a variety of responses by transmigrant households in the search for improved and secure livelihood prospects. As the constraints might be environmental, economic or socio-political in origin, the range of possible responses is varied, but in turn is restricted by a variety of practical considerations. For example, one possible response to pest problems in rice might be for farmers to opt for the cultivation of perennials. However, in practice many lack the knowledge or expertise, the technical support and a source of credit to tide them over until the first harvest. There are also important cultural factors that constrain this possible option: the central commitment of transmigrants from Java and Bali to attempt to grow sufficient rice for family consumption, for example, has been widely observed despite poor and declining yields. In a study of a transmigration site in South Kalimantan, Abdoellah (1987) observed that after the first few years of adequate yields rice production has been decreasing annually due to soil fertility decline, soil toxicity, and a lack of fertilizer, labour, capital and other innovations adapted to local conditions. Consequently, farmers' responses to constraints are conditioned by their knowledge base, learned through practice and historically derived experience, a range of socio-cultural considerations, and their economic and technical capacity to act.

Although it is important to recognize the lack of site-specific, agro-ecologically appropriate knowledge amongst transmigrants on their arrival in a particular location, it is possible to trace the development and evolution of individual agricultural strategies as farmers experiment with different crops, planting times, and other cultivation practices in the face of a changing and deteriorating resource base. This constitutes the mechanism by which farmers not only make choices about crop combinations and acquire the skills and expertise to increase food production, but also how they gradually develop a repertoire of secondary strategies to be deployed in the face of hazards and unexpected events to strengthen livelihood security. Moreover, beyond crop cultivation the search for improved household livelihoods may include experimentation with livestock husbandry and diversification into off-farm earnings. Yet detailing such strategies at the household level should not disguise the role of inter-household co-operation and institutional arrangements that contribute to livelihood security, or those exploitative or usurious activities which accentuate differentiation at the local level.

New agricultural strategies

Given that transmigrants bring with them their existing farming knowledge, which gradually adapts to their new environment, it is inevitable that such changes are influenced by neighbours performing different agricultural practices perhaps with better results. Naturally, there are also other forces for change, such as the government's intensification and extension programmes encouraging new cropping systems, as well as opportunities to pursue off-farm employment. Yet, as noted above, there remains a strong desire amongst transmigrants to create *sawah* land for rice cultivation where possible, and to plant rice on rain-fed hill land where it is not. Rice has historically received the bulk of agricultural inputs, accounting for 80 per cent of fertilizers and almost 90 per cent of pesticides (Thorbecke and van der Pluijm, 1993). With the removal of government subsidies on inputs, farmers are facing rising prices and reducing their use of fertilizers, making it difficult to sustain rice yields. This has triggered a willingness to try other strategies, such as a switch to sugar cane or to cassava. The latter is largely seen as a 'free' crop with an ability to withstand moisture stress, and which does not require applications of fertilizer. Yet cassava is one of the more exhausting crops and a 'potent factor in the process of land degradation' (Thorbecke and van der Pluijm, 1993, p. 80).

In the settlement of Ogan Jaya, which benefits from its location on superior andesitic soils, there has been, historically, a marked difference between the different ethnic groups. The Ogan farmers are primarily dedicated to perennial 'estate' crops, principally coffee and pepper but also rubber and cacao, while the efforts of Sundanese farmers have been concentrated upon rice, *palawija* and home garden crops. The Balinese, meanwhile, have moved more into tree crops with pigs playing a part in land preparation and the control of weeds. Finally transmigrants from East Java have worked to build a relationship with the Ogan, so that many are now sharecropping estate crops or using other forms of contract to build a greater perennial element in their farming systems. Such differences in agricultural production strategies among Javanese, Sundanese and Balinese transmigrants were observed by Abdoellah in his study in South Kalimantan. He found that within the home gardens the Sundanese cultivate more vegetable and ornamental plants, the Javanese more starchy food plants, while the Balinese home garden lacks diversity and is poorly developed (Abdoellah, 1987).

It is revealing that in an overall wealth ranking of Ogan Jaya, the Ogan emerge as the richest group, with many households possessing more than 10 hectares of land largely under rubber, pepper, coffee and other tree crops. They make extensive use of contract labour for tasks such as weeding while themselves pursuing other commercial interests as market intermediaries. The Sundanese, who were the first 'outsiders' to settle here in the 1950s, are, on the other hand, the poorest group within the settlement, and make extensive use of off-farm wage employment in order to meet their livelihood needs. The popular explanation for this ranking by wealth is that superior earnings are to be made from perennial crops, and this has encouraged many farmers to attempt the switch from rice and other food crops to coffee and pepper.

In Tegal Mukti and Isorejo, however, with their poorer acid lowland soils, the main escape route from the low productivity, low income trap of rice and *palawija* crops is represented by sugar cane. This option is offered under the Nucleus Estate and Smallholders scheme, where a central enterprise, combining processing plant, capital resources and marketing, acts as a 'focus' for a large number of smallholder primary producers who are contracted to produce for the nucleus. The benefits for *plasma* (smallholder) producers is that they share in the economies of scale enjoyed by the nucleus in providing inputs and services and in processing and marketing the product. In practice, however, there are many problems, some of which are briefly outlined by Sajogyo (1993), who notes the difficulties in building trust and empathy between nucleus and farmers given the history of plantation power exercised over rural people.

Some of the advantages for local farmers in undertaking a contract with PTP Bunga Mayang to grow cane as part of the expanding *plasma* operations of the company are as follows:

1 Farmers receive credit which is channelled through the local TRI (Intensified Smallholder Sugar Cane) scheme to groups comprising eight participating households. This group (called *kelompok tebu rakyat*) is the base unit recognized by the nucleus for the purposes of organizing land preparation, harvesting and payment, the latter being divided equally between the eight members irrespective of individual yields. The credit is normally provided in chemical fertilizers, all of which should be applied to the field that will be planted to cane. It is widely recognized, however, that many farmers sell part of their allocation to raise money for food, which demonstrates the scarcity of cash and lack of alternative credit.

2 Land preparation is performed by contractors with heavy machinery who are paid by the company, although the cost (around £300 per hectare) is ultimately charged to the *tebu rakyat* group and taken from the first year's payment. Ploughing to a depth of 30 cm. has the advantage of breaking up the rhizomes of *Imperata*, although there are concerns about the use of such tillage practices on poor soils.

3 The company reveals the price that will be paid to farmers half-way through the growing season, sets a date for the harvest for each group and arranges for the cane to be cut and carried to the factory. Again, these costs have to be met by the farmer but there is some reassurance in being freed from the usual uncertainties of marketing.

Notwithstanding some of these advantages, there are serious limitations to growing sugar cane. Once planted, the cane produces four harvests, of which the third and second years are the most productive. It is usual that at the end of the first year the farmer receives no payment at all from the company once the costs of land preparation and fertilizer credit have been deducted, while by the fourth year the yield is poor and income is low. There are also logistical problems regarding the scheduling of the different operations so that farmers often lose considerable time and potential revenue waiting for the contractors to arrive to prepare the land or to cut the cane.

In this regard, opting into sugar cane tends to place severe constraints on individual decision-making and flexibility, with farmers directed on the timing and performance of cultivation tasks on their own land. For many it means the displacement of food crops without the compensation of high returns to labour. This explains why some farmers talked of cane being a transitional crop on the way to an alternative perennial system comprising rubber or oil palm. In Tegal Mukti there has been a move in this direction, with a small nursery of rubber trees established, yet credit for such initiatives is minimal and the case is not helped by continuing insecurity over land ownership. Meanwhile PTP Bunga Mayang have plans to expand the area under contracted *plasma* production from 7,000 hectares to a projected 15,000 hectares by the end of the decade in order to meet the expanding capacity of the sugar factory. It is likely that more communities will be 'invited' to participate in growing sugar cane under contract, and more research is needed to examine the ecological and socio-economic consequences of such a development.

Livestock production

This has grown in importance as households seek to diversify their farm operations and develop a form of insurance against unexpected crises. Some farmers have been able to benefit from various extension schemes (such as *banpres sapi*, designed to promote the multi-purpose benefits of livestock) and to secure one or two cattle providing an important source of animal traction. Some households have invested in developing a small herd of goats, and there has been some use of the *gadu* system of share-herding with these animals. Finally, while most households have a few *kampong* hens, improved breeds are also appearing: one entrepreneur in Isorejo has a large operation with over one hundred laying hens. Nevertheless, there is high mortality for all livestock, perhaps most especially poultry, an assertion supported by government figures that show the greatest annual loss in national animal production is caused by Newcastle Disease (Thorbecke and van der Pluijm, 1993).

Wage labour

Off-farm employment takes many different forms but is a vital element for many in all three study villages. This dependence upon wage income amongst transmigrant households accords with the findings of Leinbach and colleagues (1992) in South Sumatra and Abdoellah (1987) in South Kalimantan. In the study area in North Lampung the single most important source of employment is PTP Bunga Mayang which provides field-work all year round on the 8,500 hectares of sugar cane currently planted by the company. The harvesting of cane is done entirely manually and paid on a piece-rate basis with women wielding the machetes and men bundling and loading waiting trucks. Consequently, in sharp contrast to the literature on the colonization zones of Latin America, where women's labour is undervalued, often excluded and always rendered invisible in the performance of field tasks (Townsend, 1993a; 1993b; 1995), in Lampung women are centrally involved in off-farm and on-farm agricultural activities. Although wage rates are low – the casual daily rate in 1993 was equal to £0.30 or half the official minimum daily wage – PTP Bunga Mayang offers a regular and reliable source of work. Indeed, this has had the effect of encouraging many youngsters to leave school and become more financially 'independent', often to the consternation of their parents who see education as offering the best opportunities for their children's future. For the parents, such wage work offers no possibilities for the household beyond the most basic survival.

While it is not only the poorest households which make use of off-farm employment, those with little or no access to land are especially dependent upon wage income. According to Abdoellah (1987), almost 60 per cent of the total income of transmigrants is derived from off-farm sources, while for Elmhirst (1995) an average of 30 per cent of household consumption is covered by remuneration from off-farm work on the plantation. Besides Bunga Mayang there are other large plantation enterprises in the vicinity (producing rubber, cacao, coconuts) that make use of casual labour, as well as the cassava processing plant and a local sawmill. The potential for earning higher wages, however, increases with distance that workers are prepared to travel. Respondents spoke of occasional periods of work in other parts of Lampung and in the logging industry of South Sumatra and Riau, and also referred to individuals who had apparently travelled as far as Malaysia. Although absences are timed to fit into the agricultural cycle and while women aim to perform most of the tasks that need to be done, there are inevitably consequences for on-farm production. Abdoellah (1987) observes that wage employment outside the village often leads to the neglect of farms. This creates a chain of effects in which there is an increase in the area of uncultivated land, that provides a breeding ground for pests, which in turn affects the agricultural production of those left behind. Revealingly, within the Javanese community of Ogan Jaya it was estimated that up to 70 per cent of households have members working away for some part of the year.

Within a Way Kanan community of native Lampungese in the Bunga Mayang area, Elmhirst (1995) found that several households had sent their unmarried daughters to work in garment factories in Tangerang, near Jakarta, and their remittances had become an important part of household income. It would certainly appear that wage labour had become a vital component of the economy of both transmigrant and native settlements in North Lampung.

Hunting, gathering and market trade

For a few transmigrant households, logging provides a source of income but it is amongst the native Lampung where greater use is made of the existing forest. Activities monitored by Elmhirst include cutting and selling timber for construction purposes, collecting rattan, and trapping wild animals for their skins. For those located in riverine settlements fishing can also provide an important source of income (and protein) (Elmhirst, 1995). A few individuals in the area also hunt wild pigs using packs of dogs and sell the meat to non-Moslem (Balinese, Chinese) traders. However, the continuing rapid expansion of the agricultural frontier makes such activities increasingly unrewarding.

The term 'market trading' encompasses a wide range of possibilities – from hawking of sweets, drinks and cooked food, through the sparsely stocked *warung* (more stall than shop) to an operation in which capital is tied up in a wide range of clothing or household items. Irrespective of scale it is largely women who are engaged in market trade, and their contribution to household income naturally varies but can be strategically important depending upon other income and food production activities.

CONCLUSIONS

As this chapter has attempted to outline, transmigrant farm households in North Lampung face a wide array of problems and a number of serious constraints to securing improvements in their livelihoods. While some problems are unique to each community, such as the specific combination of crop pests and disease, others are more structural in nature and affect all households in the region irrespective of their farming activities. Low and unstable farm gate prices, for example, can make future planning hazardous and price fluctuations probably have long-term consequences for soil fertility, for future levels of food output and for the sustainability of the resource base. So when farmers receive low prices for coffee and pepper, besides undermining the value of mixed perennial agroforestry systems, it also effectively eliminates the possibility of purchasing chemical fertilizers for rice and *palawija* crops. This has the effect of reducing production of food crops, leading to diminishing returns to labour and increasing the necessity for off-farm employment. This, in turn, appears to exacerbate infestation by *Imperata* grass and ultimately may lead to the abandonment of land in favour of wage labour. Further research is required in order to determine trends in levels of self-provisioning and how this relates to food security within the household.

Indeed, there is a great deal still to be learned about the circumstances and livelihood prospects of transmigrant households in North Lampung. Notwithstanding the pace of environmental change, one is inclined to agree with Soemarwoto (1991) who argues for a people-centred approach in order to improve the lives of the rural poor, which will in turn ease the pressure on resources and allow the environment to rehabilitate itself. Certainly, action is urgently required in Lampung, where 'deforestation has reached such critical dimensions that the region is now plagued by high erosion rates and frequent floods' (ibid., p. 222). The continuous in-flow of spontaneous migrants has led to a steady process of agricultural settlement involving dramatic and often irreversible land-cover conversions, including the permanent reduction in the productive potential of land, and is eliminating many possible sustainable options as the stock of resources is diminished.

Whether these include the production of agricultural commodities (such as fish, eggs, vegetables, fruits and flowers) for the export market which Soemarwoto suggests is an interesting proposition, it illustrates his argument for the intensification of agriculture to reduce the area of land needed to support a population. It is certainly the case that outside the locations occupied by the estate sector, land use is often given over to farming systems of low productivity with low labour retention leading to widespread land degradation. The *Imperata*-dominated landscape throughout the zone of *translok* settlement, where cassava production is supplemented by off-farm wage labour, is a testimony to this development. Yet, in North Lampung, the emergence of the large estates has proceeded as part of the process of land degradation. In the first place they have been responsible for dramatic environmental change, with tens of thousands of hectares of forest cleared and ploughed by bulldozers every year during the 1980s. These estates now produce a range of commodities, principally sugar cane, cassava (for the manufacture of tapioca and citric acid), pineapples, coconut, rubber and oil palm (Pain *et al.*, 1989).

However, there is a second way in which the estate sector is both directly and indirectly responsible for land degradation. This derives from their reliance upon a cheap and abundant reserve army of labour drawn from the transmigrant settlements which cannot supply their means of subsistence. The profitability of the estates is ensured by the use of labour which partly reproduces itself from smallholder production (after Meillassoux, 1977), and thus is able to work for half the official daily minimum wage as mentioned earlier. Indeed, it is no accident that PTP Bunga Mayang, which pays such wages, was developed at exactly the same time as the establishment of *translok* settlements in the vicinity (the early 1980s), and as soon as processing operations began, spontaneous migrants rushed in. The estate sector, therefore, encourages land degradation by attracting migrants to the area, and by failing to provide adequate remuneration to meet their livelihood needs. Indeed, seasonal, hazardous and badly paid work for estate workers is more likely to heighten, rather than reduce, the vulnerability of transmigrant households, and to marginalize more sustainable land use options.

For this reason it is vital that new initiatives are encouraged that begin from a concern with sustainable livelihoods. Building sustainability into livelihoods reduces the vulnerability of rural households to stresses and shocks. Returning again to Chambers and Conway (1992, p. 14), 'Stresses are pressures which are typically continuous and cumulative, predictable and distressing, such as seasonal shortages, rising populations, or declining resources, while shocks are impacts which are typically sudden, unpredictable, and traumatic, such as fires, floods and epidemics.' Sustainable livelihoods are consequently those resilient enough to avoid or withstand such stresses and shocks. Naturally, one would expect that a household's portfolio of tangible and intangible assets (stores, resources, claims and access), and their repertoire of activities are chosen and maintained in order to reduce vulnerability and to withstand stress and shocks. Crucially, however, the sustainability of livelihoods also depends upon the capability of rural people to 'perceive, predict, adapt to and exploit changes in the physical, social and economic environment' (ibid., p. 16).

It is apparent that the capabilities of the different groups discussed in this chapter vary quite widely, although many do not appear to be well equipped to cope with the difficult agro-ecological circumstances of North Lampung. Enhancing the capabilities of such transmigrants is consequently a vital task in order that they become more versatile, resourceful and better able to respond to rapid and unpredictable changes and new opportunities. Enhancing capabilities is unlikely to be achieved through the conventional top-down training programmes led by agricultural extension agents, but will require innovative and flexible initiatives to meet locally defined needs. Examples might include: encouragement and support of farmers to engage in their own on-farm experimentation; providing a line of credit for small enterprise development; facilitating community initiatives for health care, education, new employment opportunities; and so on.

Naturally, there are many other practical measures which need to be addressed if sustainable livelihoods are to be secured. Amongst these are the resolution of conflicts over land; developing more intensive and productive use of resources (exploring, for example, the potential of agroforestry, intercropping, green manuring and other options); building local safety nets to counter seasonal stress; and identifying measures that will improve the health status and well-being of the rural population (especially women and children). But the indispensable element in all of these is the participation of local people in determining the objectives, priorities and strategies that will lead to such improvements. And inevitably, such participation will require a process of political democratization in Lampung, and in the country as a whole, if livelihoods are in reality to become truly sustainable.

6

THE RACE FOR POWER
IN LAOS
The Nordic connections

Ann Danaiya Usher

The Lao People's Democratic Republic opened its borders in the mid-1980s, thus ending more than a decade of isolation from the capitalist world. For this land-locked nation which is among the poorest in Asia and suffered untold damages during the period of the Vietnam War, hydro-power has emerged as a promising potential means of establishing a foothold in the international market. Indeed, so attractive appear the prospects of exploiting this 'white gold' that as many as sixty dam projects are in various stages of consideration by the Lao Ministry of Industry and Handicrafts. A combination of political, geographic and economic conditions in Laos has spurred a race for hydro-power that is occurring on a scale and at a speed that is unprecedented in South-East Asia. Although only two projects had reached the construction stage by 1995, several others were in the planning stage. If even a fraction of these materialize, there are expected to be serious implications for the Lao environment and for people whose livelihoods are dependent on riverine and forest ecosystems.

The focus of this chapter is the Nam Theun Hinboun, a 210 megawatt hydro-power project on the Theun River, a major tributary of the Mekong (see Figure 6.1). Construction began in late 1994, and it is expected to be completed in 1997. The project illustrates the tensions that arise with most large dams between environmental costs and the potential economic benefits of hydro-power. These tensions are heightened in the Lao context for two reasons. First, the electricity to be generated from Theun Hinboun and other planned dams will be sold abroad, mostly to Thailand, which means that the only benefit to Laos that comes from the transaction is foreign exchange earnings. With no 'development' to be generated along with the electricity, this income must therefore be measured directly against the social and environmental costs. Second, unlike most dams built in the Third World up to the late 1980s, which were funded with public money, foreign private companies are investing in Laos' hydro-power development. There is thus a direct connection – in fact, an inverse relationship – between the costs of

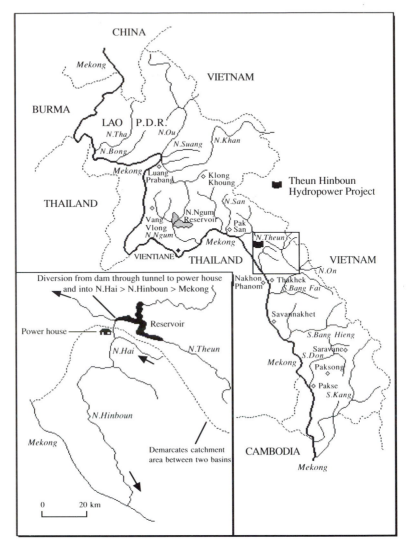

Figure 6.1 Laos, showing main hydro-power locations

avoiding or mitigating impacts in Laos and the profits to be gained abroad by these firms.

Influencing the economic benefit versus environmental cost equation of projects like Theun Hinboun is a host of actors, each with its own set of interests. The Lao government, in transition from a socialist to a market-based economy and desperate for hard currency, clearly sees hydro-power development as a quick means of generating income. But in the absence of technical expertise, strong institutions, relevant legal structures, and even a

124

sufficient number of officials with a working knowledge of English, Laos is in a weak bargaining position with respect to the other players on the scene. Northern-based construction companies and suppliers of dam-related equipment such as turbines, generators and transmission lines have seen their domestic markets shrink due to environmental concerns at home. They have responded by increasing sales, and sometimes also manufacturing, in countries of the Third World. Laos is today the scene of intense competition among these firms. To make such technology affordable, development aid institutions like the World Bank provide financing for hydro-power projects, even though pressure from environmentalists has made the Bank increasingly shy of very large projects.

Bilateral agencies typically offer 'tied aid' for dams, thereby subsidizing their national dam-building firms. So, for example, the Australian government gives aid in the form of Australian equipment, while the French, Austrians and Canadians do the same. In the Theun Hinboun project, the principal bilateral connections are with Sweden and Norway; hence the focus in this chapter on the 'Nordic connections'. Meanwhile, in the region, another important actor is the newly revamped Mekong River Commission, which has updated its historical mandate to co-ordinate dam-building throughout the Mekong basin. This river authority is represented by the governments of the four lower riparian states (Thailand, Laos, Vietnam and Cambodia), and is financed primarily by the United Nations Development Programme, as well as by bilateral donors.

Finally, private consultant firms play a crucial role as go-betweens among the above-mentioned actors. With close ties to the home dam-building industry, funding from aid agencies (both multilateral and bilateral), and well-established working relationships with the Lao government, these firms play a key role in the process because they are often hired not only to assess the social and environmental impacts, but also to design the projects. This dual role, and the potential for conflict of interest that it generates, are a fundamental weakness in the mechanism; one that practically ensures that negative effects of projects are systematically glossed over.

The Norwegian consultancy firm Norconsult that was hired by the Norwegian Agency for Development Co-operation (NORAD) to study Theun Hinboun concluded that no people's land or homes would be flooded and no serious negative environmental impacts would be expected. In fact, the height of the dam was designed so that the reservoir would not inundate any of the thirteen surrounding villages. Based on this assessment, no compensation or mitigation measures were written into the project design, and the impacts of the project would not have been discussed further had it not been for the involvement of another set of actors from civil society. These include non-governmental advocacy groups, journalists, and individuals in governmental environment agencies that monitored the decision-making process in Norway, and stimulated a public debate around Theun Hinboun.

Also monitoring the project were environmental groups and journalists in Thailand. Added pressure came from village movements in Thailand who have demanded compensation for damages caused by similar projects, such as the Pak Mun Dam in Ubon Ratchathani province (*The Nation*, 7.11.1994). As a result of the pressure created by all of these, the proponents of Theun Hinboun were forced to give more attention to environmental and social concerns than they otherwise would have done. The debate illuminated several impacts that the consultants had failed to identify: that seasonal agricultural land will be permanently flooded by the reservoir of Theun Hinboun; that natural fisheries on which local people depend for food and income will decline; and that the combined impacts of further planned dams on the same river will exacerbate these impacts. NORAD was even forced to reject the Norconsult study due to its poor quality and the vested interest of its authors. But in the end, the same firm was hired again to undertake the technical design for Theun Hinboun (*Development Today*, 6.7.1994).

Theun Hinboun bears many similarities to other ongoing projects in Laos, but it is portrayed by its Nordic proponents as something of a 'kinder and gentler' dam because of its smaller size, the fact that the Lao government maintains a majority ownership, and, not least, because of the involvement of Nordic interests. Nordic agencies and companies tend to view themselves as being less predatory than their competitors; of having the good of Laos, and other countries of the South, at heart. While this is probably something of an exaggeration, access to information laws and strong environmental movements in Sweden and Norway may ensure greater public scrutiny of Nordic development aid spending. Nevertheless, the environmental review process for Theun Hinboun was riddled with irregularities, few of which were ever resolved in spite of the extensive public debate. And if this project is in fact kinder and gentler than the others, the future for the Lao environment looks quite bleak indeed.

Sadly, the one 'group' that has so far had little influence on the hydrodam discussion in Laos is the rural villagers whose subsistence is derived from a combination of rain-fed agriculture, hunting, fishing and gathering from the forests. Hydro-power will affect such communities directly by inundating their fields and homes, wiping out their fish stocks, blocking river transport routes, and destroying their forests; in short, undermining their primary sources of food, medicine and income. But the thousands of pages of project documents are written in English (see also Lang, this volume), and so even if they could read them, they possess little political leverage to influence project design, to ensure fair compensation, to demand a share in the benefits or – as villagers across the border in Thailand (Plate 6.1) have been doing increasingly over the last decade – to say no. With no indigenous Lao non-governmental organizations yet established, and limited press freedom, the political clout of these rural people is not likely to increase for many years to come.

Plate 6.1 Hydro-power station, North-East Thailand

The dilemma then with Theun Hinboun and the many other proposed dams, as Philip Hirsch put it in 1991, is this: given that the Lao government needs foreign income, and possesses huge hydro-power potential, how can this best be developed to provide sustainable levels of income without so disrupting local people's livelihoods, and the environments on which they depend, that the benefits become meaningless for the majority of the rural populations (*The Nation*, 24.4.1991)?

THE REGIONAL CONTEXT

While Theun Hinboun will be one of the first projects in this era of privately financed dams, plans for dams in the Mekong basin are hardly new. In fact, the plans date back as early as the 1950s when a retired general of the United States Army Corps of Engineers, Raymond Wheeler, headed a mission to study the hydro-power potential of the Mekong River. Wheeler's recommendations resulted in the creation by the United Nations of the Bangkok-based Mekong Committee in 1957, a body whose mandate was to promote and co-ordinate development of the lower mainstream. Decisions were to be made by the governments of the four lower riparian countries – Thailand, Laos, Vietnam and Cambodia – with extensive financial and technical assistance from Western donor agencies such as the United Nations Development Programme (UNDP).

Wheeler and his team envisioned a great cascade of seven dams along the Mekong's mainstream from Northern Laos down to Cambodia's Great Lake that would produce more than 20,000 megawatts of electricity (Lohmann, 1991, p. 61). In the tradition of American megadams like Hoover and Grand Coulee, Wheeler's men proposed that massive structures be built along the mainstream, and dozens more on the tributaries. The first of these was to be the 250 metre High Pa Mong that would alone generate 4,800 megawatts of electricity, necessitating the removal of a quarter of a million people on the Thai and Lao sides of the river. But decades of war made the construction of such large infrastructure projects impossible. And for thirty-eight years, the Committee sat virtually idle, unable to fulfil the mandate that General Wheeler had inspired.

The leap from High Pa Mong as it was envisioned in the 1960s to the comparatively tiny Theun Hinboun of the 1990s reflects the emergence during the last thirty years of public concern about human rights and the ecological impacts of dams. In such circumstances, it is easy to understand the attraction of a project which, according to the original claims of Norwegian consultants appraising Theun Hinboun, would have no negative environmental impacts and would not necessitate resettlement. By the same measure, Western donors who must answer to environmentally conscious constituencies in times of shrinking aid budgets, are also becoming wary of very large dam projects. By 1991, the Sardar Sarovar dam on India's Narmada River, requiring the resettlement of people on a similar scale to that originally envisaged for the High Pa Mong, had become a world-wide symbol of all that can go wrong with big dams in the tropics (Morse and Berger, 1992). The international movement against the project caused the World Bank to withdraw its support. Furthermore, because of the Narmada debate, 'the Bank has been through the fires of hell to strengthen [its] environmental regulations [for dams]' (John Besant-Jones, Principal Energy Economist at the World Bank, interview, September 1994). The result, much to the industry's chagrin, is that the Bank is approving fewer dam projects, and the world's traditional financier of big dams in the Third World appears to be pulling out (Patrick McCully, of International Rivers Network, personal communication; *International Water Power and Dam Construction*, May 1994).

In the United States, concerns about the long-term ecological impacts of dams, and the effects on native communities caused Daniel Beard, the head of the US Bureau of Reclamation – an institution that has played a similar role in dam-building internationally as the US Army Corps of Engineers – to announce that 'the dam-building era in the United States is now over' (Beard, 1994). Beard says the Bureau will no longer promote dams in the Third World, but offers a candid assessment of why the practice continues:

> There is a substantial infrastructure that surrounds dam-building. There are people in the business to make money. They are spreading around

128

the globe trying to encourage dam construction. They are less interested in the problems because promoting dams is their business.

(*The Nation*, 23.11.1994)

Similar in some ways to the Bank and the Bureau, Sweden's Agency for International Development Co-operation (formerly the Swedish International Development Authority: henceforth the acronym SIDA will be used for both) has taken a firm position against dams on the Mekong's mainstream. The reasons for this are rather particular, but illustrate the tensions within the Swedish aid establishment about how to handle dams: SIDA has during the last decade been one of the major bilateral donors (along with Australia and the Netherlands) to the Mekong Committee, providing financing mostly for environment-related work in the basin. In 1990, Erik Skoglund, a Swede working at the Committee, came into conflict with the then Executive Agent of the Mekong Committee Chuck Lankester over Pa Mong. By this time, the project had been radically scaled down (and re-named Low Pa Mong), largely to reduce the size of the reservoir. Lankester wanted to push it through, as this would be the first dam on the mainstream and a clear sign of vitality for an agency whose very existence many were coming to question. But Skoglund had reservations about the resettlement of 60,000 people, and the impact on fisheries: 'fish production will go down and it will cause a lot of damage. This is not just an environmental issue. It's a livelihood issue, especially for poor people. To destroy that resource is just not acceptable' (*Development Today*, December 1994; and *Far Eastern Economic Review*, February 1991).

The row ended with Skoglund's resignation. But before he left, a SIDA mission to Bangkok made their support for Skoglund's position clear:

Without taking a stand on whether [Mekong] mainstream projects like the Low Pa Mong are economically and technically feasible and environmentally acceptable, the members of the SIDA delegation expressed that . . . it would be difficult for the [Committee] to find soft financing for such projects. Resettlement schemes involving 60,000 people constitute a major obstacle to financiers such as SIDA . . . and other financiers including the development banks.

(SIDA briefing, April 1991)

One wonders, in the light of recent developments, whether SIDA would have taken such a strong stand had the run-in not occurred.

Clearly though, neither SIDA nor NORAD has a general anti-dam policy. The two Nordic agencies have provided joint financing for dam projects in several countries in the past five years, including Pangue in Chile, Pangani Falls in Tanzania, and Epupa Falls in Namibia. Incidentally, the decision-making processes of these three projects were also criticized in Sweden, Norway and the recipient countries for downplaying social and ecological impacts in favour of industrial interests. But while NORAD is an active promoter of hydro-power – and has even suggested lowering environmental standards in Third World projects to make them cheaper for recipient governments (Grimstad, 1994) – SIDA's position is less clear. The Swedish agency is more wary of dams that require the resettlement of large numbers of people. After supporting the construction of the 45 megawatt Xeset Dam in Southern Laos, which was completed in 1991, the agency pulled out of the Lao energy sector altogether to allow Swedish resources to be used in areas like forestry and road-building. However, some SIDA officials regret this decision and argue that with its expertise in dam-building and progressive attitude, SIDA could have helped to protect Laos from less altruistic foreign interests in this era of privatized dams (Johnny Andersson, Energy Division, SIDA, interview, September 1995).

Following the Skoglund fiasco, SIDA also reduced its support to the Mekong Committee over the 1992–1994 period, partly because of its lack of a clear mandate. This was reportedly due also to concerns about the overall environmental impact of dams on both the mainstream and tributaries, as well as Thailand's water diversion projects that will decrease the flow into the Mekong from the Thai side: 'The whole river basin is a very fragile system. Any project can have consequences downstream. If you build a dam, environmentally you must be very careful, especially with the Mekong Delta,' SIDA's Mikael Bahrke said (*Development Today*, May 1995).

There can be no question, though, about the Committee's role in promoting dam-building. In December 1994, the Mekong Secretariat published a study, entitled *Mekong Mainstream Run-of-River Hydropower*, which identified eleven dams on the mainstream whose 'scale of development [was] deliberately constrained to avoid or to minimise impacts' (United Nations Development Programme, 1990, p. 1). Still, most of these are massive projects, with generating capacities of over 1,000 megawatts. Five months later, in Chiang Rai, a new Mekong River Commission was formed by the four lower riparian states, thus replacing the four-decade-old Committee. In spite of the blatant contradiction between the new Commission's mandate and SIDA's position on mainstream dams, the Swedish agency renewed and increased its three-year support for the Commission's environmental work. By all accounts, the new river authority has even less power than the Committee it replaced to deter individual states (like Thailand) from diverting water from the mainstream, and its dam-building agenda is now clearer than ever (*Bangkok Post*, 31.3.1995; 11.4.1995). Some thirty Thai NGOs and local water basin groups issued a statement in Chiang Rai, opposing the 'influence of the dam-building industry' in the creation of the new Commission: 'In this era of industrial-based economic development . . . those governments with more political and economic power have a tendency to oppress less powerful neighbouring governments and people, and therefore new conflicts are likely to emerge,' they warned (Thai NGOs, 1995).

THE NAM THEUN HINBOUN DAM

The site of the Nam Theun Hinboun dam is between Bolikhamsay and Khammuan provinces on the Theun River, the fourth-largest tributary of the Mekong, some 100 kilometres upstream of the mouth. As with most dams being planned in Laos today, all the electricity to be generated by Theun Hinboun – 210 megawatts – will be sold across the border to the Electricity Generating Authority of Thailand (EGAT). Ownership in the US$280 million project is divided among the Lao state utility, Electricité du Laos (60 per cent), through a US$60 million loan from the Asian Development Bank (ADB); Nordic Hydropower, a consortium of two hydro utilities, Sweden's Vattenfall and Norway's Statkraft (20 per cent); and MDX Public Company, a Thai real estate firm (20 per cent) (TERRA, Nordic Hydropower briefing, 1995). Profits are to be divided pro rata according to shares of ownership. The Lao government expects to earn between US$44–83 million in foreign exchange annually. This arrangement will hold for thirty years, after which, presumably, ownership will revert back to the Lao government. Theun Hinboun is the Nordic utilities' first investment outside the Nordic region.

In addition to the ADB loan, the Norwegian Agency for Development Cooperation (NORAD) has provided three grants to the Lao government for the Theun Hinboun project: in 1993, NOK10 million for a project survey, including an environmental impact assessment; in 1994, NOK38 million for technical design; and NOK3.5 million for supplementary environmental studies and a water management plan. Norconsult, the largest Norwegian consultant firm working on hydro-power projects in the Third World, was hired for the first two. The latter was divided among a number of Norwegian private consultancies and research institutes.

Theun Hinboun is a trans-basin diversion project, which will divert up to 100 m³/s of the Theun's flow into the Nam Hai, a tributary of the Nam Hinboun, which in turn flows into the Mekong; hence the name, Theun Hinboun. Project designers have made use of the 240 metre difference in elevation between the Theun and Hinboun basins for power generation. Water will be diverted by the 25-metre high dam to the power station and then through a 4 kilometre tailrace canal into the Hinboun river, and thence back to the Mekong about 30 kilometres upstream from Thakkek. Electricity will be sent to Thailand via a 230 kV transmission line, 100 kilometres on the Lao side to Thakkek, and, from the border, another 90 kilometres to the connection point at Sakon Nakhon.

The Theun river's mean annual flow is 460 m³/s, with dry season flows well below this. Diversion implies that the river bed will be 'reduced to a series of pools' over a 40-kilometre stretch downstream of the dam in February, March and April during years of normal rainfall. Upstream, the headpond will extend 24 kilometres along the Theun, 14 kilometres along the principal tributary, Nam Gouang, and there will also be smaller intrusions into two other rivers, Nam Ao and Nam Pheu. Theun Hinboun has been designed with the assumption that a much larger dam will be constructed up-stream on the Theun by the year 2000 – the 600 megawatt Nam Theun 2. If Nam Theun 2 is built, it will cut the flow into the Theun Hinboun by half, thus exacerbating downstream effects and decreasing somewhat the electricity generating potential.

Norconsult's environmental impact study concluded that the project has 'significant *beneficial* environmental impacts' (Norpower, 1993, pp. 1–7; my emphasis), while the only adverse impact of the project identified is the reduction of flow below the dam site. The project will require no resettlement of local people, and cause no negative impacts on fish, the final report assured. Their conclusions were disputed by most parties that reviewed the documents – various Norwegian government agencies, the press and non-governmental organizations in Norway and Thailand, and even the environmental adviser of Vattenfall – with the result that NORAD was forced to pay for supplemental studies.

Several key issues were absent from the consultants' report:

- Permanent flooding of seasonal agricultural land along the river banks that is planted by people living in the thirteen villages located along the reservoir during the dry season in times of low flow.
- Blocking off of migration routes for the more than 100 species of fish that are estimated to inhabit the Theun basin, and degradation of this vital source of food and income for local people.
- Impacts on forests caused by the transmission lines, access roads, and other logging activities stimulated by the project.
- Estimation of how these impacts will affect local people, mitigation measures to lessen these impacts, and compensation for their losses.
- Finally, an assessment of the cumulative impacts that will result if the Nam Theun 2 is built up-stream.

As we have seen, the supplemental studies were commissioned only in response to the criticism which followed the evaluation of the original report. However, these have had no impact on the decision-making process. Before they were completed, Norconsult was hired again by NORAD to do the technical design for Theun Hinboun.

Norconsult called Theun Hinboun a 'relatively small project', which it is by Nordic standards, and, certainly, in comparison to some of the other dams being planned in Laos today. Yet the size should be judged by the current local context. (Incidentally, the International Commission on Large Dams (ICOLD)'s definition of a large dam is 15 metres high: Theun Hinboun is 25 metres.) There are currently two dams generating electricity in Laos, the Nam Ngum (150 megawatts) and the Xeset (45 megawatts), built during the 1970s and 1980s, respectively. When it is completed in 1997, barring the completion of other projects before, Theun Hinboun will be the largest infrastructure project in Laos, alone doubling the country's installed electricity generating capacity.

DAM-BUILDING IN LAOS: THE BOOT ERA

If Laos is to increase foreign exchange earnings quickly, then on the face of things, hydro-power is an obvious option; one that the Lao government has embraced with fervour. A number of economic and political conditions have shaped the current dam-building frenzy in particular ways.

First, the Lao PDR can clearly not afford the minimum quarter billion dollar investment that a dam like Theun Hinboun would require. Laos is one of the poorest countries in Asia, with a Gross Domestic Product of just US$1 billion and exports of US$209 million in 1994 (*The Nation*, 5.6.1995). Furthermore, Theun Hinboun is only one of several dams in the pipeline. The Nam Theun 2 upstream, for example, has a price tag three times this amount (*The Nation*, 23.6.1995). Dam-building in Laos is thus unimaginable without foreign investors – whether public or private – playing a crucial role.

Second, the country consumes only a third of the 215 megawatts of electricity that its two main hydro dams can produce, selling the rest to Thailand. With no domestic market to speak of, hydro dams are being built purely to bolster the country's export economy. This is relevant in discussions about sustainable development in general, and about dams in particular, which are normally couched in terms of costs versus benefits of development within a given country. With large infrastructure projects like dams, as the argument goes, some have to suffer for the benefit of the majority. In the case of the dams being built and planned in Laos today, however, the 'costs' will all be borne in Laos by the environment and rural peoples, while the 'benefits' will be exported to Thailand. Apart from the cash, Laos will not gain 'development' in any conventional sense. And, to judge by the experience of her neighbours, and her own very pressing economic needs, it is extremely doubtful that the affected communities, and environments, will derive any direct developmental benefits from the cash windfall that the Nam Theun Hinboun will provide the Lao exchequer. (Presumably, if the numerous aid agencies supporting dam construction in Laos believed that foreign exchange earnings were equivalent to development, they would simply sign blank cheques over to the Lao government.) For this reason, assessment of the environmental and social impacts of dam projects acquires a heightened importance.

Third, Laos shares a 1,800-kilometre border with the region's most rapacious energy consumer, which will likely remain the sole purchaser of Laotian electricity for many years to come. By the end of 1993, peak power demand in Thailand was about 12,000 megawatts, representing a greater than 70 per cent increase since 1987 (Sherman, 1995, p. 7). By far the major power user is the industrial sector which in 1993 consumed 46.7 per cent, followed by the commercial sector, 26.5 per cent, and the residential sector, 20.7 per cent (*Bangkok Post*, 3.3.1994). Electricity demand in Thailand is rising by 10 to 15 per cent per year and could, by one estimate, increase seven-fold by the year 2020 (*The Nation*, 5.6.1995). Certainly, Laos has a guaranteed market. But with the Electricity Generating Authority of Thailand (EGAT) as the only buyer, Laos' bargaining power is, by definition, limited. Laos cannot play one market off against another. And although EGAT has agreed to buy 1,500 megawatts from Laos by 2000, Thailand is far from dependent on Laos for its energy (*The Nation*, 10.1.1995). Officials of Electricité du Laos (EDL) complain, understandably, that price negotiations with their Thai counterparts have been 'difficult' (*The Nation*, 5.6.1995).

A fourth condition that is shaping the politics of dams in Laos is the role of the private sector under so-called 'BOOT' – or Build, Own, Operate and Transfer – schemes. According to this arrangement, groups of investors (that in theory may or may not include the national government) finance, build and operate a dam for periods of twenty to thirty years, after which they re-negotiate the deal, or transfer it to the government. Private investors would provide the capital that once came from development aid budgets. Such schemes differ substantially from the traditional approach to dam-building in the Third World, where public money, either from the national government or from aid institutions like the World Bank, played the key role. In the name of increased efficiency, privatized energy markets and a growing suspicion of subsidies for energy production, the World Bank has endorsed the BOOT concept since 1989 (Lohmann, 1991, p. 62). Nordic investors have stressed that with a majority share (60 per cent) in Theun Hinboun, the Lao government can maintain control over how the resource is used. In comparison, the government will own only 25 per cent of Nam Theun 2.

BOOT may relieve countries like Laos of some of the financial burdens that the old system for financing dams imposed. Quite apart from the well-documented environmental and social problems associated with dams in the tropics (Goldsmith and Hildyard, 1984; Pearce, 1992), recent studies raise further questions about the economic viability of large dams. The Bank's principal energy economist, John Besant-Jones, who is conducting a financial audit of seventy dams financed by the World Bank during the past thirty years, points to the economic risks of publicly financed dams. He has observed an average of at least 30 per cent cost over-runs and schedule slips, which have made dams significantly more expensive than they were supposed to be. Moreover, there appears to be a correlation between the size of dams and indebtedness; that is, the bigger the dam, the bigger the debt (Adams, 1991; *Development Today*, October 1994). Besant-Jones specifically raises the example of Columbia whose 'hydro boom' of the 1980s contributed to the country's national debt (Besant-Jones, interview, September 1994). Such experience could have a special relevance for Laos.

The World Bank economist attributes these gross miscalculations to what he calls a 'pervasive appraisal optimism' in project documents, which tend to over-estimate the benefits while down-playing the costs (Besant-Jones, 1994, p. 42). But he concedes that the burden of these oversights rests with the client government. They still have to pay back the loans.

BOOT schemes would, in theory, lessen the investment risks for the Lao government, transferring them to the private financiers. Laos still must repay US$60 million to the ADB for Theun Hinboun, though probably under the old system, this loan would have been substantially higher. Proponents call the arrangement a 'win-win' situation that provides electricity for the buyer, and profits for the producer. A Thai commentator describes a glowing future for Laos: 'With its small population, dramatic topographical mix of high and low lands and strong-handed government, Laos is seen as near perfect dam land.' (*Bangkok Post*, 4.6.1994). Laos could, in the most optimistic scenario, become the 'Kuwait of Asia' (*The Nation*, 4.3.1994).

While it might alleviate some debt worries, the BOOT system raises other tricky questions a SIDA-financed review (de Vylder and Sonnerup, 1994, p. 38) warned:

> We . . . fear that the massive inflow of foreign capital that would be required to finance the planned hydropower development may jeopardize the Lao PDR's possibilities to maintain a minimum of national control over basic natural resources and, indeed, over the general economic development of the country.

As a government's share in a project shrinks, who will co-ordinate negotiations for electricity sales between EGAT and a myriad of foreign contractors? In this 'rent-a-river approach', what guarantee is there that full commissions on profits – the rent – is paid to the Lao government? And in what condition will plants be when they are finally handed over after thirty years? If turbines need replacing, and reservoirs are filled with silt, will national governments be able to bear these costs? And what if after several decades of use, as is occurring in Thailand, the dams no longer work? 'After all . . . there must not be just a heap of junk to be turned over [to the government after twenty to thirty years],' noted the Swedish report (ibid., p. 38).

It is also likely that environmental and social impacts will be given less attention under the BOOT system, as foreign private investors have neither a mandate to alleviate poverty nor a parliament and citizenry to which they are legally accountable.

Ultimately, the BOOT scheme raises the question of sovereignty. By renting out its rivers on the basis of BOOT, a country could lose effective control over its land and natural resources (Ryder and Rothert, 1994, p. 9). At the very worst, the Lao government could find itself in a relationship with foreign contractors that leaves it as much say over national energy policy as a vassal state in the colonial era.

Finally, Laos is attractive to dam-builders because they are unlikely to encounter the sort of public opposition and debate that have given dams such a bad name both in their own countries and in Thailand. Years of resistance caused the Thai government to announce in 1995 that no more power dams will be built in the country (*Bangkok Post*, 24.2.1995). For Northern firms that have seen their markets shrink due to environmental concerns in the industrialized world, and have watched anti-dam sentiments in Thailand rise, Laos offers welcome relief. As Vattenfall's Anders Hedenstedt remarked in 1993:

> The funny thing about [Theun Hinboun] is that there are almost no environmental problems. It's a run-of-river dam that will not use any surface that is not already flooded during the rainy season. This is the only project in the region that seems to have no opposition.
>
> (*The Nation*, 13.4.1993)

Without a free press or indigenous non-governmental organizations, there is no safe forum in which to challenge such claims, and open debate within the country remains difficult. Laotian critics of dam projects, whether government officials or private individuals, tend to word their comments extremely carefully, and more often, speak only off-the-record.

NORDIC CONNECTIONS

Viewed from the Nordic perspective, the shape of the Theun Hinboun dam that is currently under construction in central Laos has as much to do with the Mekong regional constellation as it does with the politics of aid and the dams industry in Sweden and Norway. In this sense, Theun Hinboun should also be seen as the product of the convergence of various Nordic interests.

The Nordic company that owns 20 per cent of Theun Hinboun was formed to implement this project in 1993 by Vattenfall, the Swedish state electricity utility, which has built many of Sweden's dams, and Statkraft, which is its Norwegian equivalent. Theun Hinboun marks a unique shift for these utilities because it is the first dam project that either has undertaken outside the Nordic region. Partly, this shift is a necessary response to the widespread public resistance to dams in Sweden and Norway that has stopped construction of all but the smallest projects at home (Lövgren, 1994, p. 56). Since 1987, Sweden's Natural Resources Act formally forbids the exploitation of the country's last four free-flowing rivers – Torne, Kalix, Pite and Vindel – for hydro-power. In Norway, there was a bitter struggle in the early 1980s over Alta, a dam project in northern Norway. Alta was eventually built, but it would be the last of its kind. Remarking on the political costs to the government of pushing it through, Norwegian anthropologist Terje Brantenberg comments that Alta became like a 'unit of measurement, so that today when Norwegian politicians are considering a controversial project, they ask if it is worth another Alta' (Terje Brantenberg, interview, May 1995).

In the words of Vattenfall's Senior Manager Karl-Erik Norlander: '[w]e have completed the implementation stage in the Nordic countries. There are very few new projects in Sweden and Norway. But still we need professionals in the field, so we go abroad to find possibilities to use our skills' (*Sveriges Natur*, May 1994). Indeed, as the home market shrinks, the dams industry in the Nordic and other regions has found alternative sites in the Third World. Nordic multi-nationals have been especially successful on the international dams market. Sweden's Atlas Copco, which produces rock excavation and drilling equipment, is the world leader in the field (Hans Liljeström, Atlas Copco, interview, April 1995). The same is true for the Oslo-based Kvaerner, which supplied turbines for 13 per cent of all hydro dams built in the world during the last decade (Per Berg, President, Kvaerner Hydro, interview, May 1995). Kvaerner, in fact, won the turbine contract for Theun Hinboun in 1994 (*Inside Kvaerner Energy*, 1995, p. 5). The Swedish construction firm Skanska is expanding its international operations from 10 to 25 per cent of their portfolio (*Skanska Annual Report*, 1993), while the Swedish-Swiss Asea Brown Boveri, or ABB, is one of the world's major manufacturers of electrical equipment and generators.

Nordic dam-related companies are active in the Third World, but ownership is something new. One might well ask the question: why start with Laos? Part of the answer has to do with the traditional reliance of the dams industry on development aid budgets. SIDA has a presence in Laos which stretches back several years. Sweden's anti-war policy during the 1970s caused Laotians to view the Nordics in a favourable light, an historical relationship from which Nordic dam-builders are now benefiting: 'SIDA has been working in Laos for more than ten years, and the Lao people tend to trust the Swedes,' explains Karl-Erik Norlander of Vattenfall. He says that personal contacts played a key role in putting together the deal for Theun Hinboun, describing the 'close links' between business opportunities and development assistance (*Sveriges Natur*, May 1994).

One such link is the Xeset Dam in Southern Laos (the country's second dam; a 'pre-BOOT' project), the construction of which began in 1981. Financing was provided by SIDA, NORAD and the ADB, while Norconsult and Kvaerner won contracts on the project. Gjermund Saetersmoen of Norconsult, who worked on Xeset from the beginning, lobbied actively for Theun Hinboun with Lao officials (Gjermund Saetersmoen, interview, June 1994). A second link is Zia Noorzay, a former employee of ADB who now works for Vattenfall. Noorzay worked on behalf of the Bank during the 1970s on Laos' first dam, the Nam Ngum.

Then in 1992 and 1993, Norlander himself was hired by SIDA to work as an energy advisor to the Lao government (this contract was SIDA's final contribution to the Lao energy sector). While in Laos, Norlander gave advice about how Laos might begin to harness some 15,000 megawatts of unexploited hydro-power, and to prioritize a dizzying number of projects. At the end of two years, Theun Hinboun came out on top; the Swedish consultant arguing that Laos should begin slowly with small projects where the government maintains a majority share in the ownership. This view is shared by SIDA (Karl-Erik Norlander, interview, March 1994; and Johnny Andersson, interview, September 1995). Norlander subsequently joined a private Lao-based firm representing Nordic hydro-power developers.

Similar close connections can also be identified in the Nordic region, between key officials in the aid agencies and the companies, where, to take a few examples: the Director-General of NORAD, Per Grimstad, was formerly the head of Norconsult; Alf Samuellson, the former Permanent Secretary of the Swedish Ministry for Development Co-operation (who argues that more aid should be channelled into areas where Sweden has expertise like dams) was head of Vattenfall's Finance Division until 1991; and where Anders Hagvall formerly at SIDA's Energy Division, worked previously with ABB (*Sveriges Natur*, May 1994). Following a restructuring at SIDA in July 1995, seven of the eight officers in the department of concessionary credits left to join multi-national companies like Ericsson, ABB and Volvo (*Development Today*, 27.7.1995).

Because of their role in assessing the environmental impacts of dams, the nature of the relationship between consultants and aid agencies is especially important. Agencies like SIDA and NORAD choose not to use in-house environmental expertise to evaluate the impacts of infrastructure projects like dams. Rather, they hire 'independent experts' to conduct studies, and then base final decisions on these reports (Mats Segnestam, Environmental Advisor, SIDA, interview, September 1993). These consultants occupy a fascinating grey zone between the public and private sectors. They depend almost exclusively on aid money to finance their work on hydro projects in the Third World – whether for technical, economic or environmental studies of dams. They are, in a sense, direct recipients of Nordic development aid. Lennart Lundberg, Director of Hydropower for Swedpower, for example, estimates that half his company's dam-related contracts are paid for by Swedish aid money, while international institutions like the World Bank or the Asian Development Bank fund the rest: 'There is nothing special about Sweden or Swedpower in this regard. You can debate on whether or not this is sound, but that's the way it works in Japan, Germany, England, everywhere,' he says (*Sveriges Natur*, May 1994). In this regard, they are like an extension of the aid bureaucracy, but with none of the regulations (like public access to information rules) that govern agency activities.

In theory, the studies assist decision-makers by providing them with a neutral assessment of the costs and benefits involved. The system is predicated on a belief in the political neutrality of the subject matter, and on the capacity of these individuals to maintain objectivity. In practice, the documents tend to be overwhelmingly positive. They smack of the 'pervasive appraisal optimism' that John Besant-Jones identified in his review of seventy World Bank-funded dams.

The optimism of environmental impacts studies can also be explained by the consultants' close ties with the industry. Only three companies dominate the international hydro-power consultancy market in Sweden and Norway. Swedpower is owned mostly by Vattenfall. Sweco is a daughter company of the engineering firm VBB-Viak, which also built many of the Swedish dams. And Norconsult is a consortium of Norwegian consulting firms working abroad, and is partly owned by Statkraft. When the same consultants are hired to do both the environmental studies – to decide whether to proceed with the project – as well as the technical studies – to determine how to design the project once it has been approved, conflicts of interest invariably arise. Norconsult's environmental study for Theun Hinboun, which a NORAD official described as 'asking the fox to watch the geese', was eventually discredited for precisely this reason (*Development Today*, 15.10.1993).

THE POLITICS OF 'APPRAISAL OPTIMISM'

Understanding the environmental review process for projects like Theun Hinboun, and the reasons why assessments tend to be so positive, necessitates the somewhat unpleasant task of seeing through the emperor's 'new clothes' – or rather, in this case, those of his advisors. Like the naked ruler whose subjects are made to believe he is dressed, the politics of appraisal optimism is a process of erecting and maintaining around consultants a protective wall that hides their expertise from public scrutiny. Any little boy (or girl) can clearly see that the game is rigged in favour of a positive outcome. Environmental and social impact reports are written, it often seems, as if the authors did not expect anyone to read them seriously. To function smoothly, in fact, the mechanism depends on the public – like the emperor's subjects – dutifully turning their eyes away.

The environmental review of Theun Hinboun reflects a struggle between two competing views of Nordic aid. On one side is a strong environmental profile. Historically, this has been enhanced by Sweden's hosting of the 1972 environment conference that, in turn, spawned the 1992 Earth Summit in Rio de Janeiro, and by the association of Norwegian Prime Minister Gro Harlem Brundtland with the concept of sustainable development. These, coupled with the general public perception of aid agencies helping the 'poorest of the poor', have made it politically difficult for agencies to ignore the negative effects of the dam projects they finance on rural populations and the environment. Pierre Schori, of Sweden's Ministry for Development Co-operation, told Thai journalists during a visit to the Mekong region in 1995:

When we go into energy projects, we make it a point that everything done is in harmony with the people living in the area as well as the environment. We are very sensitive on this issue, given our own environmental profile and policy.

(*The Nation*, 16.6.1995)

On the other side is the increasingly popular view of aid as a form of export credit; a way of transferring state funds to national firms seeking foreign markets. While OECD regulations define the conditions under which this is permitted, Nordic industries naturally lobby their governments for larger subsidies, and the aid agencies find creative ways around the rules in order to use aid money to support Swedish and Norwegian companies (*Development Today*, 28.4.1994; 6.7.1994; 7.11.1994).

According to the agencies' decision-making process for hydro-power projects, environmental impact assessments (EIAs) must be written by independent experts, and completed before projects are decided upon. In NORAD, the impact study must then be reviewed by at least three government agencies outside of the aid bureaucracy: the Directorate for Nature Management, the State Pollution Control Authority and the Norwegian Water Resources and Energy Administration. This rigorous external review procedure owes its existence to a relatively high public interest in the environmental impacts of Norwegian development projects. Its virtual lack of influence on the outcome of projects, however, indicates that the industrial focus of aid spending in Norway, and to some extent Sweden also, remains stronger than the environmental concern.

The appraisal optimism of Norconsult's environmental impact study for Theun Hinboun is clearly reflected in its conclusion: 'The project is highly viable and well justified. Immediate implementation, which commences with preparation of final design and preparation of tender documents for procurement of plant and civil works, is therefore recommended' (Norpower, 1993, pp. 1–8). This recommendation was received without critical comment by NORAD officials in charge of the project in 1993. It was not until the external reviewers had voiced their concerns that the public debate began.

The State Pollution Control Authority recommended to NORAD: 'the environmental assessment report is far from satisfactory . . . we cannot recommend implementing this project on the basis of the existing data', while the Norwegian water authority warned that 'there is potential for far-reaching environmental and social disturbances if these matters are not taken seriously' (*Development Today*, 15.10.1993). Meanwhile, the Directorate of Nature Management noted that of the nine claimed positive effects of the project, three were 'highly questionable', and five were 'rather insignificant, indifferent or liable to different interpretation'. The Directorate also delineated numerous omissions, the most serious of which was the potential additional impacts of Nam Theun 2 (Norway: Directorate for Nature Management statement, 1993).

The Oslo-based International Association for Water and Forest Studies (FIVAS) criticized NORAD for violating its own environmental procedures (FIVAS, 1994, p. 60). Another non-governmental environmental group, the Bangkok-based Towards Ecological Recovery and Regional Alliance (TERRA) commented:

> [the consultant] is aware that fish are a major source of protein for local people and also that fish provide people with food security in times of poor rice harvests. Inexcusably, this . . . is not matched by a commitment to conduct a careful and rigorous assessment of potential losses and corresponding impacts on local people.
>
> (TERRA communication to FIVAS, 4.8.1993)

Speaking at a public meeting in Stockholm in 1994, Napha Sayakhoummane, Vice-President of the Lao Women's Union, Khammuan Province Branch, noted further:

> The information I have is neither sufficient nor clear. There [should] be further study of this project, to analyse deeply both the possible positive and the negative impacts. And to give a chance to local government and local people to know about these plans, so that all concerned are able to express their opinions. Or at the very least, so that local people have an idea of the changes that are coming and have time to prepare.
>
> (Sayakoummane, 1994, p. 28)

A two-page brief, which she described as 'vague and propaganda-like', was all the information available from the builders to the Khammuan provincial administration in 1994.

Even Per Sjöström, senior ecologist for Vattenfall (the Swedish utility investing in Theun Hinboun), pointed to three crucial areas that were not covered in sufficient depth in the Norconsult report: water quality; fish production and fish migration; and material transport and sedimentation. He also noted the absence of concrete plans for fisheries management in the reservoir, rural electrification, irrigation, and mitigation of downstream impacts (Sjöström, 1994). He insisted, however, that his company had already decided to proceed with the project, and that these concerns could be addressed along the way (Sjöström, interview, March 1994).

The amount of information that was publicly available about the Theun Hinboun project in Sweden and Norway is a credit to the openness of these societies. It should be noted that in many countries in Europe – Germany, for example – all aid agency documents, including environmental impact assessments, are considered secret.

Given the widespread criticism, it is perhaps not surprising that NORAD eventually conceded that Norconsult should have been disqualified from the beginning because it is part-owned by Statkraft, the other Nordic owner of

Theun Hinboun. NORAD's deputy chief Sven Holmsen told the Norwegian press: 'Everyone now agrees that it was a mistake from the outset ... We agree that we don't have enough information. Relevant questions have been asked on environmental aspects of the project' (*Development Today*, 15.10.1993). The agency called for an additional environmental study to be done. As mentioned above, however, the project proceeded unchanged, in spite of all the issues that had been raised. Part of the reason for this may be the lack of direct accountability between decision-makers and those affected by the dam itself. True, Nordic-based NGOs, journalists and government agencies monitor development spending, on the grounds that they have the right to know how their tax money is being spent. But there remains no mechanism to link villagers in Khammuan province with officers at NORAD or SIDA; much less with private firms like Statkraft, Vattenfall or other companies investing in Lao BOOT dam projects.

An industry insider offers (off the record) another possible explanation for the persistence of this pervasive appraisal optimism:

> Consulting firms have a conflict of interest as long as they themselves may benefit from one outcome over another ... For example, if they find that a certain project is feasible, they are often in a good position to undertake the subsequent studies, design work and construction supervision associated with further project phases (which is often more profitable than the initial feasibility study) ... This conflict could be avoided if the evaluation was carried out by an impartial party which was aware that it would not subsequently receive any further project-related work, regardless of evaluation outcome.
>
> (anonymous, personal communication, 1994)

An essential characteristic of the current decision-making process is that it 'hides' the consultants from public view; it renders them invisible. Their assessments are at the heart of the issue, yet they – the experts – remain hidden behind the commercial nature of their contracts with the aid agencies, the anonymity of their reports, and the neutrality of their expertise. In general, they are under no obligation to deal with the public or the press, and usually refuse to release their reports without permission from the client.

Encounters with the experts reveal a strong bias in favour of hydro-power, as well as the ideologically charged nature of the subject matter. But arranging a meeting with the author of Theun Hinboun's environmental study was no simple matter. In the first place, as Norconsult's Gjermund Saetersmoen explained to me in 1993, the company had no-one with the appropriate expertise on the ecology of Lao rivers, and therefore sub-contracted the work to an Australian consultant based in the region, Charles Adamson. It was not until the following year when he visited Norway that I was able to organize an interview, during which he responded to the various critics: 'It is difficult for people living outside to understand the poverty in Laos. It is to me unethical and disappointing that people are wanting to hold back development in countries where the population is suffering', Adamson said. Such concern does not appear to extend to those who will be affected by Theun Hinboun, however. Adamson reiterated that he foresaw no need for compensation payments because no homes or paddy land will be flooded.

The forests being cleared to make way for 100 kilometres of transmission lines are merely 'secondary regenerating forests in fields that have been left fallow', and as the forest farmers working these fields have no legal ownership of the land, they are entitled to no compensation when it is appropriated. 'There may be a claim of ownership, but . . . if you want to talk about compensation, then you are referring to permanent agriculture. It would be unusual to compensate people for bits of regenerating ground that have transmission lines going through them,' he said. Similarly, farmers who practice seasonal agriculture in the 'draw-down' zone of the river banks in the area of the reservoir during the dry season have no right to compensation for the loss of this land: 'It is a very small area . . . [and] they can draw water from the headpool,' he suggested.

Though he admitted that he is not a fish expert, the consultant maintained that the Theun Hinboun will improve the situation for local people by increasing the level of water behind the dam, thus creating an ideal environment for raising fish (that is: more water, more fish). Even taken at face value, the idea was accompanied by no details in the report about how such a scheme would be organized, who would pay for it, and how the original fisherfolk of the area – accustomed to fishing indigenous species – would benefit.

The fact that the Theun River will be dried up for three months of the year, thus potentially destroying the habitat below the dam for some of the 140 species of fish, was not elaborated in the consultant's discussion. Though he agreed that, after Nam Theun 2 is completed, the river's dry period will extend to six months each year, Adamson argued that there is not enough yet known about the pattern of fish movement to be sure that the overall effects will be negative. Rather than opting for a precautionary approach, the consultant noted: 'Our attitudes have always been based on salmonid species . . . that have routes into specific areas. We are assuming that tropical fish also have definite migratory routes. The question is, would those fish that normally populate the Theun River downstream of the dam be likely to take an alternative route?' for example, up an in-flowing tributary like the Nam Mouan. The question remains unanswered by the Norconsult report.

In examining the possible impacts of the project on the Theun and Hai river basins, the Norconsult report recognizes only activities that conform with market-related, private property-based systems. It becomes evident that beyond the consultants' pro-dam bias, the environmental and social impact study was infused with an approach that undermines other forms of knowledge, and other ways of living. The lack of understanding, for example, of the various and complex farming systems practised in the area, of fish ecology, or of communities' dependence on fisheries for protein and income, are not viewed by the consultant as deficiencies. These are, rather, ignored; effectively equated to zero on the cost-benefit scale. Meanwhile, stated concerns about poverty alleviation are not matched with investigations about the potential for poverty creation as a result of the project.

LAOS'S DILEMMA

What then will be the social and ecological impacts of the Theun Hinboun dam? What costs will be borne by local people and by Laos as a country that should be weighed against foreign income earnings? What of Laos's dilemma: to earn cash, but in such a way that the impact on the rural majority is not too devastating? (Hirsch forthcoming). Clearly, the Nordic environmental impact study offers little help in answering these questions because most of the negative impacts were neither measured, nor described. Meanwhile, if Theun Hinboun is a less environmentally destructive dam than the many others currently being planned, as the Nordic builders claim, there is good cause for concern about the overall impact of the dam-building programme in Laos.

Around the world, in countries North and South, the emergence of civil society in the form of peasant movements, environmental groups, free press, independent scientists and so on, has forced a recognition of the ecological and social costs of dam projects. Ironically, it is this very environmental debate in Thailand and the industrialized world (and, paradoxically, the lack of it in Laos), that make the country attractive to dam-builders. They are engaged in a race for power in Laos – for hydro, economic and political power – taking full advantage of the country's vulnerability, and 'greenness' of the naive variety. The race for power, of course, involves actors within the country as well as outside. It would be wrong to depict Laos as being entirely at the mercy of the international community: there are many in the country who have a vested interest in seeing the dam-building programme continue, and in continuing to turn a deaf ear to muted protest from within. However, and as has occurred in Thailand, Sweden and Norway, awareness and public debate will, presumably, eventually also be possible in Laos: indeed, there is some evidence that environmental groups in Thailand are slowly helping to stimulate popular awareness and concern within Laos. Only with the development of such democratic and environmentally informed criticism is there hope that the impact and implications of projects such as the Theun Hinboun will gain their proper place on the national political agenda.

7

PLENTY IN THE CONTEXT OF SCARCITY

Forest management in Laos

Jonathan Rigg and Randi Jerndal

INTRODUCTION

In terms of per capita income, Laos is one of the ten poorest countries in the world and in World Bank terms is classified as an LDC – a Least Developed Country (see Table 7.1). Yet, from an environmental perspective, it could be argued that Laos is far richer than its more prosperous neighbours, and in particular Thailand. Laos covers a land area of 237,000 km² of which approximately 50 per cent is thought to be forested; by contrast, only 3 per cent of the country is cultivated (see Figure 7.1). With a population density of just 19 people/km², pressures on land and environmental resources would seem to be limited.

This picture of statistical poverty amidst environmental plenty has lent credence to the perspective that Laos is the last great natural treasure house in the region – a country which, through a combination of political isolation and economic stagnation, has managed to keep the more rapacious elements of the market at bay. Such a view glosses over significant loss or degradation of environmental resources in the past, but it does, none the less, highlight a series of important transformations in the resource economy of the country. No longer is Laos an isolated backwater. The effects of political rapprochement with Thailand, the economic reforms encapsulated in the New Economic Mechanism (NEM) or *chin thanakaan mai* ('new thinking'), burgeoning foreign investment – much of it targeted at the exploitation of natural resources – growing population (2.8 per cent/year, 1980–1993), and a driving increase in the pressure of needs in a country where consumerism is making its mark, are all playing a role in the nascent transformation of the economy. Taken together, these recent changes have placed considerable strain on the country's natural resources, and in particular its forest resources, creating an urgent need for greater control and management. In this chapter, we review the status of Laos's forest resource and identify what are, in our view, the main barriers to successful management.

145

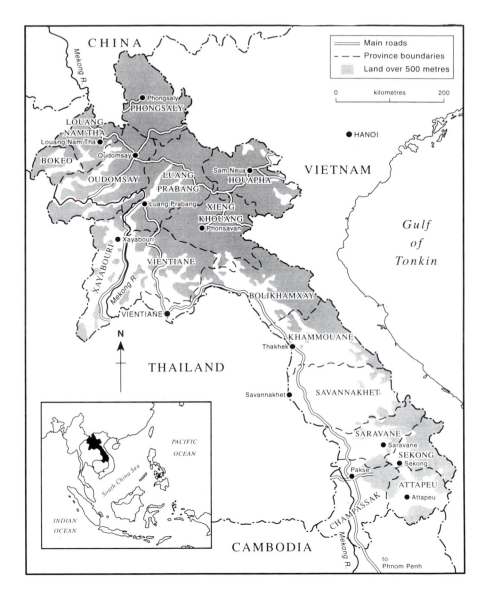

Figure 7.1 Laos

However, due to a lack of economic data – and in particular accurate data – a particular difficulty lies in identifying and weighting sources and rates of change in Laos. This is all the more acute in natural resource inventory assessments where long-run statistics are particularly important. Recent work using satellite imagery may have allowed researchers to gain a fairly good

Table 7.1 Laos: economic and social indicators

Economic indicators	
Labour force in agriculture	76%
GDP: average annual growth, 1980–1993	4.5%/yr
GDP, 1993	US$1.334 billion
GNP/cap, 1993	US$280
GDP/cap (PPP[S]), 1991	$1,760
Exports, 1992	US$92 million
Exports, 1993	US$108 million
Imports, 1993	US$250 million
Social indicators	
Population, 1993	4.6 million
Population in absolute poverty	67 %
Life expectancy	52 years
Children aged <5 underweight, 1988–93	41 %
Per cent of population with safe water	37 %
Telephones per 100 population	0.2

Sources: various

idea of the gross picture so far as the country's forest resources are concerned (Government of the Lao PDR, 1992), but the rate and nature of change can only be guessed at. In addition, the absence of micro-studies examining, for example, farmer encroachment and the role of shifting (slash-and-burn) cultivation in forest loss, represent a severe handicap in understanding the forces behind the statistics that do exist. An unpublished World Bank assessment of the country opens with the admission that the figures on which the report is based 'constitute educated guesses rather than confirmed facts', while the United Nations Development Programme highlights as one of the country's principal development constraints 'insufficient information on the country's key physical, social, economic and climatic variables' (United Nations Development Programme, 1990, p. 9). The lack of accurate information is due to a number of factors: the country's extreme poverty; a lack of skilled and trained personnel; the after-effects of the war and the long-term dislocations that it caused; the political imperative among government departments and individual bureaucrats to demonstrate progress in achieving targets, rather than any attempt to reflect reality (Zasloff, 1991); and the sheer size of the country in relation to its population and income. There is some recent evidence that economic and social statistics are becoming both more available and more accurate (Handley, 1993, pp. 28–31), but the problem none the less remains severe.

In one sense at least, this chapter is about 'sustainable development': we highlight the manifold reasons why Laos's abundant forests are being exploited

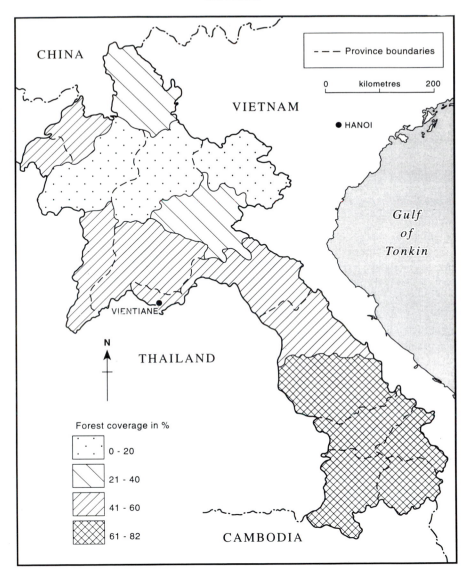

Figure 7.2 Laos: rainforest resources (by province)

without due regard to the long-term conservation of that resource. However, and like other contributors to this volume, we believe there to be multiple, over-lapping sustainabilities, not one neat concept and reality. Indeed, it is the conflicts which exist between different user groups and their different notions and practices of sustainability which give rise to many of the problems which we outline below. Shifting cultivators, settled agriculturalists, the

Table 7.2 The role of forest products in Laos's economy

Per cent of foreign exchange earnings	
1984–1987 (average)	26%
1988	36%
1991	54%
1992	38%

Sources: Ireson, n.d., p.5; World Bank, 1993; Mongkhonvilay, 1991; Stuart-Fox, 1995

state, domestic and foreign logging operations, all have distinct and often conflicting visions of how best to conserve Laos's natural resources for their own use(s). We see no reason why notions of 'sustainable development' should be restricted only to certain groups within Lao society. Indeed, Lao government efforts to persuade shifting cultivators to embrace settled agriculture represents an attempt to inculcate one group, not just with a new livelihood strategy, but with a new vision and practice of sustainability.

LAOS: THE FOREST RESOURCE

Laos is arguably the last country in the South-East Asian region which can be said to be truly 'agrarian': agriculture and forestry account for about 60 per cent of GDP, and employ over 85 per cent of the labour force. As Figure 7.2 shows, northern Laos has largely been deforested, with forest cover down to 10–20 per cent. In contrast, central and southern Laos have forest cover estimated at between 40 and 80 per cent. Pressure on cultivated land, and therefore on the forest resource, is increasing, particularly in the north and in rural areas close to towns and roads in the central and southern regions. As a result of rapid farmer encroachment, the government intends to convert 10 per cent of the country's total forest area into so-called Protected Areas (World Bank, 1993). At the end of the 1980s, total forest cover was estimated at 11.27 million hectares or 47 per cent of Laos's total land area, of which 5–6 million hectares or 40 per cent was believed to be commercially exploitable (Saramany, 1989). The rate of exploitation is indicated in a government projection which estimates that by the year 2000, Laos will have lost 44 per cent of its forest cover in fifty years. From some 17 million hectares of forest cover in 1950, by 2000 this will have been reduced to just 6.2 million hectares (Chongkittavorn, 1989).

Only a limited number of species is exploited: the frequently occurring dipterocarps provide the main raw material for sawn wood and plywood, while the rarer rosewood species and other durable hardwoods are used for furniture, parquet and weather-exposed construction components. Despite some diversification in the economy, forest resources remain of tremendous importance to the development of the country, and it is likely that many of the manufacturing enterprises that do become established in Laos will be involved in the further processing of timber products (furniture, plywood, etc.) and other natural resources. In recent years timber and timber products have accounted for between a quarter and one-half of Laos's foreign exchange earnings. In 1992 the figure was 38 per cent (Table 7.2). As in Burma (Myanmar), the exploitation and use of the country's natural resources will remain a key component in future economic growth.

FORESTS AND THE ECONOMIC REFORMS

After ten years of 'Communist reconstruction', the leadership of Laos embarked on the road to economic reform in 1986, following the adoption of the New Economic Mechanism (NEM) by the Party Congress of that year. The reform programme is more popularly known as *chin thanakaan mai* ('new thinking') or *kanpatihup setthakit* ('reform economy') and the aim of the NEM is not only to correct macro-economic imbalances, but also to give greater scope to the private sector and to market mechanisms in economic development (for a more detailed assessment of Laos's economic policy since the victory of the Pathet Lao in 1975, see: Chanda, 1982; Stuart-Fox, 1986; Zasloff and Unger, 1991; Kitiprawat, 1992; and Rigg, 1995). As former General Secretary Kaysone Phomvihane stated at the critical Fourth Party Congress in 1986:

> In all economic activities, we must know how to apply objective laws and take into account socio-economic efficiency. At the present time, our country is still at the first stage of the transition period. Hence the system of economic laws now being applied to our country is very complicated. It includes not only the specific laws of socialism but also the laws of commodity production. Reality indicates that if we only apply the specific economic laws of socialism alone and defy the general laws pertaining to commodity production, or *vice versa*, we will make serious mistakes in our economic undertaking during this transition period.
>
> (General Secretary Kaysone Phomvihane, Fourth Party Congress, 1986; quoted in Government of the Lao PDR, 1989, p. 9)

There can be little doubt that the introduction of the NEM in the mid-1980s, and its rapid and continued development, have put increasing pressure on the country's forest resources. As forest products have traditionally been one of the country's largest exports, the necessity to rapidly increase foreign exchange earnings to support the NEM has placed yet greater pressure on the industry to meet this need (see Ireson, n.d.). Forests represent an available and accessible economic resource, and one that foreign concessionaires – mostly Thai – are more than willing to exploit. Unlike other export-based industries, the lead time is short, there is little requirement for skilled and trained personnel, and the system of licensing such work is comparatively simple. The possibility of foreign involvement in the forest sector only materialized in the late 1980s, yet within the two years between 1987 and 1988, 120 joint ventures in the timber sector were agreed, mostly with Thai partners (Moseley, 1994). In an economic environment where there was – and is – a need to rapidly increase exports and foreign exchange earnings, forest exploitation became an obvious, indeed probably the only, candidate. In an interview, the Lao Deputy Prime Minister Khampoui Keobualapha and Chair of the Board of Planning and Cooperation identified the development of forest industries as one of the country's main economic objectives: 'In forestry we want to develop more processing, like making chipboard. We are still selling logs under existing contracts, but in the future we will limit raw timber exports' (*Far Eastern Economic Review*, 1993, p. 32).

Nor is it just a case of Laos's domestic economic situation encouraging such a pattern of exploitation; it is also important to see developments in Laos as an integral part of an emerging regional natural resource economy encompassing all the states of mainland South-East Asia – sometimes known as the Greater Mekong Sub-region or GMS (see Hirsch, forthcoming). Thailand, the region's greatest consumer of forest products, was effectively logged-out by the end of the 1980s, and in January 1989 a total logging ban was introduced (Rigg and Stott, forthcoming). All the countries of the GMS are experiencing severe environmental degradation. The major issues include: deforestation, soil loss and siltation, water pollution, accumulation of hazardous and toxic substances, habitat destruction, climate change, loss of biodiversity, degrading urban environment, air pollution, and threats to public health (Pham, 1994). Of the countries of the GMS only Laos and, to a lesser extent, Cambodia have abundant forest resources. In Cambodia, it has been widely assumed that Thai logging firms, with the support of the Thai army (sometimes the two being one and the same) and the permission of the Khmer Rouge, are involved in extensive logging operations. Although concern for the regional environment is often voiced within the GMS-group, national governments actively promote the business interests of companies engaged in resource extraction. Sluiter (1992, p. 18) quotes a Thai former forestry official:

> Before 1986 the possibilities for export were minimal. Then the doors opened and boom! In 1987 and 1988 120 joint ventures in the timber industry were established between foreign companies and the Laotian partners. Thai businessmen raced for concessions as the Thai government had imposed a logging ban in 1989. In 1991, when Thailand's wood industries were threatened by the logging ban in Laos, the Thai Foreign Minister crossed the Mekong to try and convince the Laotian government to lift the ban.

Thai businessmen have a reputation – well-founded or not – for dubious business practices. As journalist Paul Handley recently observed: 'Thai businessmen are their own worst enemies. When Cold War tensions eased in the late 1980s, Thai entrepreneurs rushed into Burma and Indochina. Most went into timber and gem mining, where they earned a reputation for slash-and-burn capitalism' (Handley, 1995, p. 28). A senior Thai consultant at the Thai Board of Investment admitted to Handley that '[w]e have been too predatory in our approach' adding, '[b]eing free to exploit everything here [in Thailand], lately we have been doing so in neighbouring countries' (ibid., 1995, p. 29).

THE MULTIPLE SOURCES OF DEFORESTATION

Although the role of the NEM is important, the erosion of Laos's forest resources can not be linked exclusively to recent economic reforms and the role of foreign concessionaires. Forests are also under threat due to their exploitation by various domestic groups. It is therefore important to identify the multiple sources of deforestation in the country if the process is to be fully understood and therefore effectively managed.

The Tropical Forestry Action Plan (TFAP) for Laos identifies the following sources of deforestation: commercial exploitation (logging), shifting cultivation (uncontrolled fires), encroachment by farmers and rural use (Government of the Lao PDR, 1990). However, a further important source of deforestation should be added to this list, or at least specifically identified: illegal logging. According to the TFAP report, sustainable commercial exploitation of natural forest resources in Laos allows for an annual allowable cut of commercial wood of 275,000m^3. In 1988, official (i.e. legal) logging was recorded by the government as being 350,000m^3 (Saramany, 1989). However, illegal logging is thought to contribute another 150,000m^3 to this figure, giving a total annual cut of perhaps 500,000m^3. This figure should, though, be treated only as a very rough approximation.

The main destination for illegally logged wood is Thailand. The establishment of major forest industries and log trading yards in Thailand, opposite Lao border crossing points, has been taken as evidence that log volumes are substantially in excess of officially recorded 'transit' and 'import' statistics (World Bank, 1993). In the first eight months of 1993, over 90 per cent of Lao–Thai trade with Thailand's Northeastern region was timber (*Bangkok Bank Monthly Review*, 1994, p. 26). The incidence of illegal logging operations not only has considerable implications for the management and protection of forests, but also denies the country substantial tax revenue. The TFAP report, for example, estimated that the government foregoes tax revenue of US$10–15 million per year as a result of illegal logging (Government of the Lao PDR, 1990). As total exports in 1993 amounted to just US$108 million, this therefore represents 9–14 per cent of export revenues.

The annual destruction of forest by some 250,000 shifting cultivating families is estimated to range between 70,000 and 300,000 hectares

---

Box 7.1
Population groups: division by decree

Laos's ethnically diverse population is usually divided by ecological zone into three groups: the wet rice cultivating, Buddhist *Lao Loum* of the lowlands, sometimes called the Lowland Lao, who are politically dominant and constitute over half of the total population; the *Lao Theung* or Midland Lao who occupy the mountain slopes and make up about a quarter of the population; and the *Lao Soung*, or Upland Lao, who live in the high mountains, practise shifting cultivation, and represent less than a fifth of the population. The terms were brought into general usage by the Pathet Lao who wished to emphasize that all of Laos's inhabitants were 'Lao', and to avoid the more derogatory terms that had been used in the past – such as the Thai word *kha*, meaning 'slave', to describe the Mon-Khmer *Lao Theung* like the Khmu and Lamet.

Although the words have a geographical connotation, they should be viewed, as Ovesen and Trankell argue, more as contrasting pairs of terms: *loum* and *theung* mean 'below' and 'above' (rather than hillsides and lowland), while *soung* is paired with *tam*, meaning 'high' and 'low'. These two pairs of oppositions were then brought together by the Pathet Lao into one threefold division. It is possible that *Lao Theung* in one area may occupy a higher location than *Lao Soung* in another area. In addition, economic change, greater interaction between the groups, and the settlement of lowland peoples in hill areas, mean that it is possible to find Lao Loum villages in upland areas, where the inhabitants practise swidden, not wet rice agriculture. Thus, although it is broadly possible to characterize the mountain slopes as inhabited by shifting cultivating *Lao Theung* of Mon-Khmer descent, in practice the neat delimitation of people into discrete spatial units breaks down, and as the years go by is becoming increasingly untenable (see Ovesen, 1993, pp.31–2; Trankell, 1993, pp.12–15).

(Saramany, 1989; Chongkittavorn, 1989; Mongkhonvilay, 1991; Stuart-Fox, 1995). The role of shifting cultivation in forest destruction is, though, extremely sensitive and subject to multiple interpretations. The received government line is that shifting cultivators are the major source of forest loss, outstripping that of commercial logging (see Trankell, 1993, p. 14). The former Director of Planning and Foreign Relations in the Ministry of Agriculture and Forestry, Somphong Mongkhonvilay, wrote in a paper in 1991, for example, that shifting cultivation has 'severely degraded natural vegetation and soils [in the Midland and mountainous areas]', adding that the 'situation is serious' (Mongkhonvilay, 1991, p. 35). Government policy is to persuade families to abandon shifting cultivation in favour of settled agriculture – or what has been termed in the Lao context 'fixed engagement farming' (*Bangkok Post*, 1994). This view of upland groups as environmentally destructive is linked to the 'patronizing and contemptuous attitude [of the lowland *Lao Loum*] toward the [upland] Lao Theung and Lao Sung, who are thought of as backward and less susceptible to socio-economic development.' (Trankell, 1993, p. 14; see also Ovesen, 1995, p. 14: see Box 7.1). In other words, the view not just of the government but also the majority *Lao Loum* towards upland groups in general, and shifting cultivating upland groups in particular, is informed by a disdainful attitude towards their way of life and, almost by association, their agricultural practices. When the interests of the state come into conflict with those of the hill peoples in the exploitation of the forest – the former for purposes of revenue generation and the latter for subsistence and sale – cultural preconceptions are reinforced by economic imperatives.

Agricultural encroachment, driven by increasing population pressure, allied with growing commercial forest exploitation, are undermining the role of the forest as an insurance system for many poor upland Lao. No longer is it possible for these people to use the forest as a source of food, shelter and products for sale (see Ireson n.d., p. 41).

The characterization of the upland Lao as environmentally destructive is disputed by many academics and environmentalists. They argue that shifting cultivation has always been practised by hill-living people in Laos and that in its traditional form is not destructive of forest resources on a wide scale and does not pose a serious threat to the forest resource. Sluiter's description of shifting cultivation in Laos, in a book published by two environmental groups, the Project for Ecological Recovery (PER) and TERRA, represents an example of this perspective:

> In preparing a field for cultivation, farmers do not cut down the largest trees, as they are considered sacred and home to the spirits of the forest. Smaller trees are cut and their trunks are left undisturbed in the field so that they can grow back quickly. No plough is used to break the topsoil and great care is taken to avoid erosion. Some farmers plant as many as 40 different plants at one time to make best use of soil nutrients and keep pests down. The best types of mountain cultivation aim to keep the soil fertile and the forests in balance. For thousands of years this was the gospel of life and legend in the highlands.
>
> (Sluiter, 1992, p. 31)

However, both the governmental and the academic 'ideals' suffer from overgeneralization. Different hill peoples manage the forest resource in different ways, and their impacts therefore also vary considerably (cf. the chapter by

Lang in this volume). Ovesen, for example, in his recent study of the Hmong in Laos writes that:

> [t]here is nothing in the historical experience of the Hmong that can make them understand the sentimental (and, we now know, ecologically sound) attitude of many Westerners towards forest preservation and bio-diversity: For the Hmong, wild animals are a source of food which is tapped as long as it lasts; forest produce is a potential source of cash income, and the cash is always needed; and large trees are a nuisance near human settlements, so they should be cut down.
>
> (Ovesen, 1995, p. 12)

I thought, initially, the words in parentheses in the first sentence of this passage to be intended sarcasm. However, later in the book the author writes, regarding the collection of *mak kham phaep* (?tamarind):

> The problem is that the oldest and largest trees yield the greatest amount of fruit, but since the people find it exceedingly difficult and dangerous to climb a 15–20 meter high tree, the problem is solved by simply cutting down the tree in order to get the fruit. For a westerner like myself, this no-nonsense attitude to getting what you want was tantamount to sacrilege, but there was no way I could convince the people that it might be 'better' to collect less fruit and save the trees for the years to come.
>
> (ibid., p. 63)

Important differences between groups of shifting cultivators and their variable impacts on the forest resource are compounded by a tendency to ignore the effects of the integration of hill peoples into the market economy and rapid population growth. Changing resource realities mean that methods of cultivation are continally adapting in a dynamic and complex way; studies which imply that cultivation strategies are somehow 'fixed' in time overlook this important issue.

Nor, it should be stressed, is the exploitation of the forest by the rural population restricted to the activities of shifting cultivators. Demand for fuelwood is also an important element in the equation. Rural use of wood constitutes 80 per cent of total fuel requirements in Laos today. The annual average fuel consumption is estimated at 1 m^3 per capita (Government of the Lao PDR, 1990). This is equivalent to 3.5–4.0 million m^3 annually (Saramany, 1989). However, although this may be a substantial volume of timber, wood for rural use – especially as fuel – seldom involves the cutting of trees. Most fuelwood is collected from already felled trees or from brushwood. It can therefore be seen to be substantially less damaging – despite the volume involved – to the total forest resource than the clear (or even selective) felling practised by commercial firms.

One major new source of deforestation is that associated with the construction of a series of ten new dams on tributaries of the Mekong River. The dams will supply electricity to Thailand (like the already completed Nam Ngum and Xeset dams) and will earn Laos valuable foreign exchange (see Figure 7.1). Areas are already being clear-cut for the proposed dams without, apparently, any regard to the damage caused to crucial watersheds and catchment areas. It has been alleged that the areas around the proposed dams are also being exempted from the logging ban, and some foreign observers see the dam proposals as little more than a pretext to continue logging (see below) (Lintner, 1994, p. 71). Because of a shortage of funds, the Lao government is anticipating constructing the facilities with private sector funding on a BOOT (Build-Own-Operate-Transfer) basis (see also Usher, Chapter 6). Although this will make a massive increase in generation possible in a relatively short period of time, the fear is that, as a report of the Swedish International Development Authority (SIDA) put it, Laos will be in 'imminent danger of [losing] control over the exploitation of one of its major natural resources' (quoted in Lintner, 1994, p. 70). Not only will the planned development of Laos's water resources shift control from the state to private firms, but as most of the power generated will be sold to Thailand, it means that these private firms will be negotiating with the Electricity Generating Authority of Thailand (EGAT), bringing in yet another outside actor into the decision-making and management process.

UNDERSTANDING RESOURCE DEGRADATION IN LAOS

Three key factors are associated with resource degradation in Laos: poverty, market failures, and institutional weaknesses.

Most Laotians can still be characterized, loosely, as subsistence farmers. However the market reforms and slowly improving communications are gradually integrating the rural population into the national economy. This has created new needs and, along with rapid population growth, is putting pressure on traditional systems of production. If shifting cultivation is not to expand to meet these demands, then agriculture must be intensified through the adoption of new cultivation techniques and made more productive. Poor upland farmers are unlikely to shift from environmentally damaging practices to more suitable ones like tree crop agriculture which cannot meet their basic food requirements.

In textbook terms, markets are generally thought to lead to the efficient allocation of resources so long as there are no externalities, access to information is free, and the ownership of resources is defined and secure. When these conditions are not met, however, markets fail to generate prices which reflect the full social costs (and benefits) of resource use. There is good reason to believe that Laos, as a crypto-market economy, is suffering from serious market failures and that these play an important role in the pattern of resource exploitation. Further, and despite the broad nature of the reforms already in place, it will be some time before the economy will have adjusted. Among the particular market failures which affect natural resources in Laos are:

- Open access/lack of defined property rights.
- Lack of market integration.
- Unequal access to information.

Forest resources in Laos have no clearly defined property rights attached to them, making their over-use in a context of burgeoning demand highly likely. At the same time, and despite substantial investment in roads, communications remain poor and many areas and rural people are remote from, and poorly integrated into, the market economy. The stock of roads is extremely limited and most are in chronically poor condition: of 10,000 kilometres of roads, it is thought that only about 20 per cent are sealed and most of these are in a poor state of repair (*The Economist*, 1994, p. 74; Rigg, 1995). The northern provinces are virtually cut off during the wet season, and even Route 13, the national artery which runs from the capital Vientiane south to the Cambodian border – and which has been targeted for special attention by the government and international aid agencies – is only part made-up (Stuart-Fox, 1995). The subsistence sector is very large and co-exists with a small commercial sector found in the few urban cores of the country. The lack of efficient communications also results in many farmers having limited access to information on the production technologies necessary to increase productivity and create sustainable farming of uplands. Technology development and transfer are also extremely limited in Laos, due to an absence of appropriate research, extension and marketing information systems.

Indeed, institutional capacity in general in Laos is weak and the Department of Forestry (DoF) is no exception. A lack of skilled personnel and vehicles makes it difficult for the DoF to operate effectively, while the very low salary levels leads to problems of motivation among the personnel. It is likely that corruption is widespread, although the evidence for this is largely anecdotal. Moseley writes of attempts to control illegal logging in the south that 'the authorities in Pakse are said to be at a loss as to how to police logging activities [over 10,400km^2] with the small resources they have at their disposal: 113 poorly-paid officials, one truck, two motorcycles and a bicycle' (Moseley, 1994). The weakness of the DoF also accounts, in large part, for the limited knowledge of the different types of shifting cultivation and shifting cultivators, and their relative impact on the resource base. In the past this lack of knowledge led the government to adopt unrealistic objectives and policies to restrict shifting cultivation. Finally, the overall awareness of environmental issues is very low in Laos. Unlike Thailand, there is no domestic environmental movement, and the calls for protection and conservation originate

---

Box 7.2
An estimate of the cost of inaction

- If half the area under shifting cultivation has commercially valuable forest cover;
- If the commercial volume of timber is 10m^3/ha;
- If that timber is conservatively valued at US$50/m^3, then the country is annually losing US$25 million in foreign exchange earnings.
- In 1991 export income from forest products amounted to US$33.3 million.

Source: TFAP 1990

---

from overseas-based NGOs, aid organizations, and pressure groups like PER (the Project for Ecological Recovery) and TERRA (Towards Ecological Recovery and Regional Alliance).

Institutional weaknesses in the DoF also have knock-on effects in terms of serious deficiencies in the practice of logging, even by legitimate logging operations. Methods are inefficient and undertaken without proper planning and management. This leads to a loss of timber value as trees are damaged during extraction, as well as a waste of wood. Stockpiles of old logs nationwide have been calculated to represent as much as 250,000–300,000 m^3 of timber. As much of this is high-value timber such as rosewood, losses could be at the level of US$10 million or more (see Box 7.2). Further, efforts at afforestation on plantations are of poor quality and in any case are estimated to amount to just 200–300 hectares per year (Government of the Lao PDR, 1990: another paper (Mongkhonvilay, 1991) estimates annual reforestation at 1,000–2,000 hectares). In the context of the current extent of deforestation, afforestation is effectively zero, undermining any attempt at sustainable use.

FOREST POLICY-MAKING ON THE RUN

The organization of the Ministry of Agriculture and Forestry (MAF) is based on a hierarchical structure which seems to be informed more according to political principles than functional demands. The system in place is oriented towards the twin objectives of restriction and control, with little attention paid to development. This is similar to the role that the Thai Royal Forestry Department has played in the management of Thailand's forests. Lohmann (1992, p. 87) has remarked, for example, that, 'the philosophy of the RFD [Thai Royal Forestry Department] ... [was] ... that the "tropical forest is a cash crop" under its jurisdiction'. Rigg and Stott agree, writing that 'the forest was treated [in Thailand] largely as an resource to be exploited, not protected, and notions of conservation were largely absent' (Rigg and Stott, forthcoming). There are similar parallels between Laos and other Asian countries in the hierarchy and structure of administration within the MAF: communication and lines of authority run mainly vertically through the organization; horizontal communication between provinces, for example, is rare and structurally constrained. This absence of horizontal communication and co-ordination within the MAF and DoF is reflected in the duplication of effort in the work undertaken by international agencies and donors working within the forestry sector in Laos such as Lao–Swedish Forestry Co-operation and Finnida/World Bank. But not only is there structural fragmentation leading to duplication in policy, but forest policies have also been inconsistent over time. In the case of forest policy, it has been a case of 'policy-making on the run':

- In order to control commercial logging, raw log exports were totally banned in January 1988. The objective was to make a larger volume of high quality timber available for domestic processing. But due to the poorly developed production techniques and a lack of market knowledge, it was impossible satisfactorily to manufacture at internationally competitive products. As a result, illegal logging and smuggling boomed, the planned processed wood products did not materialize, and Laos was left short of tax revenue and export earnings;

- In July 1988, to try and counteract this development, the government increased the resource tax substantially and imposed export taxes on forest products. The objective was to encourage the use of the stock of already felled logs in the forest, noted above. But this policy only resulted in the virtual cessation of all official logging due to the excessive costs and time involved in processing the felled logs. The outcome was an even larger loss in export earnings and increasing unemployment in the sector;

- In December 1990 a privatization programme was agreed for the forest industry and implementation of the programme commenced in January 1991. Yet nine months later in September 1991, a new ban on logging was introduced by the government;

- In 1992, three military companies were given the rights to carry out all logging operations in the country. Foreign companies wishing to do business in Laos are obliged to establish joint-ventures with these companies. This is a step away from the privatization programme embarked upon in 1991, since the government – through the military – now has total control over the logging business;

- In May 1994, Prime Minister Khamtai Siphandon issued a further decree to control illegal logging and stem 'systematic' corruption in the forest industry (Stuart-Fox, 1995). Companies will need to secure ministerial permission to cut above their assigned quotas and sawmills and plywood factories will not be able to cut timber themselves. Stopping the smuggling of timber sourced in Laos but labelled as imported and merely in transit to a third country will be tackled by further quotas on the trans-shipment of timber through Laos. Despite this plethora of further restrictions and controls, Stuart-Fox (1995, p. 188) is doubtful 'whether these regulations will be any more effective than previous ones'.

This inconsistent and sometimes contradictory sequence of events was not merely due to particular weaknesses within the MAF. Laos had no constitution until 1991 and the government used decrees to rule, based largely on *ad hoc* decisions. The decrees themselves were often contradictory and full of loopholes. They also did not give any legal security to those affected, since they could be withdrawn or replaced at any moment.

ESTABLISHING PROPERTY RIGHTS

The need for a Land Law, which provides secure land use rights, is perhaps the most urgent task facing the government in its attempt to control, regulate and develop forests and the forestry sector. The lack of legally established property rights is hampering the NEM-linked privatization process, the implementation of effective forest management programmes, the stabilization of shifting cultivation, and the establishment of Protected Areas.

At present a traditional community decision-making structure in the form of a village council exists to regulate the tenure of forest and agricultural land. Villagers enjoy usufruct rights to land which are inherited and can also be transferred to a third party. This traditional land ownership system co-exists within the overall principle that all land in Laos is the property of the state. The village council is empowered by the government to collect local agricultural taxes, and to plan and monitor community land and forest use. But the present system does not give villagers the authority to control or stop the activities of outsiders engaged in illegal logging and agricultural encroachment, as they do not have any legal and enforceable rights to the land and the forests (Andreasson and Markgren, 1993).

PROSPECTS FOR THE FUTURE

In decrees and programme documents, the protection, preservation and regeneration of the forest are always emphasized. 'Sustainable' forest management and development is the name of the game. In reality, though, everyday decisions and actions mitigate against sustainability. In the decree concerning 'the management and use of forests and forest land' it is stated that:

all activities pertaining to all types of forests and forest produces in Lao PDR with the aim of preserving the forests, forest land, the environment, water resources and wild life in view of meeting the requirements in the national economic development and the pluri-ethnical population's living conditions in terms of forest produces in a sustainable way without *any* impact on the environment.

(Government of the Lao PDR, 1993, Decree No. 169, p. 1: our emphasis)

It is clear that the government of Laos does not wish to repeat the mistakes of Thailand – i.e. create the conditions in which excessive deforestation can occur. However, this appears to be just what is in danger of arising, and the government seems to be powerless to halt the process for the reasons outlined above. This picture is, admittedly, one built up largely through anecdotal evidence, personal experience, and example. But given that official data are lacking and of dubious veracity, there is, arguably, nowhere else to turn:

- 'Roads on the outskirts of Pakse [in the south] are now jammed with convoys of trucks hauling processed timber and freshly cut logs to buyers in the Thai town of Ubon Ratchathani' (Moseley, 1994; also Jerndal, personal observation, has noted that many of these are Thai trucks).

- Farmers complain of loss of fertility and production due to forest loss: 'Life was better before. Nowadays there are no birds in the forest, no fish in the stream' (Mr Tang of Luang Prabang, quoted in Håkangård, 1990, B1).

- 'An interrupted chain of logs lines the road, all the way from Vientiane to southern Laos along the Mekong. Convoys of trucks are creaking on their axles under the weight of forest giants. The landscape resembles a hastily deserted battlefield, strewn with wooden corpses while shrubs hold a wake, shrouded in the red dust of Road 13' (Sluiter, 1992, p. 17).

The losers in this process of rapid exploitation are the poorest farmers and the hill groups who are entirely dependent on the land and the forest. Little money trickles down to the remoter regions. At present, it is the core areas of the country, based on the main urban centres of Vientiane, Pakse and Savannakhet, which are benefiting from the New Economic Mechanism and the process of resource exploitation. The interests of the politically and economically powerful groups in the core areas of the country are often in conflict with the poor, rural-based farmers on the margins. At present it seems clear that the forest is being exploited more for the benefit of the former group than the latter.

From 1975 through to the introduction of the NEM in the mid-1980s, the Lao economy was inward-oriented, mainly supplying the domestic market, while foreign trade (with the exception of electricity exports to Thailand) was in the form of barter with other Socialist countries. Although logging occurred, it was not as extensive as it has become since 1986. Environmental degradation and forest depletion were primarily linked to traditional agri-cultural practices, especially shifting cultivation. Since 1986, with the introduction of the NEM and the opening-up of the Lao economy, the root causes of forest loss and, more broadly, environmental degradation, have changed. Commercial logging, legitimate and illegal, under the auspices of the state or by private firms, whether domestic or foreign, has taken over as a leading source of forest loss. Laos finds itself increasingly integrated into a mainland South-East Asian regional resource economy led by Thailand whose 'increasing reliance on neighbouring countries' natural resources has potentially far-reaching environmental implications' (Hirsch, forthcoming). Hirsch continues:

> The resource relationship between Thailand and Laos is . . . illustrative of a wider asymmetry. Laos' two largest sources of foreign exchange, timber and electricity, are both largely oriented to Thailand. It may be argued that Thailand is dependent on Laos for these resources, important to its continued industrialisation, but the dependence is nevertheless skewed, with Laos dependent on Thailand to a much greater degree.
>
> (ibid.)

In 1989 the then Prime Minister, Chatichai Choonhavan, expressed the vision 'to turn Indochina from a battlefield to a marketplace'. Although this may be beneficial to both countries, it seems that Thailand will be the main winner. Thailand is the most important trading partner, and in 1993 accounted for more than one-third of all foreign investment in the Lao PDR (Ivarsson et al., 1995, pp. 25–6).

The transition to a mixed economy with an increasing number of private companies has also decreased the government's ability to control logging – especially when the institutions of control are so weak. However, it would be wrong to assume that Laos is making a transition from 'command' to 'free-for-all'. Before 1986, the Lao economy was largely on autopilot. The state had little command over large swathes of the countryside and a significant proportion of the population. The failed attempts at co-operativization after 1975 illustrate the degree to which the government could not control events even within an ostensibly 'command' system. The difference today is that the state is only marginally stronger in its ability to control events, but the number of actors operating in this weak administrative environment is far greater.

Part III
METHOD

8

ENVIRONMENTAL CHANGE IN MALAYSIAN BORNEO

Fire, drought and rain

Victor T. King

THE ISSUES

The literature on human–environment interactions in Borneo and environmental change is substantial but during the last fifteen years debates about the processes and consequences of these interactions have intensified, particularly in relation to commercial logging operations. Indeed, there has been an increasing shift of attention to South-East Asian rainforest issues since the early 1980s. A barometer of this increasing interest in Borneo is the *Borneo Research Bulletin*, which, in recent years, has published numerous papers on environmental issues. The literature both in and outside the *Bulletin* ranges from the rather popular propagandist attacks on the tropical timber industry from various of the NGOs and pressure groups to detailed scholarly studies of such matters as shifting or swidden cultivation and indigenous environmental knowledge (for examples, see King, 1993a). There is now an urgent need for informed debate based on the evidence of scientific research (see Primack, 1991).

One of the most contentious issues has been the relative effects of shifting agriculture on the environment as against commercial logging. The usual view of politicians and government bureaucrats is that this form of extensive cultivation based on the forest-fallow system and the rotation of fields rather than crops is the major factor in explaining forest destruction. For example, a highly critical commentary on tropical forest agriculture was presented in a Sarawak Department of Agriculture report (Hatch and Lim, 1978) which emerged from a workshop attended by local government officials and scientists. The report claimed that shifting cultivators destroy substantial amounts of valuable timber in that they clear primary forest and that clearing and burning forest for cultivation cause long-term environmental deterioration in tropical forest region: '[S]hifting cultivation represents probably the greatest single threat to the integrity of Sarawak's natural resources and results in totally unacceptable degrees of human suffering'

(Hatch and Lim, 1978, p. 28). The policy recommendations arising from these conclusions are that shifting agriculture should be restricted and eventually eliminated, and that, in some cases, farmers should be resettled and alternative forms of land use such as plantation cultivation and commercial forestry should be encouraged. In his evaluation of this perception of shifting agriculture, Cramb (1989) outlines the historical context of the emergence of this view in Sarawak and traces it back to culturally biased European perceptions of 'traditional' agriculture. He concludes that the charge that 'shifting cultivation in Sarawak is inherently destructive of natural resources' is not supported by much, if any, empirical evidence (ibid., p. 24).

On the other hand, there have been studies of socio-economic change in Sarawak which have been highly critical of commercial logging and government development policies (e.g. Hong, 1987; Colchester, 1989). The charge is that the penetration of a capitalist economy, the operation of patron–client politics, and top-down development planning are the main causes of both environmental destruction and cultural deterioration. Needless to say the critics of logging have provoked sharp reactions from representatives of government (Wong, 1992). But even the moderate International Tropical Timber Organization (ITTO) has criticized Sarawak's logging practices in that its level of extraction, in 1989, particularly in the hill dipterocarp forests, was not sustainable at the then rates, and that the forests would be exhausted by the year 2000 if the extraction rates were maintained (Primack, 1991; and see Aiken and Leigh, 1992, p. 126). Exports of sawlogs alone in 1989 stood at 15 million m³ in volume, and the ITTO indicated that even 13 million m³ of timber per annum would be unsustainable. Indeed, overall log production had increased from about 14 million m³ in 1988 to 19.5 million m³ in 1991 with only a slight decline to 18.8 million m³ in 1992 (*Annual Statistical Bulletin*, 1992, p. 58). In Sabah too, logging has been intense and well over two-thirds of commercial forests have now been exploited.

Concerns about forest clearance and its effects on the environment in Borneo, particularly on rates of soil erosion, water pollution and the quality of vegetation cover, have also recently concentrated on the relations between the disappearing forests and climatic change. This emphasis on climate was given even more urgency with the prolonged drought reported in eastern parts of the island and the extensive forest fires of 1983 (Mackie, 1984; Beaman *et al.*, 1985; Leighton and Wirawan, 1986). Less severe fires linked to drought were also reported again in 1987 and 1991 (Soegiarto, 1993; Wirawan, 1993) and one should note comments about the severity of the drought in 1991, even in generally wetter regions of Borneo (Salafsky, 1994, p. 374).

Kalimantan has about one-third of Indonesia's 120 million hectares of forest, supplying about 30 per cent of the country's tropical timber. Nearly half of these Bornean forests are found in the province of East Kalimantan. Having already been subject to intensive logging for some fifteen years East

Kalimantan (and the East Malaysian state of Sabah) experienced a prolonged drought commencing in mid-1982. In late 1982 fires broke out in many places in the province and by March-April 1983 vast areas of forest in the middle and lower Mahakam basin, along the coasts from Balikpapan to Sangkulirang, and in forest reserves in Kutei and near Samarinda were engulfed in flames. It has been estimated that about 3.5 million hectares of timber were destroyed, or about 20 per cent of the forests of East Kalimantan (Avé and King, 1986, pp. 85–6). In Sabah, too, fire affected a minimum of 950,000 hectares of forests, 85 per cent of it in logged areas (Wirawan, 1993, p. 242). Generally, the heavily disturbed forest suffered very serious fire damage, and shifting cultivation was also implicated in this process of forest clearance and in the increasing fire hazards (see below).

There has been speculation about the causes of drought, but scientists have linked some of these periodic dry periods in such places as eastern Borneo to 'oceanic and atmospheric fluctuations' in the Pacific known as the 'Southern Oscillation' (Nicholls, 1993, p. 154) and particularly extreme anomalies in ocean-warming processes and changes in currents in the equatorial regions of the eastern Pacific known as El Niño. The link between the oscillation and its extremes is usually referred to as the 'El Niño–Southern Oscillation' or ENSO (Nicholls, 1993, pp. 154–7). El Niño is also interspersed with another extreme of the Southern Oscillation, referred to as La Niña which brings unusually high levels of rainfall to parts of South-East Asia, and therefore the problems of severe and widespread flooding. ENSO and La Niña therefore increase rainfall and dry period variability in the areas of South-East Asia affected by them.

It is difficult to link forest clearance to these climatic changes directly, but the removal of vegetation cover has been related to the forest fires in at least two indirect ways. First, it has been suggested that forest destruction may play a role in the intensity of El Niño because of increased turbidity in the seas surrounding heavily logged islands such as Borneo. In other words, large amounts of eroded silt and debris from cleared land affects water movements, sea surface temperatures and air currents above them. It is clear that ENSO is not uniform in its patterns and effects in South-East Asia. Second, and of more immediate importance for farmers and rainforest dwellers, is that logged-over forests do not retain moisture as well as primary forest, and they dry out more quickly. An additional factor is that opening up forests produces trails and dead wood residues. The amount of plant biomass builds up near and on the ground. It is argued that logging trails can act as wind tunnels and highways for fire thus encouraging its spread, and that dried-out woody debris and ground level vegetation act as kindling for fire. Resin-bearing rainforest trees are also especially flammable.

It is clear from the variations in the areal extent of burning in different kinds of vegetation that the logged forests are especially vulnerable to fire and drought conditions. In the closed canopy of the primary forest where vegetation is more dense and damp, fires burn less well (King, 1993b, p. 301). There is a further consideration which links logging, shifting agriculture and fire. It is well known that, in various parts of Borneo, shifting cultivators move into recently logged areas because they are more easily accessible and present fewer obstacles for clearing and farming. This, in turn, might also be contributing to increased forest destruction. Furthermore, the fires in East Kalimantan were probably started by farmers who were clearing their farms prior to planting, and these fires, coupled with drought conditions, were then difficult to control.

What is more, after the great fires of 1982–83 there was serious flooding in East Kalimantan, and dwellings along the Mahakam river, for example, were under water for several months in the 1983–84 rainy season. After the devastation of the vegetation in 1982–83 the severe flooding also caused an acceleration of the rate of soil erosion (Wirawan, 1993, p. 244).

It has to be emphasized that shifting cultivators in Borneo have employed fire for clearing vegetation over a very long period of time without the devastating effects of 1982–83, though we should note that there were 'extensive' fires reported in Kalimantan in the serious drought of 1914 (Henley, 1994, p. 2). Nevertheless, more generally, changed conditions seem to be contributing to a greater likelihood of both fire risk and hazard and larger-scale fires. In addition, logged areas accessible to urban populations are being exploited not so much by poor, marginalized shifting cultivators but by urban-based merchants, government employees and others, who often subcontract the work to freelance cutters, and by inter-island migrants such as Bugis commercial farmers who are opening land for pepper and other cash crops, though initially they put in subsistence crops until cash crops are ready to yield (Vayda, 1979). This movement into Kalimantan of large numbers of people is also contributing to increased fire risks.

It is the complexity of these rapid changes in Borneo and the serious effects of these on the environment and human populations that led Mackie (1984), in her evaluation of the Great Fire in East Kalimantan, to emphasize the urgent need for interdisciplinary research in order to understand the causes and consequences of these environmental changes, and particularly human-forest interactions.

INTERDISCIPLINARY RESEARCH PROGRAMMES

Unfortunately, prior to about 1980 there is little evidence of this kind of interdisciplinary team research. It was piecemeal and usually not well coordinated. However, signs of an imminent change in the approach to the study of environmental issues were evident in the late 1970s. A model for interdisciplinary research comprises the studies arising from the UNESCO 'Man and Biosphere' Program in East Kalimantan. It was a collaborative project between various Indonesian and American institutions, including

Herbarium Bogoriense, Universitas Mulawarman (Samarinda) and Rutgers University (New Jersey). The co-ordinators were Kuswata Kartawinata, Andrew P. Vayda and R. Sambas Wirakusumah.

The general 'Man and Biosphere' Program was 'a research and training program intended to provide information and methods for better management of different types of ecosystems or human use systems' (Kartawinata *et al.*, 1978, p. 28). The project in East Kalimantan which emerged from this mission statement was specifically designed to examine the interrelations between human action and tropical ecosystems. It concentrated on human activity, its causes and effects, and the processes and systems in which it is embedded and to which it gives rise. The context of it was the rapid increase from 1968 in the production of logs for export using mechanized methods and the substantial damage which was resulting. Kartawinata *et al.* (1978, p. 30), for example, had reported cases where only 50 per cent of the trees remaining after mechanized logging were undamaged, while 30 per cent of the ground was left exposed and damaged.

In this project different types of environment were selected for detailed study by a combination of qualitative and quantitative methods (Vayda, 1981). For example, the remote Apo Kayan interior region was studied by Timothy C. Jessup and Herwasono Soedjito where the Kenyahs practised a stable, forest-fallow system. Farmers were mainly using secondary forest, some of it first cleared about 100 years ago. By careful cropping and fallowing a forest system had been sustained. This interior region had not reverted to grassland or been seriously degraded (Vayda, 1979, p. 26). Second, Carol J.P. Colfer and Soedjito investigated the more accessible Telen area, which had been affected by logging and where Kenyahs had been resettled from the Apo Kayan. What this comparison revealed was that, despite being of the same cultural group and from the same homeland, the Telen Kenyahs practised a different, less conservationist form of shifting agriculture. They had access to technology in the form of chainsaws, rice hullers and outboard motors. Thus, they used labour-saving devices and had more efficient means of transport. Colfer (1983a, pp. 3–21; 1983b, pp. 70–87) discovered that the Kenyahs in Telen could therefore get access to more primary forest and clear it quickly (Colfer, 1983a, p. 14). They therefore utilized larger areas for farming to obtain a larger crop yield from it, and they still had sufficient time to divert labour to other income-generating activities (Colfer, 1983b, p. 73). They sold surplus rice, cut timber for sale, and collected minor forest products such as rattan for the market. Furthermore, other forms of land use were expanding with the influx of commercial loggers and transmigrants. Colfer (1983b, p. 82) concluded that, 'It is clear that the land and the forest cannot long support the kinds of human activity currently anticipated.'

The third region under study within the programme was in the vicinity of Samarinda; this part of the exercise was undertaken by Vayda and students from Mulawarman University. Kartawinata visited all three locations. It was in this peri-urban area that unlicensed timber cutters and commercial farmers were felling trees and clearing land in already logged areas, especially near roads (Vayda, 1981, pp. 10–11). These operators, because they were small-scale ones, still did less damage than the mechanized commercial loggers, but such practices as clean-weeding pepper gardens for up to ten years did eventually result in *Imperata* grass infestation.

Another scientist, Nancy Peluso was also brought into the project to investigate the collection, trade and transport of minor forest products such as rattans, resins, birds nests, aloes wood, and ironwood. Peluso discovered that several activities destructive of the environment were in evidence and these were driven by short-term profit motives (ibid., pp. 12–13). Among other things, Peluso demonstrated the problem of the over-exploitation of rattan in the Mahakam area, the rapid increase of rattan traders and the consequent decline in the quality of the product. Hall (1993) has provided a similar picture for *gaharu* collection in the upper Barito. The decline in the availability of important forest products has been reported for other parts of Borneo as well, including the western areas of Sarawak (Leaman *et al.*, 1991; Burgers, 1992; King, 1993a: see also the chapter by Parnwell and Taylor in this volume).

A most important finding of this excellent range of studies in East Kalimantan was that shifting cultivators operated very variable, flexible, responsive farming systems, and that, given certain conditions such as low population density, this form of agriculture can maintain a stable forest-fallow rotation over a very long period of time. This is precisely what has been demonstrated from other detailed studies of swidden agriculture (e.g. Chin, 1985; Dove, 1985), and in the overview by Cramb (1989). Indeed, Chin (1985, p. 246), in his meticulous study of Baram Kenyah cultivation, concludes, 'The Kenyah, through their agricultural and other subsistence activities, successfully and efficiently exploit their lowland rainforest environment [and that] logging when it reaches the Kenyah territory will not only destroy the primary but also the old secondary forests.'

Nevertheless, in certain regions, where there has been population pressure, and the introduction of modern devices such as chainsaws, as well as the influence of markets, then shifting agriculture has also contributed to environmental deterioration; there has been a decline in long fallow, forest-maintaining practices, especially in East Kalimantan (Potter, 1993, p. 109; Lian, 1993, p. 331). Cramb (1990, p. 44), however, argues that the debates about shifting agriculture and its effects can only really be resolved by detailed studies which combine aerial photography, satellite imagery and local level surveys.

Sadly, up until recently Vayda's programme had not been replicated in other parts of Borneo, and Cramb's call for a combination of macro- and micro-level perspectives had not been answered to any extent (see Brookfield and Byron, 1993, for some examples). There have been team projects since Vayda's East Kalimantan programme but these have been studies mainly in the physical sciences, and have been less concerned with the consequences of environmental change for human populations (but see Cleary and Eaton, 1992). There has been some interest, however, in the changes wrought by human action. Many of the co-ordinated projects have also been undertaken in relatively unaffected reserved areas of primary rainforest.

For example, the Royal Geographical Society's (RGS) multidisciplinary expedition to Gunung Mulu National Park in 1977–78 concentrated mainly on botanical and zoological subjects, but there was some attention to human factors in the study of Penan hunter–gatherers who inhabited the region, and the problems and possibilities of integrating them into the national park area. The expedition was charged with drafting a Management Plan for the park (Jermy and Kavanagh, 1982; 1984; Hanbury-Tenison, 1992). In 1991–92 the RGS in co-operation with Universiti Brunei Darussalam (UBD) was also involved in the expeditionary phase to establish a field research centre at Kuala Belalong in Batu Apoi Forest Reserve, Temburong, Brunei (see also the chapter by Dykes and the section by Wills in this volume). The main activities were again located firmly in the physical sciences and the overall focus was on tropical biodiversity. The field centre, which was taken over by UBD in 1992, is directed to teaching, training and research as part of the country's environmental education curriculum (Cranbrook, 1992; Cranbrook and Edwards, 1994). As a result of its oil and gas wealth Brunei has not exploited much of its rainforest resources.

In 1984–85 the Sarawak State Planning Unit in co-operation with the World Wildlife Fund Malaysia undertook the Sarawak Conservation Study which focused on environmental sustainability and the provision of an inventory of environmental resources in the state (Dandot, 1992). An element of the study examined human settlement and land use, but the main emphasis remained within the natural sciences.

Cambridge University is involved in Project Barito Ulu in Central Kalimantan with the Ministry of Forestry. Again this is a multidisciplinary programme concerned with botanical and zoological research, but the only major piece of work which is concerned predominantly with human–environment interaction is that by Hall (1993). She concentrated on the important institution of house garden agriculture as a response to labour shortage among Ot Danum and Bakumpai, but also has some material on the collection and trade of forest products, particularly *gaharu*.

Finally, the most ambitious team project is undoubtedly the Royal Society South-East Asian Rainforest Research Committee's multidisciplinary programme based at the Danum Valley Field Centre (DVFC) in Sabah, Malaysia (Marshall, 1992). Other participating bodies comprise the Sabah Foundation, the Sabah Forest Department and Universiti Kebangsaan Malaysia Sabah Campus. Again the focus of the project entitled 'The Recovery

of Tropical Forest Following Disturbance: Patterns and Processes'
(1985–1993) is on the physical sciences, particularly concentrating on forest
ecology and primary forest dynamics, hydrology, and vertebrate biology. The
main work has been undertaken in an undisturbed lowland rainforest conservation area, but some research has also been carried out in nearby selectively
logged areas for which there are historical records of clearance over a twenty-
year period. Some blocks are currently subject to logging.

The proposed next five-year phase of the programme from 1994 will again
comprise predominantly physical scientists but it represents a shift in focus
in that it is recognized that areas of primary forest will continue to decrease
and the major issues will need to address the transformations of forests and
the environmental costs of these, for example, in the loss of biodiversity.
Therefore, the new programme entitled 'Restoration of Tropical Forest
Ecosystems' will examine the problems, patterns and processes of forest regeneration in areas of disturbed forest, particularly forest in Sabah which has
been selectively logged. The main research will be in the fields of meteo-
rology, hydrology and forest dynamics and its practical aims will be to assist
those responsible for forests in the management and improvement of processes
of restoration and in the reduction of logging impacts.

In this regard the Royal Society programme has contacts with the Overseas
Development Administration-funded forest management project in East
Kalimantan. The ODA has also provided some support for the DVFC work.
There has also been some collaborative research involving DVFC personnel
and a current ESRC special initiative on Global Environmental Change. In
the next phase of the Royal Society programme these links should strengthen
because of the development of a number of mutual interests. One subject
for research based at the DVFC in its next phase will be the stages and
speed of forest regeneration in relation to drought using remote sensing
imagery. There will also be a study of regional variations in dry periods and
droughts in Malaysian Borneo and Kalimantan, and the changes of these
through time going back to the nineteenth century.

It is my view that much more multidisciplinary, integrated team research
needs to be undertaken in Borneo, and particularly in Sarawak, and I welcome
the promising signs of increasing collaboration between social scientists and
natural scientists in this regard. The current research involvement of a team
from the Universities of Hull and Manchester in an ESRC-funded project
is intended to make a contribution to this social-physical science interface
and the study of processes of environmental change. Various of the team's
preliminary findings are reported below.

THE BINTULU RESEARCH PROJECT, SARAWAK

A group of social scientists and geographers, including physical geographers,
was formed and funded by the ESRC under its Global Environmental Change

initiative to bring together a multidisciplinary team to investigate human–rainforest interactions in another part of Borneo outside Kalimantan, in this case the two East Malaysian states of Sarawak and Sabah, to combine the use of satellite imagery with local field studies, and to examine the processes of environmental change, some of its consequences, and the responses of local populations to these transformations. The team comprised personnel from the University of Hull who concentrated on Sarawak and the University of Manchester who focused on the area inland of Lahad Datu and particularly around Danum Valley in Sabah.

The region chosen for study in Sarawak was that of Bintulu. The research on the Dayak Iban population of Bintulu division was undertaken over an eighteen-month period commencing March 1992; it included detailed field research and a large-scale survey of thirteen Iban communities between October 1992 and April 1993 as well as archival and library research (further information on these study communities, including a map and a table of their general characteristics, can be found in the chapter by Parnwell and Taylor in this volume). The Hull research team comprised four personnel (a human geographer, a biogeographer, a sociologist and an anthropologist) with the logistic support of two of the lecturing staff in rural development from the then Bintulu branch campus of Universiti Pertanian Malaysia and an Iban field assistant.

The Bintulu region was chosen as a field site for four main reasons. First, it is a little-studied area in Sarawak. Ethnographic research and studies of rural development have tended to concentrate on the more densely populated western and south-western regions of the state in the environs of Kuching and the basin of the Batang Lupar, and in the ethnically diverse regions of the Rajang and Baluy rivers.

Second, it is a rapidly changing region in which there is a range of processes taking place which are resulting in pressures on the environment: these include plantation development and the expansion of commercial agriculture; forest clearance and commercial logging; urbanization and industrialization in the Bintulu town and port area, particularly with the growth of oil- and gas-related activities aided by the Bintulu Development Authority, established in mid-1978; the extension of infrastructure, particularly the major Sarawak highway which cuts through the region and connects Bintulu town with Miri and Sibu; and population growth both from natural increase and particularly in-migration to take advantage of the employment opportunities in and around Bintulu town as well as in the logging industry and in market-related activities along the main road. In fact, the region shares many similarities with the boom areas of East Kalimantan.

It must be remembered that the Bintulu area and the adjacent Miri region to the north-east have witnessed dramatic and rapid changes only during the last twenty years. Thus, the environmental pressures, the effects of which we are currently witnessing, have been squeezed into a very short time period.

This too has implications for the various coping and adaptive responses of local populations.

Third, the Bintulu region provides us with a range of ecosystems for investigation, from communities in coastal swamplands cultivating swamp rice, coconuts and sago to those in the hilly interior still practising forms of shifting cultivation of hill rice, with some hunting and gathering in the remaining forests; from peri-urban communities near Bintulu town which are heavily involved in the money economy and undertake a range of market-related activities to those living close to plantation developments and finding their main employment, for both males and females, as estate workers. Nevertheless, the dominant population comprises the Iban ethnic group, and therefore the study could, for most purposes, hold the ethnic factor constant, while examining a variety of community responses. What is more, despite the variations between the study villages in such features as population density and size, location, accessibility and environmental circumstances, all the communities, though to varying degrees, are involved in market activities and are facing and having to come to terms with rapid market- and state-generated transformations.

Fourth, the location in the region of the then Bintulu campus of Universiti Pertanian Malaysia provided logistical and practical support. What is more, the University has excellent links with government departments and with rural communities in the hinterland as a result of its own research in to rural development issues.

FIRE, DROUGHT AND RAIN

A broad issue which the research team had to address was that of the relation between commercial logging, shifting cultivation, climatic change and in particular the possible increase in the length and intensity of drought and the increase in fire hazards and risks. As I have already noted, the importance of investigating these complex relations was brought into sharp relief by the events of 1982–83 in eastern Borneo and the widespread destruction of forests by fire in eastern Sabah and the province of East Kalimantan. Furthermore, connections have been made between periodic droughts in parts of South-East Asia and El Niño–Southern Oscillation (ENSO) events. ENSO is produced by changes in and the interactions between ocean currents (particularly in the eastern Pacific) and atmospheric temperatures and circulation. The details of these complex physical processes are still being studied by natural scientists, but we know that rainfall, temperature and wind variations in parts of the Indonesian archipelago are connected to them. The severe drought in eastern Borneo was certainly linked to an ENSO event, and was part of a global pattern which also saw similar droughts in Australia, India and areas of Africa. These phenomena are also long established, and, as Nicholls (1993, p. 158) points out, research has demonstrated that 'several

El Niño events (notably 1877–8, 1864 and 1804) were associated with Indonesian drought'; Henley (1994, p. 2) indicates the droughts of 1902 and 1914 are also linked with El Niño. It is also important to note that ENSO is connected to climatic changes over much of South-East Asia, although its effects in terms of rainfall variation and timing are variable across the region. Nevertheless, the duration of both El Niño and La Niña phenomena are typically about twelve months. As yet there are no predictive models of sufficient reliability to suggest the likely effects of global warming on ENSO behaviour.

El Niño events can variously give rise to tree deaths, and other vegetational changes, an increase in wildfires and crop failure, and, with La Niña, accelerated soil erosion. Increased susceptibility to fire in tropical rainforests is also linked to other effects including acid rain and deteriorating levels of health among forest populations. For example, after the Kalimantan fires of 1982–83 the acidity of the water increased there, which, in turn, apparently affected fish populations in the Mahakam river (Wirawan, 1993, p. 246). Furthermore, research suggests that 'combustion in tropical rain forests and savannas generates at least as much CO [carbon monoxide] as in fossil fuels' (ibid., p. 246). Nicholls (1993, p. 159) argues sensibly that these climatic variations associated with ENSO 'should be considered when sustainable development strategies are being developed'.

RAINFALL STATISTICS

It is clear from research on rainfall statistics in Borneo that drought frequency and severity are very variable across Borneo. A recent study by Walsh (n.d.) on drought frequency changes in Sabah since the late nineteenth century confirms our team findings that eastern Borneo is generally more prone to severe drought and associated fire hazards than areas to the west. Thus, our field sites in the Bintulu district of Sarawak in north-west Borneo, and the Danum Valley and coastal regions of eastern Sabah presented some marked climatic variation and associated environmental differences.

In general, Borneo enjoys a wet tropical climate, but there are considerable variations in annual rainfall from over 5,000 mm in parts of the interior to a little less than 2,000 mm in some east coast regions. With regard to Sarawak, by world standards the state has a large volume of water per inhabitant. Nearly all of Sarawak has a mean annual rainfall exceeding 3,000 mm, and averaged across the whole of the state the mean is just under 3,700 mm (Walsh, n.d., p. 5). However, rainfall is generally heaviest in the western and south-western regions of Sarawak and somewhat lower in the north-eastern coastal regions (Bruenig, 1969, p. 128). In Sabah the mean annual rainfall is lower, ranging between 2,000 and 3,000 mm, and falling below 2,000 mm in a few places on the east coast and in the interior Keningau valley of western Sabah (Walsh, n.d., p. 5).

Table 8.1 Drought frequency and intensity in Northern Borneo

Station	Period	Years	Number of droughts recorded in different intensity (CRD) categories				Most intense drought	
			200–299 mm	300–399 mm	400–499 mm	500–599 mm	CRD mm	Year
Sandakan	1879–1992	103	7	2	1	0	443.7	1903
K.Kinabalu	1889–1992	81	13	6	0	0	384.7	1991
Labuan I.	1855–1992	79	5	0	0	0	287.9	1973
Kilanas	1937–1988	52	3	0	0	0	229.3	1987
Pontianak	1879–1988	86	0	0	0	0	168.0	1967
Kuching	1876–1991	105	0	0	0	0	95.4	1914

Source: After Walsh, n.d.

Table 8.2 Interannual variability at Borneo stations, 1961–1970

Station	Maximum yearly total (mm)	Minimum yearly total (mm)	Mean annual rainfall (mm)	Mean number of wet months >200 (mm)	Mean number of dry months <100 (mm)
Quoin Hill Cocoa Research Station	2747	1960	2403	7	0
Kuching	4908	3352	4210	11	0
Sandakan	3730	2193	3062	7	1
Sibu	3597	2805	3255	11	0
Bintulu	4355	3645	3818	12	0

In plotting the frequency and intensity of drought in northern Borneo, Walsh (n.d., p. 9) uses monthly rainfall data series, and defines a dry month as receiving less than 100 mm of rainfall 'on the basis that monthly potential transpiration of lowland tropical forests is of that order and that soil water deficits and plant water stresses will start to develop if monthly rainfall falls below 100 mm'. A 'dry period' comprises one or more consecutive months with rainfall below 100 mm per month and 'drought intensity' is calculated by adding together the total amount of the deficit in a dry period below the 100 mm level ('the cumulative rainfall deficit' or CRD). Walsh used statistics for various stations across northern Borneo, from Kuching (Sarawak) and Pontianak (West Kalimantan) in the west to Kilanas (Brunei) and Labuan Island and on to Kota Kinabalu and Sandakan in Sabah.

The results of Walsh's investigations are presented in Table 8.1. The data for Sandakan and Kuching span over 100 years from the 1870s. However, some of the statistics do not provide a continuous series. This is especially so for Labuan.

Walsh's main findings (n.d., p. 2; pp. 25–6) are as follows:

1   There have been two broadly identifiable periods 1878–1915 and 1967–92 within which droughts have been more frequent and intense, while the period 1916–67 was drought-free. (At Sandakan drought years were 1885, 1903, 1905, 1906, 1915 and 1969, 1983, 1986, 1987, 1992).

2   The period 1967–92 is characterized by a marked increase in the frequency and severity of droughts. For example, at Kota Kinabalu four out of six droughts of CRDs over 300 mm have occurred in the last ten years.

3   Given the Sandakan data, the drought of 1982–3 which resulted in considerable environmental damage was not as long and as intense as the droughts of 1903 and 1915.

4   (a) Dry periods are short and rare in Sarawak, south-west Sabah, Brunei and central and western Kalimantan.
     (b) Droughts are relatively short but seasonal in north-western Sabah.
     (c) Droughts are not so frequent but are more intense in eastern Sabah and areas of eastern Kalimantan.

5   Several droughts were linked to ENSO events ranging from very strong to moderate, but there were other ENSO events which did not give rise to major drought. Importantly, Walsh notes that '[t]he link between ENSO events and drought does not appear to be as strong for Sabah as reported by Leighton (1984) farther south for 16 stations in East Kalimantan' (n.d., p. 18).

In our recent research, data have been taken from various stations in Sarawak (Sibu, Bintulu) and Sabah (Quoin Hill), in addition to Kuching and Sandakan, which again demonstrate the wetness of the western regions of Sarawak and the drier areas of eastern Sabah. Table 8.2 gives data from the ten-year period 1961–70 indicating interannual variability of rainfall and mean number of wet and dry months.

Let us now examine the Bintulu rainfall statistics in more detail from 1915 to 1991, but note, several years are missing from this series (see Table 8.3). The mean monthly precipitation at Bintulu town is 315 mm; the maximum monthly figure ever recorded was in January 1963 at 1281 mm and the minimum figure was 23 mm in February 1985. Usually November, December and January are the wettest months and February, April and May the driest. The mean annual rainfall is 3,778 mm, and the frequency of drought months is 0.45 per year. Since 1915 there have never been more than two drought months (i.e. a drought month has 100 mm or less rainfall) in the same calendar year; this has occurred on only seven occasions (1918, 1920, 1929, 1930, 1931, 1987, 1990 (see Table 8.3). On only two of those seven occasions have the drought months occurred consecutively in any twelve-month period: in December, January and February 1930 when there was a three-month drought, and again in February–March 1987. However, given that we are dealing with calendar monthly figures, we can

Table 8.3 Drought months at Bintulu (1915–1991)

Year	Drought months [and precipitation levels in mm]	Annual total (mm)
1915	February [61]	3480
1918	March [78], July [100]	3672
1920	March [89], May [92]	3264
1926	February [85]	3461
1927	February [82]	3662
1928	June [94]	3602
1929	July [83], Dec [77]	3526
1930	Jan [97], Feb [25]	3191
1931	Feb [34], Aug [31]	3301
1938	Feb [39]	4766
1955	May [74]	4149
1956	April [80]	3866
1958	July [26]	3376
1966	May [59]	4030
1972	April [95]	3047
1977	April [73]	3849
1978	Feb [86]	3534
1979	May [78]	3311
1982	June [75]	3301
1985	Feb [23]	3391
1987	Feb [79], March [68]	2901
1989	June [76]	3676
1990	March [77], Aug [55]	2994

only provide a rough-and-ready indication of drought, because a 30-day or more dry period cutting across two months would not necessarily show in the calendar month data.

Variation in annual precipitation levels is quite marked from a maximum figure of 5,111 mm in 1925 to a minimum of 2,901 in 1987 (Figure 8.1). There have been very few years when the annual rainfall has fallen below 3,000 mm and the deficit has usually been relatively small: 2,997 mm in 1965, 2,901 mm in 1987 and 2,994 mm in 1990. If one compares this with Lahad Datu in eastern Sabah, one of its driest years was in 1983 when only 1,115 mm of rainfall was recorded.

Although the frequency of drought in Bintulu since 1975 is above the mean (0.59) and 1987 and 1990 recorded two rather low annual precipitation levels, it is still less than the mean of 0.78 for the years 1918–1938. There was a series of relatively low annual rainfall totals in the 1920s and 1930s and a severe drought in 1929–30.

However, these data appear to contradict, in part, the analysis of Salafsky (1994, p. 378), who on the basis of a forty-one-year series of rainfall statistics from Pontianak, in particular, suggests that even in the wetter western regions of Borneo, 'ENSO linked dry periods occur [and] over the last two

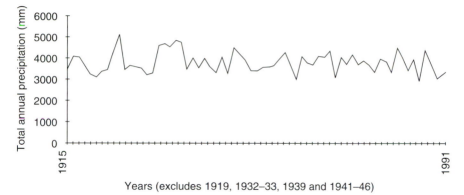

Figure 8.1 Variations in annual precipitation levels, Bintulu (1915-1991)

or three decades they have been increasing in intensity.' One should note that the statistics demonstrate a correlation between the intensity of the normal dry periods at Pontianak and ENSO events in 1972, 1982, 1986 and 1991. Of course, one would need a much longer time series to demonstrate this proposition definitively.

Again if one takes the same time series as Salafsky and applies it to Bintulu, then the rainfall data seem not to be so clear-cut. There was no noticeable drought at Bintulu in the ENSO years 1951, 1953, 1957, 1963, 1968, 1969, 1976, 1983, 1986, 1991 in which there was no month below 100 mm; the ENSO year 1965 did record a much lower annual rainfall overall of 2,997 mm, although without any Cumulative Rainfall Deficit (CRD).

CRDs of 74 mm, 5 mm, 25 mm and 53 mm were recorded in the ENSO years 1958, 1972, 1982 and 1987 respectively. Furthermore, in 1987 there were indeed two consecutive months of drought. However, in the non-ENSO years of 1955, 1956, 1966, 1977, 1978, 1979, 1985, 1989 and 1990 there were CRDs of 26 mm, 20 mm, 41 mm, 27 mm, 14 mm, 22 mm, 77 mm, 24 mm and 68 mm respectively. One might conclude from this that drought months have increased in frequency, particularly since 1977, when, out of a span of fifteen years, eight years have recorded CRDs, and the intensity of these has increased since 1985. However, only two of those fell in ENSO years, and if one took a similar span from 1917 to 1931 (with data for 1919 missing) there were also eight years of CRDs, six of them occurring in sequence from 1926 to 1931 and culminating in the relatively high CRDs of 78 mm in 1930 and 135 mm in 1931.

Therefore, in contrast to Salafsky's findings, it would be difficult to argue for Bintulu that there has been a recent trend towards an increased intensity of drought associated with ENSO events. Furthermore, the more frequent occurrence of CRDs at Bintulu during the past fifteen years is replicated in an earlier period in the 1920s and 1930s. However, comparing the Bintulu

179

material with Walsh's data for Sabah, then the latter do reveal a marked increase in the frequency and severity of drought and far higher CRDs.

Certainly rainfall levels and times remain relatively unpredictable, but Bintulu is not prone to severe droughts. Total rainfall amounts across any given year are usually at such a level that substantial reductions would be needed for drought to become a real problem.

Thus, we seem to be dealing with a quite variable drought phenomenon which has significant impacts in parts of eastern Borneo, but is not a problem generally in north-western regions like Bintulu.

IMPLICATIONS OF DROUGHT

Given the experience of 1982–3 in eastern regions of Borneo, it is clear that the impact on the environment of drought associated with fire is considerable (Beaman et al., 1985; Leighton and Wirawan, 1986). Furthermore, the temporary disruption of the tropical forest biomass by selective logging, which under Borneo conditions often damages up to 50 per cent of the canopy, makes it vulnerable to further degradation, particularly during drought. As Walsh (n.d., p. 19) has said 'Regenerating forest following logging is more susceptible than primary forest to fire accompanying drought, and man's activities (including the proximity of logging and agriculture) also increase the chances of fire . . . in adjacent primary forest'.

In surveyed forest in Sabah affected by the drought and fires of 1982–83, it was found that 5.6 times more of the logged forest burned than the unlogged areas (Beaman et al., 1985). Furthermore, in areas of logging the mortality rate of trees which had suffered from both drought and fire was greater than that in unlogged forest (Walsh, n.d., p. 20). Specifically 85 per cent of the forest in Sabah affected by fire had been subject to logging (Wirawan, 1993, p. 256). Having considered the evidence for Sabah, Walsh (n.d., p. 25) concludes that 'minor droughts that would have been otherwise of little ecological significance are rendered very damaging if they lead to fire'.

In East Kalimantan too, the estimate of 3.5 million hectares of timber destroyed in 1982–83, representing about 20 per cent of the province's forests, demonstrated connections between logging, fire and drought damage. About 70 per cent of the fire-damaged timber occurred in logged areas (Wirawan, 1993, p. 256). Although one should be wary of attributing forest clearance directly to the increased risk of fire, once started, the fires burned longer and more intensely in logged forest, and areas of adjacent unlogged forest also suffered (Mackie, 1984, p. 65). According to reports, the East Kalimantan fires spread rapidly through the thick organic deposits of the logged peat-swamp areas, which had dried out to a depth exceeding 0.5 of a metre (Avé and King, 1986, pp. 85ff.). Fires also travelled along timber trails littered with wood debris. Aerial surveys of the affected areas estimated that 800,000 hectares of unlogged dipterocarp forests were severely damaged, 550,000 hectares of swamp and peat forest, 750,000 hectares of secondary forest and farmland and 1.4 million hectares of logged forest (King, 1993b, p. 301).

The fire may well have been started by shifting cultivators during the burning period prior to sowing, but, it would seem that its devastating effects were in no small way the result of logging. Furthermore, Dayak farmers are being augmented by settlers from other parts of Indonesia and transmigrants who are slashing and burning the forest for farming, and putting down cash crops such as pepper. Population increase and the escalating pressures on land and forests are apparently leading to an increased incidence of fire, although this has to be placed in the context of the exposure of eastern Borneo to drought and a much more variable rainfall regime than areas to the west.

Although the effects of fire, drought and human action are complex and, in some cases, large canopy trees in East Kalimantan were unaffected by fire, but died directly from drought, nevertheless, damage did tend to correlate with the degree of disturbance prior to the fire. In lightly disturbed forest, lower and middle layers of the vegetation were affected; in moderately disturbed forest these layers were seriously damaged as well as some of the upper canopy; and, in heavily disturbed forest all layers were very badly damaged (Wirawan, 1993, p. 250).

IBAN ACTIVITIES AND RESPONSE IN THE BINTULU DIVISION OF SARAWAK

Let us now turn to a more detailed consideration of environmental change in Bintulu. The division had a population of approximately 107,700 in 1991 (*Annual Statistical Bulletin*, 1992), with a population density of 8.5 persons per km², about 60 per cent of it located in towns. In 1947 there was a population of just over 21,220 people in Bintulu district, which had increased slowly to 27,436 by 1960, and then more than doubled in size by 1980 to 58,549, and then nearly doubled again in the decade to 1991. During the last twenty years the increase has been due principally to in-migration because of the rapid economic growth in and around Bintulu town and port, and such other activities as logging.

The population is ethnically mixed, but the majority comprises Ibans (c. 50 per cent), with smaller numbers of Chinese (c. 20 per cent), Melanaus (c. 14 per cent), and Malays and other indigenous groups (c.16 per cent). The Ibans have moved into the area during the past century mainly from the overcrowded Batang Lupar basin to the south-west. They were attracted by the prospects of forest-product gathering and by large tracts of sparsely populated land for shifting agriculture. These Iban migrations into Bintulu, as well as within the region, sometimes sponsored or permitted by government, sometimes illegal, are a constant theme in the history of this region (King, Taylor and Parnwell, forthcoming). Therefore, Ibans were involved early on in market-related activities through the collection of jungle products such as rattan, jelutong, gutta percha, illipe and ironwood for trade. Subsequently they showed an increasing interest in moving to areas suitable for swamp rice agriculture, though they were traditionally shifting cultivators. Presumably this was a consequence of the government promotion and support of wet rice farming as well as difficulties of sustaining shifting agriculture in certain areas.

Therefore, the Ibans are quite recent resource-hungry migrants into a region which was very thinly populated at the turn of the century. They do not represent a long-established traditionally oriented society; they responded to market-incentives and were willing to move in search of new opportunities. Some of our study communities had only settled in the area some thirty to fifty years ago. It is also clear from the historical evidence that Ibans were not necessarily conservation-minded both with regard to forest gathering for the market – on occasion they employed extraction methods which damaged or killed trees – and to the over-exploitation of land and vegetation for agricultural purposes. We should therefore beware of seeing them as necessarily and generally in harmony with their environment. However, prior to large-scale population increase and in-migration since the 1970s their effects on the environment were not a major cause for concern.

FIRE AND RAIN IN BINTULU

We have already seen that the Bintulu region falls within the wetter regions of western Sarawak and that there is no evidence of the relatively prolonged, and sometimes intense drought events experienced by eastern Sabah. In the thirteen survey longhouses those individuals interviewed were asked to address the issue of shifting agriculture, particularly whether or not it is now more difficult to predict the best time of the year to burn, and to control the burn. The questions attempted to elicit information on perceptions of climatic change.

Six of the thirteen longhouses studied no longer cultivated hill rice using forest-fallow methods. There was therefore little concern here about the use of fire for burning slashed forest or deciding the time for burning prior to planting. These villages ranged from riverine upland locations to non-riverine and riverine lowland sites.

Of the community-level responses only four of the seven still involved in shifting agriculture indicated that it is now more difficult to predict the best time of the year to burn the slashed debris. This was perceived to be because of a longer rainy season. The remaining three respondents thought that it was not a problem, or at least no more difficult than it has always been in tropical swidden systems. These are some of the responses:

> 'Yes, it is now more difficult [to predict the best time of the year to burn] because the rainy season [is] longer.'
> 'Yes . . . sometimes when we want to burn in July, it rains now, so we have to wait until August.'

'Yes, because the weather is always changing. Now it is always raining.'

With regard to controlling the burn, there were comments as follows:

'It's only difficult if we burn near someone's cash crops, otherwise we don't bother controlling it.'
'[We] don't control the burn. The burn does not affect the environment . . . [it is] not a threat to the forest.'
'No . . . if land is far from another's field we do not bother controlling the burn [we] only control [it] if near another's [field].'

Taking responses from the individual households, of which there were 222 in our survey, we got a rather firmer view of perceptions of climatic change. Only 9 respondents considered the climate/seasons to be 'always very reli-able'; while 21 felt it to be 'sometimes rather variable', and 122 thought it had become 'increasingly unpredictable'. What is especially clear is that much of the unpredictability is blamed on increased rains (83 respondents) or a longer rainy season (32 respondents).

This does not quite square with the overall rainfall statistics available to us, but these latter do not give us precise information on the exact timing of rainfall in relation to burning and planting. However, what is clear from the responses is that the prolonged drought and the extensive fires of the eastern regions of Borneo are not a problem in the Bintulu area. Fire hazards from uncontrolled burns for those still practising swidden agriculture seem not to be an issue. On the contrary, the concern for some communities is that there is perceived to be too much rain at certain periods, not too little. On the basis of our research it is difficult to conclude that periods of rain-fall have become more erratic and more difficult to predict than previously, despite the perceptions of various of our respondents.

OTHER ENVIRONMENTAL CONCERNS

If fire and drought are not at issue, what then are the important environ-mental concerns in Bintulu which preoccupy local people? From our surveys, the main ones are water pollution, the decrease in wild game and fish, and the increasing difficulty of obtaining timber and non-timber forest products for local use. However, we attempted first of all to obtain a more general macro-level view of the environmental changes which had taken place over the last decade (Taylor et al., 1994).

The changes in the areal extent of different land uses between 1980 and 1991 are provided in Table 8.4. The difference in the total area of just under 3,000 hectares between 1980 and 1991 is due in part to the increase in urban areas by 1991 (which are not included as one of the four land use categories in the comparison), as well as differences in collecting the primary data. However, this represents less than 1 per cent difference between the

Table 8.4 Areal extent of different land uses (1980 and 1991)

Land use	1980 Area (ha.) [a]	1980 area (% of total)	1991 area (ha.) [b]	1991 area (% of total)	Direction and % of change 1980–1991
Mixed Dipterocarp hill forest	530,357	65	504,082	62	Decline in area (by 5%)
Swamp forest	96,721	12	79,580	10	Decline in area (By 17%)
Settled agriculture	8,700	1	11,196	1	Increase in area (by 29%)
Shifting cultivation	175,643	22	213,821	27	Increase in area (by 22%)
Total	811,421	100	808,679	100	

two sets of data which is considered tolerable for the broad comparisons to be made.

By far the largest area of over half a million hectares in 1991 was covered by mixed dipterocarp hill forest, which from 1980 had only declined by about 5 per cent in area. Swamp forest had declined by about 17 per cent, while appreciable increases were evident in the extent of areas devoted to shifting cultivation (22 per cent) and settled agriculture (29 per cent). Areas of settled agriculture have probably increased further since 1991 because of recent oil palm plantation development within the region, particularly along the Miri-Bintulu road. These figures may seem somewhat surprising since the area has been subject to substantial logging operations particularly in the hill forests, and, based on information from our longhouse surveys, to a decrease in the activities of shifting cultivators. However, the extent of forest cover from land use maps tells us little about changes in the quality of the vegetation. It was, in part, to secure information on this matter that the longhouse-based surveys were conducted. Therefore, while the areal extent of forest cover has changed very little, its composition may well have been altered considerably.

Furthermore, the total area of land under shifting cultivation tells us nothing about the level of active agriculture. In other words, some of the land will be in various stages of regeneration, following cultivation. The overall increase in land area subject to shifting agriculture is mainly due to Iban incursions into Reserve Forest and along the Kemena river. However, the vast majority of farming is in secondary rather than primary forest areas, close to rivers and roads, and various areas are supporting established secondary forest no longer worked by shifting cultivators. Indeed, our longhouse surveys suggest that there has been a decline in shifting agriculture during the last ten to fifteen years. This is the result of declining rice yields and labour shortages arising from exhausted soils, competition from paid employment and government support for permanent agriculture. Where hill rice cultivation is practised, the length of the fallow cycle has been decreased to a point at which vegetation and soils will not attain sufficient robustness to provide good yields.

PERCEPTIONS OF THE ENVIRONMENT

According to most of the respondents in our survey the major agent of environmental change has not been climatically induced but is the result of commercial logging, which was already well established in the area from the early 1960s. Some forest areas have been logged three to four times since the 1960s and in the case of one of our study longhouses, Rumah Anchai, the environs had become so badly degraded that it was decided to establish a rattan plantation there to assist in the rehabilitation of the land and the local economy. Logging in the upper Jelalong basin commenced fairly recently but already valuable species such as ironwood have been exhausted and the two longhouses of Rumah Ugal and Rumah Jerangku were commenting on its scarcity. Further down-river, the logging camps around Rumah Utan were running down their operations because the area had already been substantially worked out. This context helps explain some of the responses to questions about perceptions of the surrounding environment. Significant numbers of our respondents thought the environment to be 'marginal', 'unpredictable', 'unstable' and 'degraded' (see also the chapter by Parnwell and Taylor in this volume). Some of our longhouses such as Nyandang, Jerangku, Utan and Anchai particularly voiced these opinions; these interior riverine communities have been especially badly affected by logging. The effects of logging were specified as a general deterioration in environmental quality, water pollution and poor health, declining soil fertility and soil erosion, and declining availability of forest and river products.

Two particular indicators of this environmental change which we used were (a) health and nutrition, and (b) the availability of non-timber forest products.

Health and nutrition

Table 8.5 provides a summary of the health problems which respondents identified. There would appear to have been a decline in certain significant nutritional elements in local diets since various items are no longer available from the surrounding forests and rivers – there has been a decrease in the availability of animal proteins from deer, boar, fish, fruits and vegetables. Some of this decline has been made good by animal husbandry, fish farming and gardening, but it seems that it has not bridged the gap completely. Obviously food can also now be purchased in the market, and, as we have seen, local

185

Table 8.5 Principal health problems in the study longhouses, identified at the community level

Health problem / Longhouse	Diarrhoea	Fever	Malaria	Muscle ache	Head ache	cough/TB flu/asthma	Stomach ache	Lice/skin ailments	Heart problems	High blood pressure	Eye problems	Age groups	Reasons given	Change in incidence	Reasons given for change
Rh. Jerangku	✓	✓	✓	✓	✓	✓	✓	✓				All	Dirty drinking water	Increased	Inconstancy of the weather
Rh. Utan				✓	✓	✓	✓		✓			All	Poor personal hygiene	Same	
Rh. Bana		✓			✓	✓	✓			✓	✓	All	Food: consumed MSG and additives	Increased	Change in diet
Rh. Nyandang	✓	✓				✓		✓		✓		All	Dirty drinking water;	Increased	Air pollution from wars overseas
Rh. Saba	✓				✓	✓	✓					All	Dirty drinking water; diet	Increased	Dirty water in dry season; pollution
Rh. Assan		✓		✓						✓		Old	Diet (MSG), Sugar	Decreased	Better hygiene, cleaner food
Rh. Galan		✓				✓	✓			✓		All, esp Old	Don't know	Increased	More diseases around; diet
Rh. Lunyong	✓	✓			✓	✓	✓					All, esp Young	Dirty food and drinking water	Same	
Rh. Anok		✓			✓	✓	✓	✓			✓	All	Pesticides in the food	Decreased	Medicine and medical services available
Rh. Anchai	✓	✓				✓	✓					All	Dirty drinking water	Increased	Imported food
Rh. Mesa	✓	✓				✓	✓					Old and Babies	Dirty drinking water	Increased (fever)	Pollution, dirty compound and water
Rh. Bunsu	✓	✓		✓		✓						All	Dirty drinking water	Increased	Less fresh food; tinned food
Rh. Ugal		✓	✓		✓	✓	✓					All	Dirty water, dirty compound	Decreased	Have piped water

communities are increasingly turning to cash and purchase. But there has also been a noticeable increase in the consumption of less nutritious processed foods or those which have been grown using artificial inputs such as pesticides.

However, a major cause of increased illness is related to water pollution. Rivers are carrying large amounts of sediment and debris from logging activities. Diesel fuel and oil from machinery and transport are also finding their way into water-courses. Increases in population, and human settlement as well as sewage and other waste from logging camps have also contributed to water pollution. In many areas, there are few, if any, sources of clear drinking water especially in the drier season. Health problems such as diarrhoea, stomach aches, skin disorders and fevers were frequently reported.

Non-timber forest products

Despite an increasing orientation to the market, many households in the survey still depended on forest resources for food, medicines and a source of raw material supply (see the chapter by Parnwell and Taylor in this volume). Just under one-third of households practised hunting and fishing, while about two-thirds of them continued to collect products in the surrounding forest. However, only about a tenth of these households still undertaking hunting and collecting continued to do so regularly (that is, more than once a month), and just under a fifth of them regularly fished. Less than a fifth of respondent households considered that hunting made a 'vital' or 'very important' contribution to their economy, though about a quarter considered fishing significant and just under two-fifths of households were of the view that collecting was still of some importance.

The effects of commercial logging have clearly led to a depletion of the availability of forest products, whilst the extension of the market economy has also, in some cases, raised awareness of the rising commercial value of increasingly scarce forest resources, leading to an intensification of exploitation from within the communities.(On the other hand, male labour migration has in other instances led to a reduction in the number of people available to engage in the pursuit of forest products (see Parnwell and King, 1995)). The longhouses of Sabah, Utan, Jerangku, Mesa and Bunsu indicated that forest hunting and gathering had virtually ceased. In particular, fish stocks in many areas of intensive logging have been severely affected by river pollution and sedimentation. The presence of large numbers of logging camps has also resulted in a marked increase in hunting with guns. Apparently local longhouse residents have also become much less conservation-minded; with the rapid destruction of natural resources taking place all around them, they too have stepped up their exploitation of the remaining forests in order to secure a share of the proceeds before the resources disappear altogether. Some local Iban are now applying modern technology – electricity generators, chemical poisons, chainsaws and guns – to exploit the environment.

Forest products also yield income for local people. For example, 35 respondents indicated that selling planks cut from forest timber was their main non-farm activity. A small number, 15 respondents, also sold other forest products in the market. What is significant is that about 70 per cent of the households were engaged in small-scale handicraft production for use, for which the forest yields essential raw materials. Overwhelmingly the products required were rattan and bemban. It is disturbing, however, that about half the respondents reported increasing scarcity of these materials during the past fifteen years, and that many had reduced handicraft production or stopped altogether. Substitute products are also now purchased in the market, though some people cultivated their own rattan and bemban.

SUMMARY AND CONCLUSION

It is clear that the incidence and impacts of drought vary across Borneo, and there are regions further west like that of Bintulu which have not been subject to long dry periods, nor would it seem that this is likely to be a problem in the foreseeable future. Climatically induced change does not seem to be an issue there in contrast to eastern regions. However, environmental change arising from human action, particularly forest clearance, is having a major effect on human livelihoods. There has been environmental deterioration with water pollution as a particular issue, and for some households and communities a decline in physical and material well-being. Nevertheless, there are processes at work which are both contributing to environmental degradation and enabling this condition to be transcended for some local Ibans. More and more people are now involved in the market-place, and livelihoods are becoming increasingly dependent on cash. To this extent there is less need to seek support from agriculture and forest-based activities. For example, shifting cultivation has shown a significant decline. This does not mean, however, that market dependence is necessarily a good thing; what it does mean is that some rural dwellers are less directly affected in economic terms at least by forest clearance. Indeed, they are themselves contributing to it as employees in logging camps, as freelance providers of timber and forest products and as competitors in securing individual benefits from the exploitation of natural resources.

However, a major issue is that, despite an increasing orientation to the market, many rural communities are still involved for the time being in agriculture and forest-related activities, and dependent on natural water supplies. To some extent, they have overcome declining resource availability by, for example, turning from 'capture to culture' – in other words by cultivating previously naturally growing products like rattan, and by constructing fish ponds for aquaculture (Parnwell and Morrison, 1994). These are welcome adaptations to change, but they could be encouraged further by government support. The provision for supplies of clean drinking water is also essential.

Overall, government measures to assist rural economies would be welcome; they will assist local coping efforts and ease the transition to a market-based economy. Fortunately the communities in Bintulu do not have to face the additional problems of drought and fire which are the lot of their Dayak cousins to the east.

9

MAPPING THE ENVIRONMENT IN SOUTH-EAST ASIA

The use of remote sensing and geographical information systems

Duncan McGregor, Julia McMorrow, John Wills, Helen Lawes and Mark Lloyd

INTRODUCTION

The major environmental problems afflicting South-East Asia at the present time include rapid deforestation, accelerated land degradation, degradation of water resources, and the rapid growth of urban centres. There is a growing awareness and recognition of such problems throughout the region, as exemplified by many of the contributions to this volume. But the nature and extent of these problems are imperfectly documented, and regional synthesis is dogged by data uncertainty. The linkages between human activity and progressive environmental change have been explored elsewhere, and are summarized in this volume for South-East Asia. However, there have been relatively few systematic longitudinal studies of the nature and dynamics of the processes involved in environmental change in South-East Asia. Attempts to form strategies for sustainable development of the region have thus frequently been hampered by imperfect knowledge of the nature and rates of recent change.

An examination of the nature and rates of environmental change over recent decades is essential to a proper understanding of why present environmental problems have arisen; and is also necessary to allow the construction of accurate predictive models of future environmental change. In this respect, progressive changes of land use have been a significant factor in many of the region's environmental problems, and also an indicator of the nature and rate of change.

Systematic monitoring of environmental change and of the progress of environmental management is therefore urgently required, and has been given impetus recently, for example, by international and regional initiatives

(International Geosphere-Biosphere Programme, 1992; Singh, 1994) and national studies (for example Global Resource Information Database, 1990). The immediate need is for appropriate methodologies for systematic data collection and manipulation. Ground surveys are expensive and time-consuming, and, particularly in forested and upland terrain, difficult to achieve with accuracy and in appropriate depth.

A vast body of suitable information does already exist, but this largely relates to discrete, often site-specific, research exercises; government land use survey data in tabulated or mapped form; and remotely sensed data such as aerial photographs, radar and satellite imagery. The progress of environmental change over the last fifty years or so largely awaits documentation from this data bank.

There are three principal problems relating to this data bank: the collation and integration of spatial data relating to different times and held in different forms; the display and manipulation of this data in a form which aids interpretation and prediction; and the sheer volume of data. A further consideration relates to the rapid rates of environmental change which are presently taking place in South-East Asia: deforestation and urbanization being perhaps the most high-profile of these. Techniques are required which enable the rapid collection of data at up to regional scales, the easy assimilation of these data into existing databases, the ability to handle large amounts of data in different formats, and the facility to present them in a broadly comparable format.

The application of remote sensing and Geographical Information Systems (GIS) technology is appropriate for such investigations (Malingreau, 1991). Their utility for environmental investigations in South-East Asia has been demonstrated in specific cases (for example, Kuchera, 1986; H.R.H. Sirindhorn et al., 1990; Susilawati and Weir, 1990; Government of Malaysia: Department of Agriculture, 1991; Karnchanasutham and Wongwantanee, 1991; Lam, 1992; Laili and Darus, 1993). The principal framework of much of this research is the investigation of land use change over time. Implicit in this is both the nature and rate of change in the past, and the construction of models of land use conversion which can be used to project and manage such change.

This chapter illustrates for South-East Asia three aspects of the use of GIS technology in the building of databases and their management. It also examines the conversion of environmental data at different spatial and temporal scales into broadly comparable formats, suitable for analysis and management of land use change.

REMOTE SENSING

Systematic aerial photographic coverage of South-East Asia dating from the 1940s at scales of between 1:4,000 and 1:50,000, and satellite imagery dating from the early 1970s, constitute an archive of land use pattern which has only recently been properly utilized in the determination of the nature and rates of regional and site-specific land use change.

Early black and white aerial photographs include regionally extensive, if imperfect in terms of coverage, over-flights dating from the Second World War. For regions such as Thailand and Malaysia, these are supplemented by subsequent over-flights at varied time intervals to provide a record of progressive land use change over the last fifty years. The accuracy of interpretation of such imagery depends, *inter alia*, on image quality, image scale and verifiable modern analogue for tonal contrasts and textural changes on the photographs. Where particular over-flights were supplemented by contemporary ground survey, such as was undertaken in the 1950s and 1960s land use surveys of Thailand, direct comparisons of photograph and ground information may provide a key for wider photograph interpretation.

Similarly, satellite imagery may be manipulated and interrogated to provide a record of land cover change. Here, the potential accuracy of interpretation has improved markedly from the early Multi-Spectral Scanner (MSS) imagery with pixel (effective ground resolution) size of about 80m x 80m, through Thematic Mapper (TM) from 1982 (pixel size about 30m x 30m) to Système Probatoire de l'Observation de la Terre (SPOT) imagery from 1986, with pixel size down to 10m x 10m.

By sensing different parts of the electromagnetic spectrum, the spectral reflectance from differing surfaces may be measured and manipulated to provide environmental information on the nature of ground conditions. Accurate interpretation, again involving ground verification surveys, of the nature of, and changes in, land cover may be attained. With the above systems, frequent satellite passes enable seasonal changes in ground cover to be assessed, cloud cover permitting, and that change to be monitored through time.

It is important to distinguish between land use and land cover. Land use is abstract because it emphasizes the functional role of a land parcel, whereas land cover is the physical expression of land use and cannot always be directly observed on remotely sensed imagery (Campbell, 1987). It is land cover which is more readily mappable from digital satellite images, whereas land use can be interpreted from aerial photographs mainly because of their better spatial resolution.

GEOGRAPHICAL INFORMATION SYSTEMS

The arguments for using a GIS approach are well documented (Burrough, 1986). They include the ability to use disparate data sets concurrently, ease of up-dating, flexibility of map output with different scales, flexibility of legends and cartographic styles and, most importantly, the analytical ability to create new information by selectively interrogating existing data sets. Applications are wide-ranging, from simple inventory and monitoring to complex environmental modelling and management (see, for example, Goodchild *et.al.*, 1993; Haines-Young *et al.*, 1994).

A Geographical Information System is a computer-based system which captures, stores, manipulates and analyses all forms of geographically referenced (spatial) data. Within a GIS, distinct classes of data are stored in different layers. If the data layers relate to the same geographic area and projection, and are of comparable scale, they can be overlaid in any combination. Whilst Computer Aided Design (CAD) systems are capable of placing different types of geographic features on individual layers and linking these layers with attributes in a database, they are merely graphic systems (Cowen, 1988). GIS systems represent an advance on CAD systems in that they allow complex manipulation of geographic data

CASE STUDIES

The utility of remote sensing and GIS in relation to the evaluation of environmental change over time is demonstrated here through three case studies. Each differs in its focus, scale and methodology, but all three seek to provide, through GIS, a basis for recording environmental change and a potential tool for environmental management.

CASE STUDY 1

MONITORING AND MEASURING LAND COVER CONVERSION IN SABAH, EAST MALAYSIA
Julia McMorrow and Helen Lawes

Background

The 500,000 hectare study area lies in the Lahad Datu and Kinabatangan districts of eastern Sabah (Figure 9.1). It includes commercial production forest and virgin Forest Reserves in the south with logged-over forest excised for agriculture in the north (Douglass *et al.*, 1995). Analysis has concentrated on four 1:50,000 map sheets lying at this forest–agriculture boundary, namely Lamag, Bilit, Sungai Bole and Mensuli (sheets 1 to 4, Figure 9.1). The principal people–forest interactions in this area are commercial logging (Marsh and Greer, 1992), small-scale cultivation in and around villages (Yapp *et al.*, 1988) and forest clearance for large-scale agriculture (Mustapha, 1991).

Forestry in Sabah, formerly British North Borneo, began over 100 years ago. Revenue from forestry and related products contributed between 50 and 70 per cent of Sabah's total revenue in the 1980s and now employs some 25,000 people (Sabah Forest Department, 1989; no date). With the current policy of management of the forest resource, timber output from natural permanent forest estate is expected to be maintained at 6 million cubic

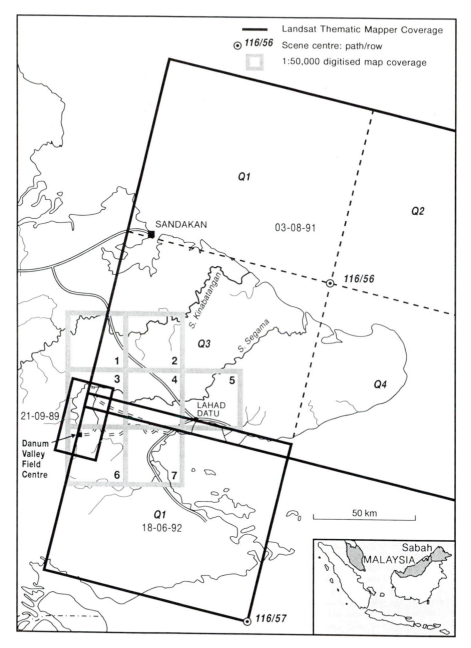

Figure 9.1 Location of digitized map sheets and TM satellite images for Sabah case study

Note: For clarity, the coverage of the Landsat 3 MSS image is not shown: it lies slightly to the west of Landsat 5 116/56

metres. However, as this figure is only 45 per cent of the 1984 level, soft-wood plantation forests, commercial tree crop agriculture and value-added timber industries, together with other secondary industries and the service sector, will contribute an increasing proportion of the state's revenue.

Against this background of a reduced timber resource, it is likely that increasing amounts of logged-over forest will be excised from Forest Reserve for clearance and subsequent conversion to commercial agriculture. Such agriculture is both in the form of privately owned estates and Federal or state land settlement schemes, for instance those run by the Sabah Land Development Board (SLDB) (Sutton, 1988). Both almost exclusively grow permanent tree crops in response to government subsidies and world market trends (Pletcher, 1991). Cocoa, the favoured crop in the 1980s, has been overtaken in the 1990s by oil palm.

Land conversion has potentially significant local and regional environmental consequences which justify a study of its rates and location. Exposure of bare soil during clearance temporarily increases soil erosion risk, and degrades water quality, thus adversely affecting fish stocks for local communities. Oil palm processing similarly exerts demand on water resources and has potential consequences for water quality (Khan and Lim, 1991). The albedo and water requirements of tree crops differ from those of the forest which it replaces, although the precise effect on the water and energy balance and its regional climatic implications is not well known. Land conversion increases potential fire risk, especially in drought years such as 1982/83 when some one million hectares of Sabah's forest were burnt (Beaman *et al.*, 1985; Malingreau *et al.*, 1985: see also the chapter by King in this volume). Oil palm estates increase ignition sources during the two clearance burns, even though good precautions are taken, and from accidental fires along access roads (McMorrow and Douglass, 1995). Weed cover crops on oil palm estates and leaf litter under cocoa seem to have been more susceptible to desiccation and fire than was forest.

In and around *kampungs* (villages) small patches of forest are cleared for hill rice, fruit trees, cocoa and mixed horticulture. The amount of land mapped as shifting cultivation by the Sabah Department of Agriculture in 1989 was minimal and had declined since 1970, but because plots are small and inter-cropping and mixed cropping are common, changes in character and extent are difficult to map from aerial photography and satellite imagery.

The aim of this case study was to create a GIS of eastern Sabah capable of analyzing the type and rate of land cover change from maps and remote sensing, and its location in relation to other map layers.

Method

Both spatial and non-spatial data sets were used. Spatial data were of two types, (a) paper maps stored in the GIS as digital maps, notably 1:50,000

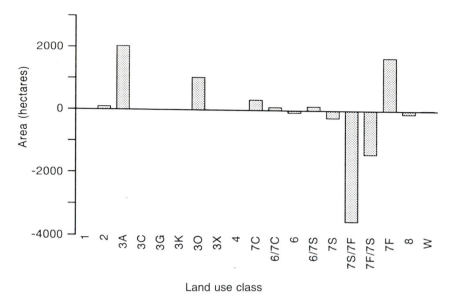

Figure 9.2 Net change in land use (1970-89): Lamang, Bilit, Sungai Bole and
Mensuli, Sabah

Source: Department of Agriculture, Sabah, Malaysia: 1:50,000 land maps published
1970 and 1989

land use maps and topographic maps; and (b) digital satellite images with
acceptable cloud cover ratings, including one Landsat 3 Multispectral Scanner
(MSS) image (path 116 row 56, 3 August 1982) and the three Landsat 5
Thematic Mapper (TM) images shown on Figure 9.1. Non-spatial data were
of two types: (a) lists and tables compiled from documentary sources, such
as the bi-annual statistics of crop area obtained from Sabah Department of
Agriculture district offices; and (b) selected information from questionnaire
surveys of *kampung* households. These data were stored as attribute tables
linked into the GIS by the geographic co-ordinates of the places described.

Data collation was a major part of the research effort, complicated by the
very real issue of restricted access to large-scale maps and aerial photographs.
A permit from the Federal government is required for any research conducted
in Malaysia and collaboration with a Malaysian academic sponsor is essential. Work began in the UK with 1:250,000 land capability maps and
1:500,000 Tactical Pilotage Charts but a minimum scale of 1:50,000 was
needed to relate to the scale of changes reported from village surveys. Official
permission to use recent 1:50,000 topographic and land use maps was granted
after several months, access being restricted to use within Malaysia.

Six thematic layers were digitized from seven 1:50,000 map sheets by the
project's academic sponsors, Universiti Kebangsaan Malaysia, Sabah campus
(UKMS) using PC ARC/INFO (registered trade name, ESRI) (Figure 9.1).

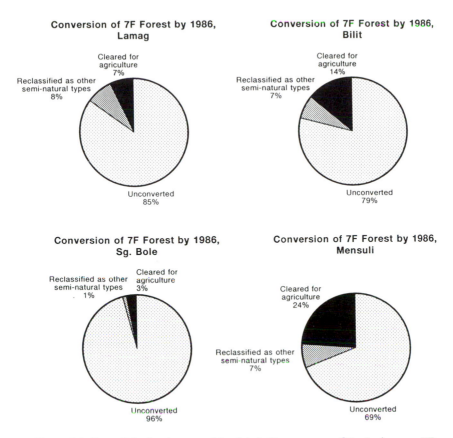

Figure 9.3 Fate of the land mapped by Sabah Department of Agriculture as 7F forest in 1970

Note: Produced by overlay of 1970 and 1989 land use maps and selective query of the resulting map

Maps were transformed into the Universal Transverse Mercator (UTM) projection and converted into a format compatible with GENAMAP (registered trade name, Genasys), the GIS in use at Manchester University, where the researchers were based.

Each map layer was edited, cleaned and verified to minimize errors in digitizing. For example, digitized lines defining each land use area were checked to ensure that they formed a closed polygon and had been given the correct land use label. Care had to be taken to avoid the effective migration of boundaries which can occur when tolerance thresholds are set too large.

Analysis can be at varying degrees of sophistication, from map analysis to spatial modelling and scenario prediction, the so-called 'what if' scenario used by policy-makers. The present work concentrates on map analysis, using overlay and buffering techniques.

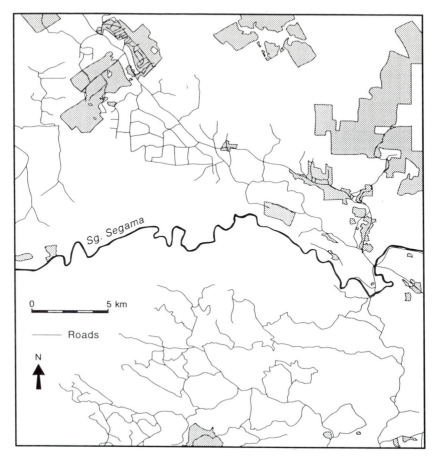

Figure 9.4 Cocoa converted from forest classes (1970–1989) Mensuli, Sabah

Note: Map sheet 4 on Figure 9.1: produced by overlay of 1970 and 1989 land use maps and selective query of the resulting map

Land conversion analysis is based on two sets of 1:50,000 land use maps published in 1970 and 1989 by the Sabah Department of Agriculture, the maps being compiled from 1:25,000 photographs dated 1969/70 and 1986, respectively (Siew, 1973a; 1973b). The classes of this hierarchical scheme found in the study area are:

1 Settlements
2 Horticultural lands
3 Tree, palm and other permanent crops
    3A   cocoa
    3G   rubber
    3K   coffee

198

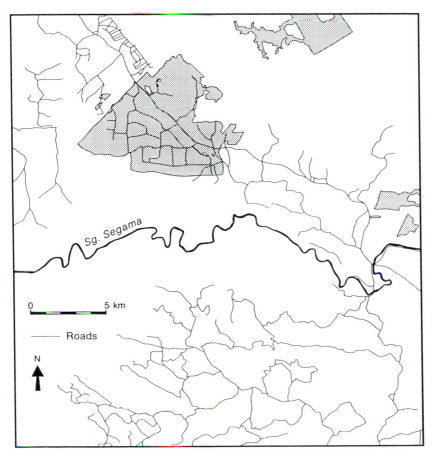

Figure 9.5 Oil palm converted from forest classes, Mensuli, Sabah

    3O   oil palm
    3X   orchards
4 Cropland
6 Grasslands
7 Forest lands
    7F   forest (closed forest)
    7S   scrub
    7C   cleared forest
8 Swamps, marshlands and wetland forests

Additional information on the road and drainage network was derived from the 1:50,000 topographic maps and Landsat images.

199

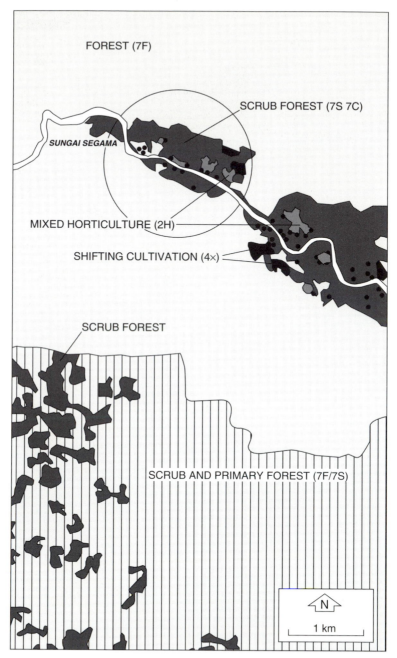

Figure 9.6 Land use around Opak village, Sabah, 1970

Note: Circular 1km buffer

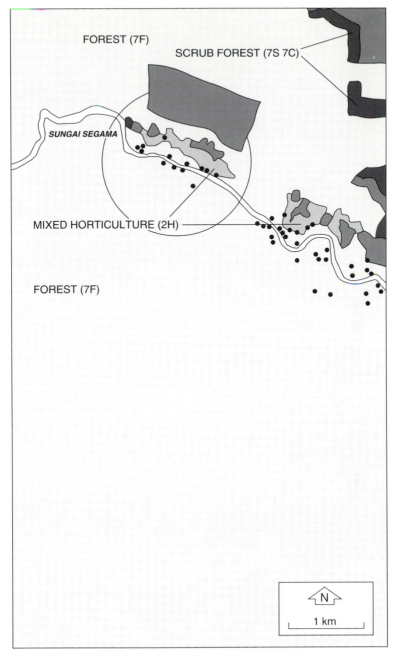

Figure 9.7 Land use around Opak village, Sabah, 1986
Note: Circular 1km buffer

Results

The area covered by each land use category for the two dates was calculated within the GIS. Net land use changes between the 1970 and 1989 maps, as mapped from 1970 and 1986 photography, were calculated (Figure 9.2). Cocoa (class 3A) and oil palm (class 3O) have increased. Surprisingly, so too has 7F forest, whilst scrub forest categories (7F/7S and 7S/7F) have declined. Such figures appear to suggest that it is degraded scrub forest rather than the less disturbed 7F forest which is being converted to agriculture, though this was not borne out when inter-conversions between classes were investigated by overlay analysis. The figures also suggest that scrub forest has regenerated to 7F forest in the sixteen intervening years. This could, however, be due to inconsistency in mapping between the two surveys. It is notable that overlay with logging 'coupe' maps showed that the 7F 'forest' category includes both logged-over forest and primary forest. This implies that land use maps cannot be used to quantify degradation of the forest resource by selective logging, but can determine deforestation where this is defined as clear felling for conversion to other land uses.

The two sets of land use maps were overlaid to investigate inter-conversions between categories (Figures 9.3, 9.4 and 9.5). This produces a new map whose polygons are combinations of the 1970 and 1989 mapped class together with an attribute table listing the original 1970 class and the new 1989 class for each polygon. Inter-conversion matrices can be produced which show the fate of each 1970 class and the origin of each 1989 class. The new map can be queried to yield a map and table of cocoa or oil palm on land converted from forest (Figures 9.4 and 9.5, for the Mensuli map, sheet 4). The analysis shows that in Mensuli 94 per cent of the oil palm area has been converted from 7F forest and the remainder from scrub.

The fate of land mapped as 7F forest in 1970 is shown in Figure 9.3. Between 1 per cent and 24 per cent of 7F forest has been converted to agriculture. Mensuli, the area closest to the town of Lahad Datu and bisected by the main road to Sandakan, has the highest rates of forest conversion to agriculture.

Losses are least for the Sungai Bole map sheet because around three-quarters of the area lies within the Ulu Segama (commercial) Forest Reserve in which timber harvesting is the designated land use. Designation is effective in controlling the rate of conversion and rates of forest-to-agriculture conversion will depend on the Sabah Forest Department's policy on land excision.

A one kilometre buffer around settlements is being used in an attempt to quantify the mapped land use changes for comparison to changes reported in the village surveys. The figure of one kilometre was chosen because respondents in the village survey reported that most of their agricultural lands lie within this radius. Figures 9.6 and 9.7 illustrate a simple circular buffer around the centroid of Kampung Opak on the Segama river, Mensuli, map sheet 4. Irregular buffers around individual houses are also being used.

Data issues

In addition to the issue of data access, there are important issues of data quality in any GIS project. First, there is the question of map currency. Some of the topographic maps were compiled from the 1969/70 aerial survey with only selected revision so that settlement and road networks were sometimes out of date and had to be up-dated from the Landsat images.

Second, there is the multi-faceted issue of accuracy, a function of map purpose and scale. One aspect of this issue is interpretation accuracy, arising from the problem of deciding the class to be assigned. Inconsistency in interpretation between dates and interpreters may explain why land use maps from 1986 photography show less subdivision of forest classes than 1970 maps. For example, patches of 7S scrub forest mapped in 1970 (southwestern half of Figure 9.6) have apparently regenerated to 7F forest by 1986 (Figure 9.7). It is difficult to know whether these are genuine vegetation changes or artefacts of a possible administrative decision to concentrate mapping effort on the, by then, much extended agricultural areas. A further problem is positional inaccuracies which enter both in plotting the position of 'fuzzy' intergrading vegetation boundaries and in capturing their position during digitizing, especially at small map scales. Subsequent overlay operations may create thin 'sliver polygons'.

Remote sensing

The 1982 MSS image was used to depict land cover status at an intervening date to the published land use maps, whilst the 1991 TM image is being used to produce a more up-to-date picture of land cover. An extract from the MSS image was geometrically corrected and geo-referenced to the UTM co-ordinate system used in mapping. Vector files of 1970 land use, roads and drainage were imported into ERDAS from GENAMAP and overlaid on the image to assist image interpretation. The drainage overlay was valuable in helping to differentiate between rivers and roads, allowing a more up-to-date map of the road network to be produced. The vector map was exported into GENAMAP for use in buffer analysis. A land cover classification was produced, based on unsupervised clustering of natural spectral classes in the data, and was refined by reference to Agriculture Department estate returns, giving the location, area and crop type for estates. MSS and SPOT images have been successfully used to produce broad forest, non-forest and cleared land classifications (Amad, 1988; Thang, 1993). The Sabah image similarly used these three classes plus cloud, cloud shadow and mixed (agriculture plus cleared). The raster, or grid cell, classified image was filtered to remove most of the single misclassified pixels. A major cause of misclassification proved to be topography, which causes sunlit forest slopes to have similar reflectance properties in the four MSS bands to that of agriculture. Future work will use a digital terrain model derived from digitized contour data to attempt to correct for topographic effects prior to classification. The classified image was digitized for input into GENAMAP.

TM image extracts were similarly geometrically corrected and UTM geo-referenced. The better spatial and spectral resolution relative to MSS images enables a more detailed and accurate land cover classification. Spectral separability of classes based on the Agriculture Department records and the 1989 land use vector map show that swamp, forest, cocoa and oil palm of varying degree of maturity can be differentiated, but that topographic effects are a greater problem. Progressive stages of forest regeneration after logging, illegal logging and an agricultural incursion into Forest Reserve were detected.

The four most significant problems in using Landsat imagery within a GIS for land conversion studies in Sabah are: (a) the difficulty obtaining acceptably 'haze- and cloud-free' scenes, hence the current use of airborne synthetic aperture radar for 1:100,000 land use/cover mapping in Sabah; (b) the need for access to 1:50,000 topographic maps for geometric correction; (c) the need for a digital terrain model to attempt to correct for very problematical terrain effects, again needing access to 1:50,000 topographic mapping; and (d) inconsistency between land use classes derived from aerial photography, and land cover classes from MSS and TM, requiring amalgamation of categories prior to overlay operations.

Conclusions

The GIS has allowed rates of land conversion and specific inter-class conversions, such as the type of land being lost to oil palm and cocoa, to be quantified. Issues such as data access and data quality were encountered with both the GIS and remotely sensed data sets. Data capture normally consumes 80 per cent of the research effort in a GIS project. In this case, data collation was another bottleneck which has reduced the time available for analysis. Despite these problems the potential of GIS for investigating the rate and location of land cover change in eastern Sabah has been demonstrated. It is hoped that the GIS will continue to be developed in collaboration with Malaysian partners and used to assist environmental management decisions in Sabah.

CASE STUDY 2

AERIAL PHOTOGRAPHY AND GIS: A BASELINE RESOURCE FOR ENVIRONMENTAL STUDIES IN SOUTH-EAST ASIA
Duncan McGregor and Mark Lloyd

This case study outlines the preliminary stages of research into aspects of land use change in South-East Asia, and illustrates the utility of aerial photographs and GIS technology for this purpose. Basic data on land cover are obtained here from the Williams-Hunt aerial photograph collection, held at the School of Oriental and African Studies (SOAS), University of London.

Table 9.1 Williams-Hunt collection inventory

Location	Number	Principal dates of photographs
THAILAND		June 1944-May 1946
North	187	
Northwest	602	
Central	396	
Bangkok	486	
Total	1,671	
MALAYSIA		August 1947-August 1949
Northwest (N)	901	
Northwest (S)	286	
Northeast	273	
Southwest	330	
Southeast	826	
General	16	
Total	2,632	
BURMA		October 1943-January 1944
Upper Irrawady	73	
Shan Plateau	75	
Central: West	100	
Central: East	153	
Rangoon	282	
South Coast	138	
Total	821	
SINGAPORE		Various 1947–1948
Total	240	
CAMBODIA		January 1946-February 1946
Angkor	128	
Phnom Penh	11	
Miscellaneous	12	
Total	151	
FRENCH INDO-CHINA		December 1945
Saigon	58	
Various	82	
Total	140	
MISCELLANEOUS		October 1944
Pacific various	65	
Nicobar Islands	84	
Total	149	
TOTAL IN COLLECTION	5,804	

Source: After Moore, 1986

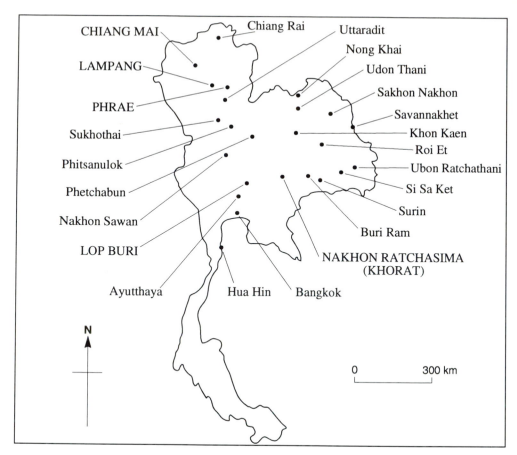

Figure 9.8 Locations represented in Williams-Hunt collection of aerial photographs

Peter Williams-Hunt served as an aerial photograph interpreter during World War II. He was posted to the Far East in 1945, and built up a collection of aerial photographs of archaeological and general interest from the archives held in the Combined Photographic Interpretation Centre in Singapore. Williams-Hunt remained in Malaysia until his accidental death in 1953 (Moore, 1984; 1986). The Collection now consists of just under 6,000 photographs, together with maps (at a scale of about 1:250,000, covering parts of North-East Thailand), some flight plans and various documents of Williams-Hunt's. Table 9.1 indicates the geographical spread and main dates of the photographs. A full description of the inventory, carried out by Elizabeth Moore at SOAS, can be found in Appendix 1 of Moore (1986).

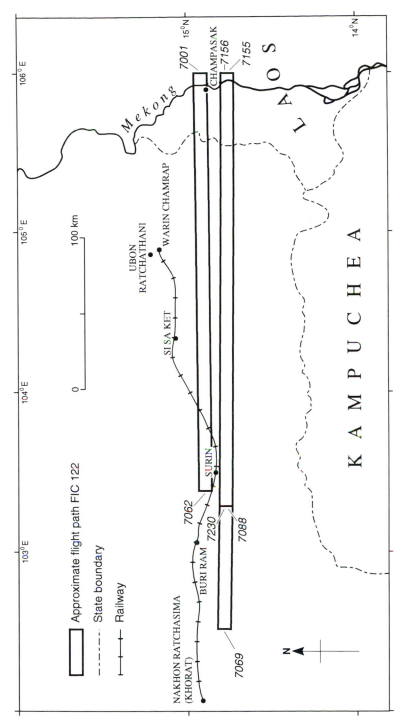

Figure 9.9 Williams-Hunt collection: North-East Thailand transects

Geographical coverage

Most of the Collection consists of vertical aerial photographs, though about 20 per cent of the Collection consists of lower-level oblique photographs (notably of Bangkok, Ayutthaya, Singapore and Saigon). Territories significantly represented in the Collection include Thailand, Peninsular Malaysia, Burma, Singapore and Cambodia. In most cases, the vertical photographs were taken with stereoscopic overlap, permitting detailed three-dimensional viewing and accurate plotting with appropriate equipment. Together, the Collection represents the oldest known freely available regional aerial photographic record of South-East Asia.

The most comprehensive cover in the Collection exists for Peninsular Malaysia (Table 9.1). Much of this is high-quality imagery at a nominal scale of approximately 1:6,500. Moore (1986) records that nearly all of this was flown by the British Royal Air Force (RAF) during the Malaysian Emergency in the late 1940s as part of Operation Firedog, designed to identify possible activity of Communist insurgents. These photographs provide a particularly good record of land use, but are difficult to locate in the absence of flight plans and recognizable settlements or transport arteries.

Coverage of Thailand dates from 1943 to 1947, with particular concentration between June 1944 and May 1946, and was taken by the RAF during the occupation of Siam. Cover exists for a number of Thailand's major centres of population and their immediate environs. Figure 9.8 shows some examples. The Collection also contains two major transects of North-East Thailand (Figure 9.9). Photograph scale of urban areas varies from about 1:8,000 to 1:50,000, and a number of photograph mosaics have survived. The North-East Thailand transects are at a scale of about 1:50,000, although of variable quality.

The earliest photographs in the Collection are from Burma. Mostly vertical at 1:50,000 scale, they were taken in 1943 and 1944. Nearly 300 photographs exist of Japanese-occupied Rangoon. The Singapore photographs are all obliques, together with about half of those of Angkor.

Pilot study: Thailand

A pilot study of the use of the Williams-Hunt Collection for investigation of land use change is under way at present at Royal Holloway, University of London. Aerial photographs of the towns of Chiang Mai, Lampang, Phrae, Lop Buri and Nakhon Ratchasima (Khorat) (Figure 9.8), together with one of the transects across North-East Thailand (photographs FIC122, numbers 7001 to 7062: Figure 9.9) are being examined.

Table 9.2 Land use percentages in the six sample areas

	N.E. Thailand Photo 234				Khorat Photo 140	
Location	(FIC122/7056)	Phrae	Lampang	Lop Buri	(Siam 6/4110)	Chiang Mai
Closed urban				3.5		
Open urban	0.3	11.6	16.5	6.8	3.7	26.5
Dispersed settlement			0.4		17.2	
Isolated settlement		0.3		0.3	0.2	
Industrial site		0.3				
Airfield						0.9
Agriculture	72.4	42.2	56.8	57.9	62.4	52.5
Rice fields			2.7			
Farm factory		0.6			2.2	
Scrub	5.9	0.4	5.2	14.8		2.2
Closed forest		4.4	0.1	3.1		0.1
Open forest	21.3	31.1	6.2	5.3	8.8	7.6
Water feature		0.9	8.4	0.8	0.6	
River		2.2		0.9	1.5	2.3
Canal				0.2	2.5	1.3
Reservoir	0.1					
Wetland		4.7		0.3		
Idle land		1.5	3.7	6.1	0.8	6.6
Total per cent	100.0	100.0	100.0	100.0	100.0	100.0

Table 9.3 Land use percentages for sectors on North-East Thailand flight transect

FIC122: Land Use (%)	7001– 7005	7006– 7017	7018– 7029	7030– 7042	7043– 7055	7056– 7062
Open urban						0.2
Agriculture	0.8	3.7	16.7	53.2	50.6	67.2
Scrub	4.1	13.0	27.2	4.1	31.8	7.1
Closed forest	89.0	83.0	56.2	42.1	12.6	12.7
Open forest						12.8
Water feature	5.9					0.1
Lake	0.2					
Reservoir						0.1
Idle land		0.3		0.6	5.0	
Total %	100.0	100.0	100.0	100.0	100.0	100.0

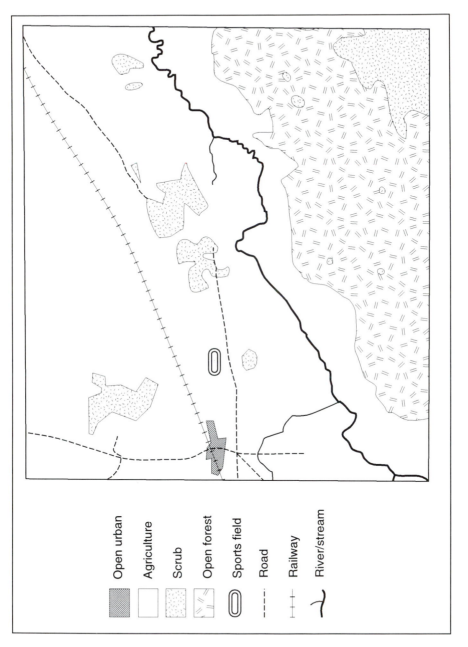

Open urban

Agriculture

Scrub

Open forest

Sports field

Road

Railway

River/stream

Figure 9.10 ARC/INFO map of NIC122/7056 (North-East Thailand) (1946)

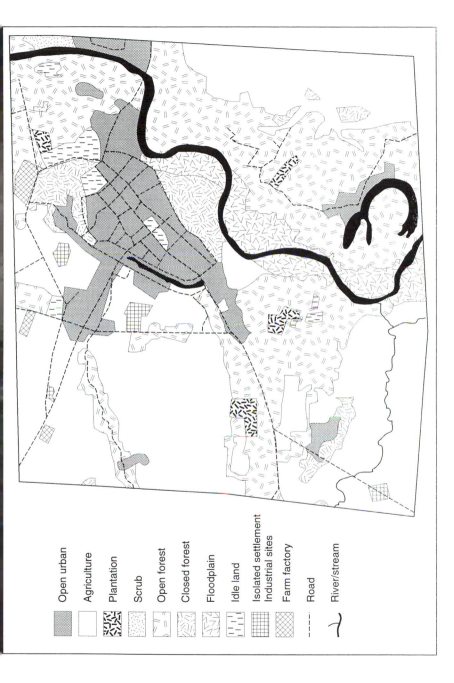

Open urban
Agriculture
Plantation
Scrub
Open forest
Closed forest
Floodplain
Idle land
Isolated settlement
Industrial sites
Farm factory
Road
River/stream

Figure 9.11 ARC/INFO map of Phrae area, Thailand, from 1945 aerial photographs

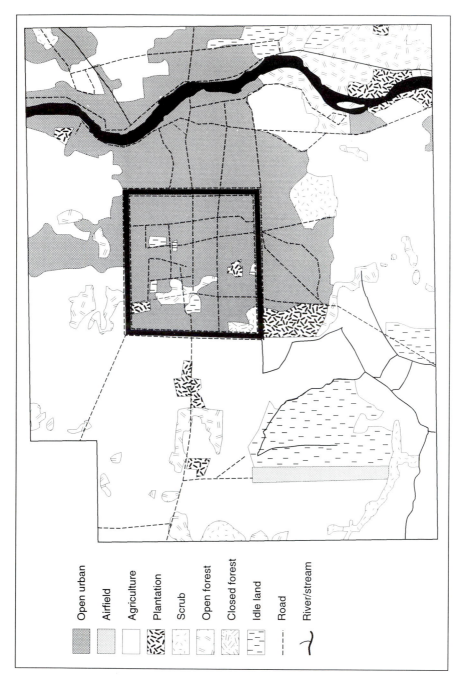

Open urban

Airfield

Agriculture

Plantation

Scrub

Open forest

Closed forest

Idle land

Road

River/stream

Figure 9.12 ARC/INFO map of Chiang Mai area, Thailand, from 1944 aerial photographs

Selected aerial photographs are presently being digitized, using the Royal Holloway Geography Department's CALCOMP 9100 digitizer. Digitized information on land use and land cover is being input into the Department's PC-based ARC/INFO GIS system.

A preliminary land use classification has been devised from study of the range of tones, textures and patterns observed on the photographs, and a table of land use codes has been set up (Tables 9.2 and 9.3). Individual units of land which have a homogeneous appearance on the photographs are demarcated on individual photographs, digitized into ARC/INFO, labelled, and assigned a land use code. ARC/INFO is then able to display the attributes of each mapped area, singly or in combination as required, and also produce statistics of percentage cover within each area.

Photograph analysis

It must be stated at the outset that these are preliminary results, and are subject to refinement. Table 9.2 shows the percentages of land use recognized on the imagery of five sample urban areas and their immediate surroundings, together with one sample photograph from the North-East Thailand transect. These were chosen to test the range of image quality present within the Collection, and include examination of individual photographs at scales of about 1:10,000 and 1:50,000, together with photo mosaics at scales of about 1:50,000. Figures 9.10 to 9.12 illustrate the output from ARC/INFO for some of these areas.

The principal feature noted within the urban areas is their 'open' nature. In all cases, the urban areas contain a significant amount of open space, usually associated with individual dwellings, and therefore presumably small agricultural plots, kitchen gardens, or similar. There is no 'closed' urban area identified, other than a small area at the centre of Lop Buri. A comparison of the amount of closed urban area, as well as overall urban expansion, will prove to be instructive.

Even in these 1940s photographs, there is almost no closed forest left, and relatively little open forest except in the vicinity of Phrae (Figure 9.11). Rice fields are identified only on the Lampang mosaic, although this may well be a function of season (that is, time of year when particular photographs were flown) or incorrect identification.

Table 9.3 shows the percentages of land use classes recognized in six sections along the transect of North-East Thailand. The photographs used are at a scale of about 1:50,000, and are generally of good quality. However, at this scale it is difficult to subdivide agriculture, or indeed to interpret minor variations within particularly designated land use polygons.

Figure 9.10 represents the output from one photograph area along the transect, FIC122 number 7056. This is located to the east of Surin, with the Surin to Si Sa Ket railway running southwest to northeast across the mapped area. The land use statistics are shown in Table 9.2.

The gradual shift from closed forest near the Laos border to increasing agricultural land use is clearly demonstrated (Table 9.3). The transect over-flies the border mountains just to the north of the site of Champasak in Laos. This is a sensitive border area, and obtaining a modern analogue in Thailand may prove problematical. However, it is this area which is of considerable interest in terms of deforestation, as clearance of such hilly terrain will lead to severe land degradation unless appropriate preventative measures are emplaced. There are numerous examples of this in similar topographic situations elsewhere in the humid tropics (for example, Jamaica: McGregor and Barker, 1991).

Discussion of the analysis

There are a number of problems associated with the analytical procedures. These are detailed below, and their implications for the use of the Williams-Hunt Collection discussed.

Transfer of information from photograph to digital format

Normally, digitized information is derived from tabulated/published spatially referenced data, or from data sources such as maps which are planimetrically correct. The uncorrected aerial photograph is not a true plan representation of the ground. Distortion arises from a variety of sources, but generally increases away from the centre of the photograph. Accordingly, only the central portion of the photograph can be digitized with any certainty. The alternative is to register ARC/INFO co-ordinates for land use units on both photograph and map, and then to rectify the photograph co-ordinates to fit the map. Maps of a suitable scale (to match the nominal photograph scale) are available in Thailand, but not as yet to this project.

It has proved difficult, due to inherent photograph distortion, to register suitable control points on the photographs to allow the building of continuously digitized maps over large areas. For example, the North-East Thailand transect has been presented here as six separate sections. The problem is technically solvable; but in the absence of topographic map sheets at the appropriate scale, errors will be present in the shape, and hence area, of the polygons.

Recognition of land use patterns

There is always an element of subjectivity in the recognition of land use on aerial photographs, particularly where this is being 'remotely sensed' from an old set of aerial photographs. Errors of correspondence will inevitably be present in interpreting different photographs of the same area. This is exacerbated with varying photograph scales and image quality, where proper 'quality control' is difficult to achieve. Features which appear distinct on a photograph of scale 1:10,000 may be all but invisible at 1:50,000. At the 1:50,000 scale, it is often difficult to determine the exact boundary between land use units; indeed, a sharp boundary may not exist on the ground. This has a knock-on effect in the analysis, in terms of placing polygon boundaries exactly. The hard copy output, showing clearly defined polygons, implies both an exactitude of interpretation and a sharp transition between land uses on the ground. Neither of these may exist in reality.

In this case, there is added complexity of unfamiliarity with these areas of Thailand (although one of the authors has considerable experience of aerial photograph interpretation in other humid tropical areas). It is necessary, therefore, at the next stage of the research, to liaise with Thai experts in aerial photograph interpretation.

Further, the process of analysis is only appropriate to the scale of digitization (and hence the scale of the original photograph). Minor land use variations within particular polygons may not show up at 1:50,000, but will be clear and significant at 1:10,000.

The digitization process

Various strategies were tested in the initial stages of the aerial photograph analysis. Tracing on to overlays was attempted, but this was found to be too time-consuming, and of no greater accuracy than conventional digitizing. A photogrammetric approach was tested, whereby stereoscopically overlapping photographs were set up, and the resulting three-dimensional model studied. This has the advantage of magnification (the mirror stereoscope used has a x4 binocular attachment) and of some correction of photograph distortion. But this proved to be both time-consuming and difficult to digitize accurately.

Available maps at scales of 1:250,000 (dating from the late 1940s and present in the Williams-Hunt Collection, and from the early 1980s) were enlarged in an attempt to provide ground control. However, significant xerographic errors were introduced, and the enlargement of detail such as roads and railways tended to mask the actual line of the feature on the ground. In most cases names, settlement patterns, and other attributes on the maps did not correspond sufficiently closely to the photographs to be of use. This was equally true of the 1940s and the 1980s maps.

It was concluded that straightforward digitizing of the photographs was the only solution at this preliminary stage. It follows that errors of correspondence discussed above will be present. These can only be quantified and eliminated by field checking with appropriate map sheets.

The land use shade set used requires refinement, both from the point of view of changes to the land use codes, and also from the difficulty of drawing patterns in very small polygons which are recognizable from the key. This will be facilitated by collaboration with Thai counterparts.

Use of multi-spectral satellite imagery would assist in the refinement of land use classification, although this does not extend back in time further than the early 1970s, and would not therefore be of direct use in analysing these 1940s photographs. The use of such imagery for comparative purposes is, however, of critical importance to the development of this project.

Research directions

The wider aims of the research are to analyse the nature and rates of land use change in Thailand and Malaysia since the 1940s; to examine environmental policy, particularly as it refers to land use change and land degradation; and to provide an empirically determined GIS upon which to base future environmental policy. The principal outcome of the research will be the provision of better information on the nature and rate of land use change in particular, as this will provide an input into the formulation of regional policy on environmental management.

The preliminary classifications of land use and land cover types require refinement, the present analysis providing the baseline data for checking by collaborative partners in Thailand (and subsequently, following parallel analysis, in Malaysia). The first stage in this process is to provide research partners in Thailand with the output from the present analysis. This will then be checked against the current situation, and a gross picture of land use change from the 1940s to the present day will be constructed. Where possible, checking against appropriate archive material to determine particular land use patterns will reduce the probability of significant errors of interpretation. This includes modern aerial photograph image analogues, together with any available record of land use patterns in the sample areas from the 1940s.

Time-sequence analysis of aerial photographs, land use maps and possibly satellite imagery will be an essential part of project development, and in addition, assessment of land degradation will be initiated using multi-criteria methods developed (for example, Stocking and Elwell, 1973; 1976; Morgan, 1986.

CASE STUDY 3

A GIS FOR THE RAINFORESTS OF BRUNEI
John Wills

This case study outlines the aims, methodology and results of a GIS developed for the Kuala Belalong Field Studies Centre in Temburong District, Brunei Darussalam (Figure 9.13). The possible applications, limitations and issues associated with using GIS in tropical forest environments are also discussed.

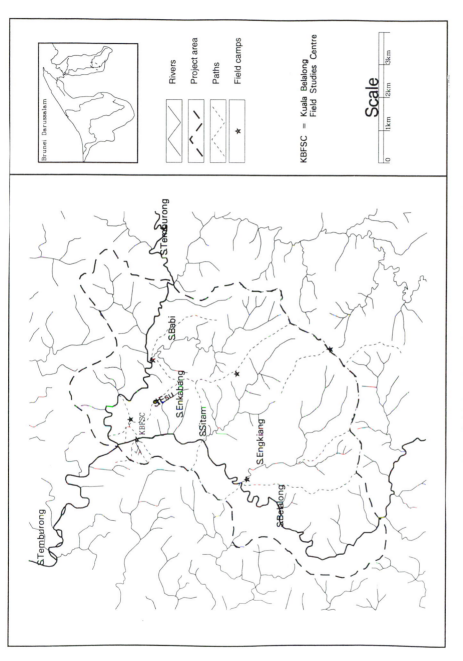

Rivers

Project area

Paths

Field camps

KBFSC = Kuala Belalong
Field Studies Centre

Scale

0   1km   2km   3km

S.Temburong

S.Babi

S.Enkabang

S.FSU

KBFSC

S.Sitam

S.Engkiang

S.Belalong

S.Temburong

Figure 9.13 Location of Brunei rainforest GIS study

Research was conducted within the context of the Brunei Rainforest Project 1991–92, a fifteen-month joint expedition between Universiti Brunei Darussalam and the Royal Geographical Society (see also the chapter by Dykes in this volume). The Kuala Belalong Field Studies Centre, located in the primary rainforest of Temburong district, provides facilities for daily and residential visits by school pupils, university students and visiting scientists. The aim of this study was to provide a resource of baseline information on the flora and fauna of a 5,000-hectare area around the field centre, within the Batu Apoi National Park. This was achieved by creating a GIS to be used as an educational tool for students and members of the public visiting the field centre, as well as providing a data archive of previous research work and an information base for detailed spatial analysis by other researchers.

Methodology

The project used PC ARC/INFO, with digital elevation, slope angle and slope aspect models produced in IDRISI.

Topographic and related data

1:50,000 scale Ordnance Survey maps on the Rectified Skew Orthographic (RSO) grid system, detailing contours, rivers and streams were digitized. Data layers for the paths, sub-camps and project boundary (based on contour interpretation) were created by drawing in on the Ordnance Survey maps and re-digitizing.

Existing data

After completing the topographic coverages a search was made for existing data relating to the project area. Much of Brunei has been covered by detailed geological and soil surveys. However, due to its remoteness, the area around Kuala Belalong has not been included in these surveys, the only data available being of a very coarse resolution. The 1983 Temburong Renewable Resources Study (Anderson and Marsden, 1983) was found to be the only comprehensive data-set available. This survey covered the whole country and provides detailed information on vegetation type in Southern Temburong district. The relevant 1:50,000 maps were digitized and exist as a separate data layer.

Survey data

With over seventy scientists conducting research in the survey area, the volume of data produced was substantial and spans a wide range of scientific disciplines. To enable a variety of users to make effective use of the

Table 9.4 Brunei rainforest GIS: coverage list

General	1	Map of Brunei showing the project area in Temburong District
Topographic data	1	Project boundary
	2	Contours: 100ft intervals
	3	Rivers and streams
	4	Paths: marked trails used by project members
	5	Pondoks: the field centre and sub-camps in the forest
	6	Grid system: Rectified Skew Orthographic grid system
Existing data	1	Vegetation type: Temburong Renewable Resources Study 1983
Project data		
A: *Plants*	1	Plants of Belalong: extracted from field log records
	2	Plant growth environment study: Dr D. Mitchell
	3	Pteridophyte collection: Dr D. Edwards
	4	Ethnobotanical survey: Dr K. Abu Salim
B: *Invertebrates*	1	Invertebrates of Belalong: extracted from field log records
	2	Phasmid survey: P. Bragg & I. Abercombie
	3	Stalk eyed fly study: Prof D. Burkhardt & Dr I. de la Motte
	4	Brachyceran Diptera study: Dr M. Drake
	5	Microlepidoptera study: Dr G. Robinson
	6	Woodlice study: Jason Medway
C: *Fish*	1	Fish survey: Dr S. Choy
D: *Herps*	1	Frog survey: Dr C. Reading
	2	Reptiles of Belalong: Dr I. Das & field log records
	3	Amphibians of Belalong: Dr I. Das & field log records
E: *Birds*	1–4	Birds of Belalong: Dr C. Mann, Dr D. Wells & field log records
F: *Mammals*	1	Mammals of Belalong: Dr J. Charles & field log records
	2	Bat survey: Dr C. Francis
	3	Mammal transect survey: Dr L. Bennett
G: *Physical data*	1	Water-level records of S. Belalong: manual records
	2	Rain gauge records at Field Centre: manual records

system it was therefore necessary to record the data produced in a concise and understandable form.

Some data were received in the form of written completion reports, whilst the remaining researchers were interviewed. Once it was decided that a particular research project would require GIS input, a specific request for data was made: for example, the points or areas sampled and the species found there. These data were entered into specific subject database files containing spatial data (for example, species present) and attribute data (for example, an identification number linking the data to a particular sampling point or area).

Some files relate to the work of individual scientists while others are the combination of results from two or more scientists working in the same field. In addition, some data were extracted from field observation records which relate to sightings of birds, plants, mammals, invertebrates, reptiles and amphibians along paths and streams.

Each database file was stored as a different data layer to allow clear viewing and analysis of the results. There are six background coverages relating to basic topology and the project's infrastructure. Twenty-four coverages relate to particular scientific investigations.

GIS output

Table 9.4 indicates the completed coverages with a description and the name of the scientist involved if appropriate. Because these coverages relate to the same base area it is possible to overlay them in any desired combination and perform spatial analysis. For example, it is possible to show which species are present in any sampling point or area on any of these coverages; and conversely, in which sampling points or areas a particular species is found.

Discussion

The use of GIS in tropical forest environments is set to increase. The Royal Botanical Gardens at Kew are currently involved in three GIS projects. A GIS has been developed for the Kew seed bank, combining geographical locations for seed collection sites with seed bank records. A GIS containing digitized geographic data is being constructed for Madagascar; descriptive and locational data for legume, orchid and palm specimens will be added in the near future. In Brunei, an integrated specimen and taxonomic data-base has been developed. More than 15,000 fully verified specimen records and a large number of specimen images have been entered into the system, although problems have been encountered in locating the position of some collection sites. Brunei Shell Petroleum, with extensive GIS experience from its oil operations, have developed a system covering the whole country, which includes forest type maps and complete LANDSAT coverages. In Australia, the Environmental Resources Information Network (ERIN) is involved in a project using ARC/INFO GIS to link the databases in regional herbaria for easy access and display. In addition, the British Museum is involved in creating a broad-scale GIS which indicates levels of biodiversity.

Applications of the present GIS

The main use of this GIS is educational and presentational, at the local scale. School groups are currently visiting the field centre fortnightly, with residential visits by students from the UK and Australia planned for the near future. The GIS is being used to demonstrate the flora and fauna of the area and to educate students about the environment around the field centre. A local teacher is presently being trained to take over the demonstration and maintenance of the system, to ensure that its continued use is sustained by local resources.

In the educational context, GIS offers an innovative means to present complex information and introduce students to rainforest issues. Along with multi-media, its use in education is expanding. In the UK, for example, the National Curriculum at time of writing included an explicit reference to the use of GIS in Geography teaching. The non-statutory guidance notes state that GIS can 'provide schools with a means of using IT [Information Technology] to handle spatially referenced data' (Cassettari, 1991). In the USA, the National Center for Geographic Information and Analysis is also developing a role for GIS in the Geography curriculum (Kemp, 1991). Secondary education in Brunei follows the UK model and the educational role for GIS is therefore likely to expand.

Limitations

The majority of GISs relate to areas for which considerable spatial data already exist; for example, data on land-use classes, geology or soil type. With the exception of vegetation data and base maps, no detailed spatial data for the area around Kuala Belalong existed prior to the Brunei Rainforest Project. Whilst significant spatial data now exist as a result of the project, this relates to buffered areas around paths and streams (for 16 coverages with streams buffered at 15 metres and paths at 60 metres) and points or sampling stations (7 coverages). As research in the project area continues and more data become available, it may be possible to extrapolate them to produce species location maps or species density predictions. However, this process is problematic, especially where the existing species data are poor and the species habitat parameters wide. Where the habitat parameters are narrow, species presence estimates are more feasible. For example, in the case of fish, a strong correlation was found to exist between stream order and altitude for the species found at thirty-nine sampling stations in the project area, allowing species presence predictions to be made and validated. In the case of plants, as data are mainly derived from passive sightings, it is not possible to conduct accurate species presence or frequency estimates.

Issues and recommendations

Data unification

Ideally the data pertaining to one subject area (for example, plants) would be combined as a single coverage to make the storage and presentation of data more logical. This would necessitate the construction of complex database files and would only be possible where all data had a comparable degree of locational accuracy and related to approximately the same geographical area. For example, in the present case, it would not be desirable to combine the data on Microlepidoptera, which relate to precise sampling locations, with the other datasets on Invertebrates which relate to much broader sampling zones.

Data protocols

The establishment of data protocols prior to the start of a GIS project would greatly simplify data entry. For example, a GIS data entry *pro forma* would facilitate collection of varied data in an appropriate form. Establishing data protocols therefore formalizes the system of data input and ensures that all data are compatible with the requirements of the GIS.

Spatial accuracy

The issue of spatial accuracy is particularly important; the success and utility of any GIS depend on it having a level of spatial accuracy which is appropriate for its intended use. A GIS used, for example, to calculate the home ranges of hunter–gathering populations will require a far higher degree of accuracy and detail than a GIS with a purely educational use. Unless the level of accuracy is sufficient for its intended use, the GIS will not perform its intended tasks well, wasting valuable effort and resources. Conversely, a GIS which has a level of accuracy and detail beyond its initial brief may have additional applications. This implies that the area covered should be mapped in detail before any data collection begins, thereby increasing the spatial accuracy of the data produced.

Ease of use

It is essential that a GIS for non-specialist use is easy to operate. This enables the maximum number of people to use the system with the minimum training and specialist knowledge.

Conclusions

This project created a large number of coverages and provides a good overview of the area's topology. In view of the educational aims of this GIS the issue of spatial accuracy was not of paramount importance. ARC/INFO proved to be a very effective package and had no significant limitations. Maximizing spatial accuracy, establishing data protocols, following closely defined aims and ensuring ease of use are considered to be the important factors in creating an effective GIS.

The main applications of the system described above are educational and presentational. The main limitations are the lack of spatial data and the uncertainty of spatial accuracy, while the main issue is the need to combine closely related data-sets and spatial accuracy. Through future repeated survey, such a system would enable the identification of change over time, as well as having the potential to build in the agencies of change and to examine the implications of such change.

GENERAL CONCLUSION

Rates of resource exploitation and associated environmental change are high throughout South-East Asia. The growing awareness of the consequences of environmental damage has not been matched by appropriate regional efforts to quantify and map the progress and processes of change. Attempts to address this problem have often been hampered by a lack of suitable data on the nature and rates of change. The introduction to this chapter pointed to the urgent necessity for setting up data collection, manipulation and display systems which would enable the monitoring and modelling of rapid environmental change. A solution to the data processing problem, if not directly to the environmental change itself, is offered by remote sensing and geographical information systems technology.

The first of the three case studies presented in this chapter illustrates the potential of GIS to create a regional picture of land cover change, with particular reference to the conversion of lowland dipterocarp forest to commercial agriculture. It uses a combination of existing maps compiled from aerial photography and ground survey together with satellite imagery and government statistics to investigate land use and land cover changes since 1970 in part of Sabah, East Malaysia. This analysis has allowed rates of land conversion and specific conversions between one land use and another to be quantified.

The second case study examines the use of aerial photographs from the 1940s as a baseline for examination of the nature and rates of land use change in selected locations of Thailand. It deals with the problems of interpreting archive aerial photographs of areas for which there are no direct ground truth data or appropriate maps. Although the preliminary classifications of land use and land cover types require refinement, this analysis provides a baseline GIS for the examination of over fifty years of environmental change at both regional and local scales.

The final case study illustrates the potential of GIS to create a systematic environmental inventory at the local scale, here in Brunei Darussalam. It illustrates how GIS technology can bring together topographic, ecological and hydrological data from existing maps and a wide variety of ecological field surveys. The main current applications of the system are educational and presentational, but it is sufficiently flexible to allow detailed monitoring of the progress of environmental change at the local scale. Additionally, a GIS such as this could be modified to provide species predictions in areas which have not been sampled, based on existing data and known habitat parameters.

Although varied in their nature, scale of research and areal coverage, these case studies demonstrate the utility of remote sensing and GIS technology for synthesizing environmental information and presenting this information in an accessible format. The varied data sources employed here illustrate the flexibility of GIS in unifying disparate data sets, and presenting them on a common spatial base. GIS can therefore provide an important predictive tool, avoiding the high cost, practical difficulties and methodological problems associated with field sampling in tropical forest environments. Equipped with such techniques, practitioners and policy-makers may be better able to identify the incidence and extent of resource and environmental degradation, and use this information to provide the basis of appropriate, and ultimately more sustainable, ecological practices.

10

PROBLEMS IN THE MAKING

A critique of Vietnam's tropical forestry action plan

Chris R. Lang

INTRODUCTION

This chapter aims to examine the Tropical Forestry Action Plan (TFAP) for Vietnam, and to assess its likely effectiveness as a tool of forest conservation in this rapidly developing and changing country. The following discussion focuses particularly on certain large-scale development projects and their likely impact on Vietnam's remaining forest areas. The enquiry is critical of the process whereby the TFAP was drawn up, particularly the widespread use of foreign consultants, and a number of the inherent contradictions and inconsistencies which are found within the Main Report.

According to the Government of Vietnam, the most serious ecological problem facing the country today is that of deforestation (Government of Vietnam, 1993, p. 27). The area still covered by good quality primary forest is less than 10 per cent of the country, and the quality is decreasing faster than the coverage (Biodiversity Action Plan Planning Team, 1993, p. 11). The TFAP views forest destruction as being caused principally by increased population, which means more land is needed for agricultural subsistence, which in turn leads to encroachment of forest land. Poverty and ignorance of 'western' modes of conservation on the part of local people are also cited as causes. The solution proposed is further economic development and increased control of forest by outsiders. More dams have to be built to supply more power to industry, which will attract people away from forests to urban areas.

An alternative view sees outsiders as either direct or indirect causes of deforestation. Logging, commercial plantations, resorts, dams and migration from overcrowded or degraded lands all result in forest loss. The key to forest conservation in this view is to restore local people's rights to manage and use local resources (see also the chapter by Braganza in this volume). As these people have been and still are dependent on the forests, they are much more likely to preserve them than outsiders who do not suffer the direct consequences of deforestation (see Permpongsacharoen, n.d.).

The aim of this chapter is not, however, to suggest specific solutions, but to raise issues for debate both within Vietnam and internationally.

THE TROPICAL FORESTRY ACTION PLAN

The TFAP has been widely criticized by NGOs, environmentalists and local communities. The main criticisms can be outlined as follows (see also: Lohmann and Colchester, 1990; Brunton, 1990; Sattaur, 1991; *The Ecologist/World Rainforest Movement*, n.d.):

- It employs a top-down approach.
- There is a lack of consultation with community-based organizations.
- There is a lack of attention to forest-dwellers' needs and traditional ownership patterns and agricultural practices.
- It places a heavy emphasis on commercial forestry.
- It fails to address the root causes of deforestation, and carries an implicit assumption that deforestation is caused by the rural poor, not logging (see Lohmann and Colchester, 1990; Winterbottom, 1990, appx. 2, p. 45).
- There is a lack of criticism of large-scale development projects which displace villagers.
- Foreign experts out-number local professional staff, they spend a short time in the country, and they have a vested interest in not criticizing either the TFAP or forestry practices within the country.
- The TFAP has failed to reduce deforestation (see Sattaur, 1991).

While the Vietnamese Forestry Sector Review was being conducted, the TFAP process was under review. This led to a meeting in Geneva to 'revamp' the TFAP in March 1991, yet this review process is not mentioned in any of Vietnam's TFAP documents.

Consultations for Vietnam's Tropical Forestry Action Plan were carried out at the end of 1990 and early 1991, after an initial visit by Chuck Lankester (then of the United Nations Development Programme, and later to join the Mekong Committee, the agency responsible for producing plans for almost 100 dams on the Mekong and its tributaries) and R.D.H. Lowe (of the World Bank). These two agencies provided a convenient way for international consultants to obtain work in Vietnam without upsetting the then US embargo of that country. The Main Report was published in December 1991. The Vietnamese TFAP documentation consists of a main report, an executive summary, project profiles and a number of field documents and technical reports. All the field documents and technical reports were written by foreign consultants, most of them men. The Vietnamese appear to have been allowed hardly any role in the process. There is hardly any consultation with villagers or people living in or near the forests. In 1991, moreover, travel for foreigners in Vietnam was quite restricted, and large areas were not accessible due to the political situation at the time. Only one TFAP consultant mentions this. The Main Report is a summary of the consultants' reports, but there is no cross-referencing. There is also no discussion of apparent conflicts between various consultants' reports. The Main Report simply selects some paragraphs from these and omits others. For a document that cost around US$700,000 to produce, it is a rather shoddy piece of work (Sattaur, 1991, p. 4).

SHIFTING CULTIVATION

From an FAO document entitled *Tropical Forestry Action Plan*, produced in the early days of TFAP in 1987, comes the statement:

> Eleven million hectares of tropical forests are lost annually and forest industries have often been marked as the main culprit. However, the fact is that all of this forest land is lost to shifting cultivation, permanent agriculture and built up areas.

Following this, shifting cultivators have been favourite scapegoats of TFAP consultants, and have been blamed for deforestation around the tropics. As evidence of this, an underlying theme of the Vietnamese TFAP is that shifting cultivation and the rural poor are the main causes of deforestation. For example, Sharma's (1988) report starts by stating that 'the major causes of watershed degradation today are deforestation resulting from agricultural expansion and shifting agriculture along with associated fires on the uplands'. No evidence is given for this claim.

Ohlsson presents some data on shifting cultivators, but does not mention where or how the study was carried out. His methodology is summed up in the following statement: 'As observed during the consultancy, *bona fide* shifting cultivators appear to clear a maximum of 1 hectare of forest each year and use it for three years. The shifting cultivators return to the same land after 9–15 years' (Ohlsson, 1988). However, he explains neither what a '*bona fide*' shifting cultivator is nor how he knows that they return after 9–15 years, except that he 'observed it during the consultancy'! He then extrapolates the data to give figures on shifting cultivation for the whole country. It is not at all accurate to visit briefly one area of an immensely diverse country, and then generalize to the remainder of the territory.

MacKinnon (1988) describes the Da River watershed as heavily threatened by the shifting cultivation practices of the local Hmong minorities. His report has several references to ethnic minorities and their destructive farming practices. For instance, he states that Cuc Phong National Park has suffered badly from the uncontrolled shifting cultivation of Hmong villagers. At Kong Tai, he reports that 'the plan to remove three villages is sound but it would be better to remove them right out of the reserve'. There is no mention in his report of how the villages came to be in the park, or how long they have been there. Sargent (1988), on the other hand, describes two Hmong co-operatives in which 'sophisticated systems of crop rotation to protect the soil were practised'. She recommends more research into shifting cultivation 'in order to determine where inputs really are needed, and whether it is possible to introduce a land use which is socially, economically and environmentally sound'. MacKinnon and Sargent are looking at ethnic groups in different zones (i.e. protected areas and 'bare hills' respectively), and yet there is no discussion of these very different views of shifting cultivation.

In reality, the shifting cultivation situation in Vietnam is extremely complex. In Son La province, for example, there are twelve ethnic groups, each practising a different form of agriculture (An Van Bay, personal communication, December 1993). The TFAP totally fails to address this complexity and generalizes both problems and solutions.

CONSULTATION BETWEEN CONSULTANTS

Both Davies (1988, p. 5) and Armitage (1988, p. 52) make the following statement: 'The natural forests will continue to comprise the single most important resource for production of logs and fuelwood well into the 21st century.' This is repeated on page 91 of the Main Report. Four pages later comes the statement 'the natural forest resources ... are not able to produce the logs needed by the wood processing industry in a sustainable fashion even if managed properly' (Government of Vietnam, 1988, p. 95). Nowhere in the TFAP documentation is this contradiction either resolved or even discussed.

Later in his report, Armitage explains that the lack of roads in forest areas has led to heavy logging in comparatively small areas. He suggests spending US$250 million on road building in forests in order to access a further 2 million hectares (Armitage, 1988, p. 7). As MacKinnon points out, 'as soon as there is a network of roads leading into a forest it will be cleared out, burned and settled by local people. This the ministry is demonstrably unable to control' (MacKinnon, 1988, p. 12). There is little evidence from anywhere in the world that building roads through forest will help to preserve it; meanwhile, there is a great deal of evidence to the contrary. In Vietnam, the first forest to be declared a National Park was Cuc Puong in the north of the country. Before Cuc Puong was declared a National Park in 1962, there were no roads, and only 200 Muong minority people lived in the park. A 15-kilometre road was built through the park to allow visitors and scientists access to the forest, which also opened the park to settlers until eventually there were 2,000 people living inside the National Park.

The fact is that, in 1991, when the TFAP consultants were in Vietnam, the Vietnamese logging industry was out of control. In order to attempt to preserve some of the remaining forests, bans were introduced in 1992 on logging and timber exports (*Bangkok Post*, 27.3.92). However, even this has not proved completely successful. In August 1993, Prime Minister Vo Van Kiet ordered immediate cancellation of all timber export contracts and a

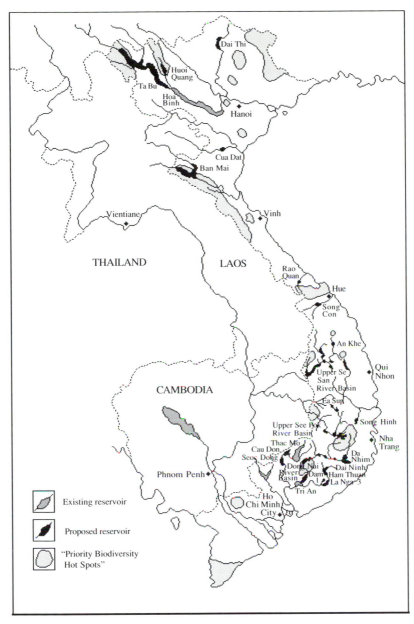

Figure 10.1 Vietnam: existing and proposed hydro-power projects, August 1995

probe into why sawn timber continued to slip out of the country. US$3 million worth of timber was stranded in Qui Nhon port as a result (*Bangkok Post*, 31.8.93). Vietnam's TFAP fails to address the problems posed by a logging industry out of control.

THE PAPER AND PULP INDUSTRY

Another threat to Vietnam's forests is posed by the demands of the paper and pulp industry (see also the chapter by Lohmann in this volume). Vietnam currently has only one large paper and pulp mill, at Bai Bang in Vinh Phu province (see Figure 10.1). None of the TFAP consultants consider the social and environmental impact of this project in any detail.

The mill was built in the 1970s with Swedish aid, production beginning in 1982. The paper mill's capacity is 60,000 tons, which although modest by Swedish standards is enormous for a country with Vietnam's small level of paper consumption. The mill required 500,000 tons of bamboo and wood per year, which was simply not available, so pulp was initially imported from Sweden. Vietnam did not have enough qualified technicians to run the mill, and spare parts and chemicals had to be imported, which the Vietnamese government could not afford. The paper was not even needed in Vietnam, and Sweden therefore bought up the paper and exported it as aid to Ethiopia (Collins *et al.*, 1991, p. 239).

Highly paid consultants and technicians moved into the province, and were provided with housing for their families, schools for their children, and a hospital. The Swedish aid agency, SIDA, was advised by consultants from Interforest AB, a subsidiary of the Finnish Jaakko Poyry Group. These 'experts' recommended huge eucalyptus plantations in order to supply the mill. Widespread planting of eucalyptus has exacerbated soil erosion and led to land use conflicts as land designated for plantations was in many cases already in use by local people (Usher, 1993). Some areas I saw around Vinh Yen town planted with eucalyptus have metre-deep erosion gullies running between straight rows of trees.

The production process led to encroachment into natural forest and into neighbouring provinces. Since 1990 the development of the market economy has created a climate in which people can sell bamboo and wood to the mill at a competitive price, which has led to an uncontrolled cutting of bamboo and trees and still further deforestation. In ten years more than 80,000 hectares of mostly natural forest lands have been cleared to supply Bai Bang (Le Thac Can *et al.*, 1993). In addition, the mill has polluted local rivers and been forced to pay out compensation to local communities. In 1991 US$1,800 was paid to villagers for rice loss and damage to fish stocks as a result of pollution from the mill (ibid.).

In spite of this record, one TFAP consultant, K.M. Gray, argues that an expanded paper and pulp industry is necessary in order to solve the future marketing problem that will be created by the government's reforestation plans (Gray, 1988). The argument goes something like this: it is uneconomic to grow plantations purely to meet fuelwood needs, as fuelwood is traditionally either collected free or sold cheaply. So only 50 per cent of the products from plantations can be used for fuelwood – the rest must be sold at a higher price to cover the costs of the plantation. The answer to this 'major marketing dilemma' is to sell the wood for poles or pulpwood. This clearly illustrates TFAP's commercial bias in its approach to forestry, as opposed to a serious attempt to meet the needs of the rural population.

Although there are a small number of extremely worthwhile reforestation projects in Vietnam, the vast majority of the reforestation that has been carried out consists of monocultures of eucalyptus, acacia or pine. The state is now investing in plantations, and offering low-interest credits for plantations supplying the paper and pulp industry (Government of Vietnam: Ministry of Forestry, 1995). International investment in plantations has been encouraged; for example, the Japanese multinationals, Oji Paper and Nissho Iwai, have a 10,000 hectare concession to plant eucalyptus, which will supply the Japanese paper and pulp industry (Inkaku Tomoya, email forest. plantat conference, 9.10.1992).

LARGE-SCALE HYDRO-POWER DEVELOPMENT

Another extremely serious threat to forests, both directly and indirectly, is that of large-scale dam construction. Once again this is not considered, or even mentioned, in the TFAP. The impacts of dams are varied. I intend here only to look at the impact on forests of three dams, one of which has now been completed.

The dam at Hoa Binh in north Vietnam is the largest dam in South-East Asia. Construction began in 1979, with Russian aid and technical expertise. It displaced 58,000 people, and flooded 11,000 hectares of productive agricultural land. Some forest was inundated. Displaced people cleared steep hillsides along most of the reservoir edge through need of land to cultivate which led to deforestation and soil erosion. The new roads associated with the dam facilitated access and therefore led to increased forest loss (Hirsch *et al.*, 1992).

The relocation of displaced people was not well managed. Even Chuck Lankester, former head of the Mekong Committee, said that Hoa Binh 'is exactly what we shouldn't be doing' (Wallace, 1992). The TFAP utterly fails to consider the impact of the Hoa Binh dam on the forests of the region. Sharma mentions Hoa Binh, but only in relation to the erosion of the Da river watershed, stating: 'the people who were displaced by the reservoir have moved uphill without applying any upland development and conservation practices' (Sharma, 1988, p. 16). The point is that people had no choice but to move uphill. They were very skilled in farming the fertile soils in the valley bottom. The only land available to be farmed after the dam was built was steeply sloping. The dam has led to land use conflicts and to extensive subsequent clearing of forest.

A further project is being planned, the Ta Bu or Son La project, which would centre on a 150-metre high dam at the end of the Hoa Binh reservoir. This would displace 130,000 people. Much of the remaining wet-rice land in Son La and Lai Chau provinces would be inundated (Hirsch *et al.*, 1992, p. 25). The impact of this dam on an already seriously deforested region would be disastrous. I visited Son La in December 1993, and found a very dry, mountainous region, with generally deforested hills and only small patches of forest remaining. On the dusty red roads were lines of women with water buckets, and occasional rusty water tankers which were wheeled from village to village selling water. An Van Bay, a forester from the Son La Centre for Agroforestry, told me it had only been so dry in the last ten years since the forests had gone. Much of the forest was removed during the Vietnam War partly for the timber to provide money to fund the war and partly by settlers from the lowlands who needed more agricultural land to feed soldiers and to replace land elsewhere in Vietnam that had been bombed, napalmed and sprayed with Agent Orange. Bay also told me of Hmong tribes who had managed to protect areas of forest and who were still completely dependent on that forest for their livelihood.

The Hoa Binh dam will be connected to the 500 kV transmission line, which is currently under construction, and will run 1,500 kilometres to serve Ho Chi Minh City. The line passes through forested areas, creating a corridor of cleared land. Politically the line is important as it will physically link the north with the oil-rich south, making the south dependent on the north for electricity. The line has met serious problems due to corruption, and has received criticism from within Vietnam (*Far Eastern Economic Review*, 25.3.93).

Another project linking in to the 500 kV line is the Yali Falls dam, in the Central Highlands. It is part of a six-dam scheme proposed by the Mekong Committee for the Se San river. Although the area is already under pressure from deforestation due to logging, chemical defoliation during the war, and increasing agricultural pressure through migration from the lowlands, there are several important protected areas. The Mom Ray Nature Reserve is particularly important. Elephants, tigers and kouprey are among the ten endangered mammals in the catchment area (Electrowatt Engineers and Consultants, 1993, pp. 8–9).

Forest will be flooded, and destroyed through construction sites. The construction will involve a labour force of 10–20,000, for a period of four to six years (ibid., pp. 7–16). This will lead to an increase in consumption of firewood and timber. The construction process itself will require a large amount of timber. New roads associated with the dam will improve access and therefore lead to further deforestation. Increased development will generate further pressure and demand for forestry products (ibid., p. 2).

Resettlement of around 3,000 people will, unless carried out extremely sensitively, lead to still more deforestation as land is cleared for agriculture. The influx of an additional 10–20,000 people could lead to a substantial amount of illegal hunting (ibid., pp. 9–15). The Environmental Impact Assessment on the project states that 'the general pressure on wildlife habitats, protected or not, will increase as a consequence of this project' (ibid., p. 14). This statement can be contrasted with the Biodiversity Action Plan (currently being produced for Vietnam), which comments that hydro-power provides the most obvious alternative to reduce the use of wood and charcoal as fuel (Biodiversity Action Plan Planning Team, 1993, p. 41). (At the Yali Falls, the villagers are not expected to use electricity immediately, as it will be too expensive compared with their income: Electrowatt Engineers and Consultants, 1993, pp. 15–39). The Biodiversity Action Plan argues that dams have some disadvantages as far as forest preservation is concerned, but also some advantages, such as 'the importance of protecting the forested catchment becomes a higher priority for funding' (Biodiversity Action Plan Planning Team, 1993, p. 41). The argument that dams help protect forest is simply not supported by the evidence, and on the contrary they typically present a serious threat to forests.

THE TFAP IN VIETNAM TODAY

Caroline Sargent, one of the consultants involved in Vietnam's Tropical Forestry Action Plan, has retained an interest in Vietnam's forests. She works at the International Institute for Environment and Development (IIED), and since returning to London she has re-written her TFAP report. Among the changes are references to other TFAP consultants' work, and the implication of a greater degree of collaboration between consultants than actually occurred. It also refers to the TFAP review that took place in Geneva, in 1991.

Since then she has co-edited (with Elaine Morrison and John Palmer) a report entitled 'National Seminar on Setting Priorities for Research in the Land Use Continuum in Vietnam.' The report contains criticisms of the TFAP draft report, such as: '[i]t is clear that the draft report should be evaluated to test whether the conclusions and recommendations are logically derived and reflect biological and socioeconomic reality' (Sargent *et al.*, 1991, p. 249n.). However, whether any such evaluation has taken place is not clear. The rest of the book consists of the Vietnamese contributions to the seminar, which were presented in (generally rather poor) English.

This kind of book, and the seminar upon which it was based, raises many questions (as do several similar instances and events in other parts of South-East Asia, however well intentioned they may have been). Who is it written for? Did anyone anywhere benefit from this seminar, or from the publishing of the book in English, in London? Why was it not conducted in Vietnamese and translated to English – after all, it was held in Vietnam and most of the participants were Vietnamese (if this still doesn't seem odd to you, imagine a seminar held in London about British forestry practices, conducted in Vietnamese, and published in Hanoi in Vietnamese, at a cost equivalent to two weeks' wages for the average British forestry worker). Furthermore, were any forests preserved, or are any likely to be saved, as a result of the seminar?

CONCLUSION

Unfortunately, none of these contradictions seems to matter in the bizarre world of the Tropical Forestry Action Plan. The FAO reports that there are around thirty TFAP projects in the pipeline or currently being implemented in Vietnam. Details are hard to come by, as the projects are the property of the Government of Vietnam, so the FAO is unable to provide further information. However, the German government is funding a project through GTZ, and the European Commission is funding a project in Nghe An province. Meanwhile, Jaakko Poyry, the firm behind the Bai Bang paper and pulp project, were given the consultancy for a participatory rural appraisal project by the FAO in 1992. Another TFAP project consisting of reforestation, logging, and reallocation of land is proposed for Son La province, funded by the Japanese agency JICA. The reports produced for the project still make no mention of the Son La dam.

Thus, in this writer's opinion, the Vietnamese TFAP was a shambles. Consultants flew in, produced their reports, took their fees, and hundreds of copies are sitting in the Ministry of Forestry in Hanoi. The failure to address the real issues of deforestation remains. Large-scale development, displacement of people and logging still represent a serious threat to Vietnam's remaining forests. The complexity of the problem must be addressed properly. Solutions need to be small-scale and locally based (cf. the chapter by Braganza in this volume). Different solutions are required in different areas, involving different people and different environments. Factors such as history, culture, vegetation and the existing situation must be taken into account, while people's rights to self-determination and control over their land, livelihood and destiny must be placed high on the political and policy-making agenda. Such a view is consistent with others expressed in this volume, as it is seen as fundamental to progress towards sustainable development.

Part IV

OPTIONS

11

THE SUSTAINABILITY OF ECOTOURISM IN INDONESIA

Fact and fiction

Janet Cochrane

Traditional forms of tourism, particularly mass tourism to resorts and beaches, have well-documented detrimental effects on the regions and countries visited, ranging from marine pollution and coastal erosion to cultural dislocation of host populations. In the last decade sustainable tourism and ecotourism have been vaunted as answers to the problems of mass tourism and as a contribution to conservation of the world's wild places and the welfare of indigenous peoples. This chapter will examine the theory and practice of ecotourism and sustainable tourism with particular reference to national parks and other protected areas in Indonesia.

TOURISM IN NATIONAL PARKS

In nineteenth century America, John Muir and other conservationists believed in 'the revitalising powers of wild landscapes in an increasingly complex society' (Pigram, 1983, p. 151), and thought that, if nature's resources could be protected and opened up to serve the recreational needs of the general population, the public would in turn become motivated to support the parks. Similar arguments are used to support the establishment of protected areas in developing countries nowadays. 'Some [Indonesian] planners believe that, over the long term, domestic tourism may be the most effective means of developing broad-based social awareness and support for nature conservation and parks' (Robinson and Sumardja, 1990, pp. 210–11).

The first North American parks were remote and difficult to reach, as many Indonesian parks are today, which meant they received few visitors. It was only much later, in the second half of the twentieth century, that carrying capacity for both the environment and the visitors themselves began to be exceeded. This was due to a variety of influences and lifestyle factors for people in industrialized countries, including the development of mass transportation systems and the availability of greater disposable income and more leisure time, and an increasing desire to spend this leisure time in natural areas. The management of visitor flows in national parks in industrialized countries therefore became a pressing issue.

The 1969 definition of national parks drawn up by the International Union for the Conservation of Nature (IUCN) stressed that the protected ecosystems should not be 'materially altered by human exploitation and occupation' (IUCN, 1985, cited in Stankey, 1988, p. 12) and that steps should be taken 'to prevent or eliminate as soon as possible exploitation or occupation in the area' (ibid., p. 12). Tourism was also one of the aims: 'Visitors are allowed to enter, under special conditions, for inspirational, educational, cultural, and recreational purposes' (ibid., p. 12). But as North American, European and Australian parks became more crowded, tension between conservation and tourism emerged and strategies were sought for dealing with the problems. In the Kings Canyon National Park in California sensitive upland meadows were being damaged by the horses and mules used to carry visitors and camping equipment, so the park authorities restricted the use of certain trails and camp sites to protect these areas (Edington and Edington, 1986). Over a nine-year period in the 1980s the National Trust in Britain spent £1.25 million on repairing erosion caused by walkers to paths in the Lake District (English Tourist Board, 1991). By the late 1980s the valley of Dovedale, in the Peak District National Park in Britain, was attracting around one million visitors per year and experiencing considerable damage, particularly erosion of footpaths. The car park most people used to reach Dovedale was reduced in size by almost half, which resulted in a drop in the number of visitors (ibid.).

The tensions between tourism and conservation have been exacerbated by the failure of the tourism industry and the parks managers to understand each other. The world of tourism has tended to view the national parks in terms of a tourist attraction rather than as a means of protecting increasingly rare ecosystems, while the national parks services have traditionally understood the parks in terms of biota and animal behaviour but not in terms of tourists' needs. Objections by field biologists to tourism in protected areas are that any form of tourism will jeopardize the integrity of the ecosystem the national park is designed to protect. As Gunn (1994, pp. 95–6) says,

> Parks and conservation areas have been established, planned and managed as isolated oases of special natural and cultural resources. Commendable as has been their protection of these resources, handling visitors has been treated mostly as non-conforming and only to be tolerated.

At worst, communication between the parks authorities and the public consisted of 'a Big Brother act comprizing signs telling the users what not to do and occasionally roaring past in a four-wheel-drive vehicle to let them know they're watching' (Robertson, 1988, p. 31).

In the 1980s conservationist philosophy altered course towards a more 'people-friendly' approach. The 1982 World Congress on National Parks (held in Bali) was in many ways a watershed. While still stressing the need to preserve ecological processes and genetic diversity, the emphasis was no longer on 'eliminating exploitation' of natural resources to protect them from change and interference, but on ensuring that through wise management utilization of these resources is sustainable (MacKinnon *et al.*, 1986). These 'benign' principles of people-oriented conservation are codified in the tenets of Integrated Conservation Development Projects (ICDPs), which recognize that people around the parks often bear substantial costs to their income potential while receiving few benefits in return (Brandon and Wells, 1992). ICDPs were designed to involve local people and win their co-operation in protecting the area near which they live by providing them with alternatives to exploiting the plants and animals within the reserve: in other words, social or economic development was specifically linked to conservation. However, in an analysis of twenty-three ICDPs in fourteen countries, it was found that:

> while many projects received a great deal of attention from international donors and conservation organizations eager to claim 'success', few projects could in fact match the claims put forward about them or had met their stated goals of linking development to protected area management.
>
> (ibid., p. 561)

One of the problems was that poor households did not automatically switch from illegal, unsustainable activities such as poaching or cultivation of protected land to the legal activities offered by the development programme: they simply used the provision of legal activities as an additional way to increase their income. Other problems were that there was a failure to involve local people in designing projects, and that project managers failed to ensure that local people understood the link between the development projects implemented and the goals of conservation. In the case of tourism, the benefits tended to accrue to local élites rather than to the community at large. Another problem with ICDPs is that the income-generating projects act as a magnet for poor people from outside the project area, with a resulting inflow of population and increase of pressure on resources in the protected area.

By the 1990s the philosophy of the conservation movement had evolved to the extent that, in their policy document on ecotourism, the WWF went so far as to say that 'biological diversity of natural resources can be preserved only if populations who are dependent on these resources for their livelihood are offered viable alternatives to use the resources in sustainable ways' (Boo, 1990, p. 3). One of these alternatives is tourism: at the 1991 Pacific Area Travel Association conference, Prince Bernhard of the Netherlands said on behalf of the World Wide Fund for Nature: 'Tourism in parks and reserves emerges as an alternative that can provide a variety of local economic incentives for sustainable protection, while simultaneously offering substantial national income' (*Jakarta Post*, 13.4.1991). It is by now largely accepted that the visiting public and people living around the protected areas must be welcomed and even embraced if the aims of protecting particular species and biological diversity are to be achieved.

There are risks, however, in over-reliance on economic arguments. If an area is to be protected because it generates revenue through tourism, then if greater economic benefits can be gained from exploiting the resource in other ways, protecting the wildlife would be a low priority (Tudge, 1991). The National Conservation Plan for Indonesia (Food and Agriculture Organization, 1982, p. 4) warned that 'if a heavy investment is made dependent on developing tourism through wildlife and the investment predictably never shows a profit, this could have severe counterproductive effects on the attitude of central government budgetters towards the conservation department in future.' But a decade later it seems that the 'conservation through tourism' reasoning may be holding sway. An analysis of funding to seventeen national parks in Indonesia between 1988–91 suggested that:

> The correlation between funding allocation and parks with biological and/or ecological value is negative. This also holds true for parks with an assessed conservation threat. In contrast, more funding seems to be allocated for those parks with many visitors.
>
> (Sugardjito, 1995, p. 19)

Whatever the drawbacks, the reality is that in the struggle for development priorities, the economic arguments are strong ones. 'Tourism has become one of the largest global economic activities – a way of paying for nature conservation and increasing the value of land left natural' (Lindberg and Hawkins, 1993, p. 7). The accepted practice of tourism in national parks has now become known as ecotourism.

ECOTOURISM: PRINCIPLES AND PRACTICE

Ecotourism is estimated to be the fastest-growing sector of the tourism industry. Visitors to one of the best-known wildlife sites, the Galapagos islands in Ecuador, have increased from 17,000 in 1980 to 46,500 in 1993 (McWilliam, 1995). Tourism receipts have increased markedly in several countries where wildlife is a primary attraction, and form an increasingly significant part of the Gross Domestic Product (Table 11.1).

But what is ecotourism, and how does it differ from conventional tourism? Hector Ceballos-Lascurain, a leading proponent of ecotourism theory, described it in 1987 as:

> Travelling to relatively undisturbed or uncontaminated natural areas with the specific objective of studying, admiring, and enjoying the

Table 11.1 Tourism receipts in selected ecotourism destinations

	Tourism receipts as a % of export earnings		Tourism receipts as a % of GDP	
	1981	*1990*	*1981*	*1990*
Belize	7.8	42.3	4.5	24.8
Costa Rica	9.7	18.9	3.6	4.8
Ecuador	5.1	7.1	1.0	1.8
Kenya	15.3	42.9	2.5	5.9
Madagascar	1.5	12.8	0.2	1.6

Source: Cater and Lowman, 1994

scenery and its wild plants and animals, as well as any existing cultural manifestations (both past and present) found in these areas.

(cited in Boo, 1990, p. xiv)

A subsequent definition by the Ecotourism Society ascribes it a more pro-active role: 'ecotourism is responsible travel to natural areas which conserves the environment and improves the welfare of local people' (Lindberg and Hawkins, 1993, p. 8). The six years separating these two definitions saw a shift in emphasis from the idea that ecotourists should simply pass through an area, enjoying it somewhat passively and leaving no impact at all, to the view that ecotourism should have positive impacts. It is notable that rather than focusing narrowly on nature, both definitions embrace culture and local people as an integral part of ecotourism. As will be shown later, in most cases it is impossible (and now considered undesirable) for ecotourism not to involve local people.

In the wider world, where tour companies and destinations compete for market share, there is a *de facto* use of 'ecotourism' to describe almost any holiday based on a natural attraction. Critics say that this is one reason the industry appears to have expanded so fast: holidays are labelled 'ecotours' because ecotourism is fashionable, regardless of whether or not the holiday contributes to conservation and people's welfare in the destination. Wheeler (1993, p. 123) lambastes ecotourism as 'ego-tourism' and is of the opinion that 'in a symbiotic conspiracy, the mature tourism industry in the west is now consciously cultivating, wooing then seducing the pseudo explorer, the adventurer, the careful eco-traveller, those regarded to be the higher echelons of the tourist continuum'. A glance at almost any tour brochure aimed at these 'higher echelons' reveals the language of this seduction in alluring phrases about 'the preservation of unique cultures and fragile environments' (Exodus Discovery Holidays, 1994/95) which appear to confer an aura of unassailable virtue to the enterprise.

This labelling, however, may bear no relation to reality. The Annapurna region of Nepal (a major destination, incidentally, of Exodus) has suffered deforestation, erosion, pollution and cultural and economic disruption as a result of the influx of trekkers (Gurung and de Coursey, 1994) and is now the focus of a major programme designed to rectify the damage caused by what is effectively mass ecotourism. In Belize, ecotourism has been heavily promoted by the government and tour industry, focusing on the country's coral reefs, variety of terrestrial habitats and its archaeological sites. However, Cater (1992) contends that many of the problems of conventional tourism are already manifesting themselves: for instance much of the revenue from tourism remains in the generating country as payments to tour operators or 'leaks' out for imported purchases needed to service tourism, and considerable environmental damage has resulted from hotel construction along the coasts. The problem is that 'ecotourism may aim to be more environmentally sensitive but, as the numbers of tourists increase, it will inevitably generate similar infrastructural demands to traditional tourism' (Cater, 1992, p. 20).

One of the arguments of developing ecotourism is to divert people's energies away from dependence on the more destructive uses of the protected area's resources. To be successful in this, tourism has to be large enough to employ substantial numbers of people. But any medium- or large-scale tourism activity is bound to have environmental impacts – so it is extremely difficult to achieve both of the aims of ecotourism at once, i.e. to conserve nature and to improve the welfare of local people. This rather fundamental point was borne out at a 1993 conference on ecotourism in Italy which 'found it almost impossible to cite specific parks and reserves where tourism had produced local economic benefits' (Healey, 1994, p. 138). Where such examples do exist, they tend to be very small scale. At the Tavoro Forest Park and Reserve in Fiji the local people manage and benefit from ecotourism, having chosen to develop this as an alternative to logging on their land. The project is deemed a success as it provides indirect benefits through funding small development projects for a community of 150 people, but in its first year of operation it employed only three full-time and one part-time workers (Young, 1992).

SUSTAINABLE TOURISM: PRINCIPLES AND PRACTICE

Most observers agree that any form of tourism – not just ecotourism – should not exceed the capacity of the physical and human environment to withstand it without undergoing serious changes. In other words, it should be sustainable. Certain criteria have been developed for assessing the sustainability of a tourism development (de Kadt, 1990; Cater, 1992; Eber, 1992; Gunn, 1994). These criteria are essentially as follows:

● Sustainable tourism should make careful use of environmental resources, reduce wasteful consumption, and promote conservation.

- Tourism developments should be small scale and integrated with other industries or land-use strategies.
- Sustainable tourism should be of long-term economic benefit to the local community, who should be stakeholders in any development and have a decision-making role.
- The local culture's contribution to the tourism experience should be recognized, and tourism should not undermine the host society and culture.
- Sustainable tourism should be capable of attracting increasing numbers of tourists and continuing to provide them with a fulfilling experience.
- Marketing campaigns should respect the host culture and environment, and tourists should be given full information on the destination.

These principles are relevant in the context of any tourism development, and particularly so in the case of wildlife or nature-based tourism in a developing country. Unfortunately they are subject to a number of inherent conflicts which render their practical application very difficult, if not impossible.

The principal problems of sustainability in tourism lie in four areas: the inherent nature and organization of tourism; the inevitability of market forces in a global, competitive economy; the tourists themselves; and the characteristics of the product life-cycle. These problems as they affect the tourism industry in the developing world generally will be summarized here.

Tourism is now estimated to be the world's largest industry, employing 112 million people world-wide in 1991 (Eber, 1992) and growing 23 per cent faster than the world economy as a whole (Wheat, 1994). It is also a highly fragmentary industry, subject to the policies and needs of a huge variety of agencies and individuals, from government departments concerned with national employment levels through transnational airline and hotel corporations to souvenir sellers on the beach. Wheeller (1991, p. 92) contends that if tourism is to answer the policy aims of governments (increasing foreign exchange earnings, creating employment, promoting regional development) it has to be big, and therefore 'at best [sustainability] is a micro solution to what is essentially a macro problem'. In any case, keeping tourism within tolerable limits means determining carrying capacities, and this has been likened to the quest for the Holy Grail. 'Limiting tourist numbers . . . presupposes those critical levels to be identifiable in advance and such that can be agreed upon. Even if identification and agreement are achieved . . . the question remains as to who would set the limits and enforce them' (Butler, 1991, p. 205).

The size and fragmentation of the industry have led Butler (1991) to draw a parallel with Hardin's Tragedy of the Commons, in that the lack of specific responsibility for common goods and the desire of individuals for short-term gains tend inevitably to cause overuse of these goods. In the case of ecotourism, such goods are essentially natural or 'exotic' cultural manifestations (in the sense that tourists would view them as exotic), which can be over-exploited by, for example, regular intrusions into animals' territory or regular, commercialized performances of formerly infrequent cultural rituals – both of which will contravene the tenets of sustainability.

Further problems arise when the demands of a competitive market meet the criteria of ensuring that the local community benefit economically, have a decision-making role, and are stakeholders. Most tourists come from the industrialized world, and their travel arrangements are facilitated by tour operators based in the generating countries. Only around one-third of the cost of a typical mid-market package tour is spent in a developing country, with the rest going on flight costs, advertizing, wages and other overheads. If the hotels used during the tour are foreign-owned, an even lower proportion of the total expenditure remains in the destination country. It has been estimated that 90 per cent of all coastal tourism developments in Belize were foreign-owned, despite the government's policy of small-scale ecotourism (Cater, 1992).

The general service principle that 'the customer is always right' ensures that the standard of facilities and services will be dictated by the tourists' needs. Many developing countries lack the skilled manpower to provide services to the international standards required by tourists, particularly in areas where ecotourism is an option, which are often distant from cities and centres of higher education. This usually means that English-speaking guides and other staff have to be brought in from abroad or from outside the region. The ability of poor farmers and fishers to become involved in tourism is limited: they generally lack both the skills needed to work directly with tourists and the capital to invest in good quality facilities. There are a few examples where this has been done: in the Annapurna Conservation Area, in Nepal, a sense of ownership of the area and a good understanding of the benefits to be gained from continuing tourism have been developed amongst lodge owners and other people. Healey (1994) contends that there are frequently opportunities to develop indigenous handicrafts around national parks for sale to tourists. Such developments are unlikely to be spontaneous, however. As in the Annapurna area, a great deal of outside organizational and funding assistance is necessary.

The exigencies of international tourists themselves mitigate against the careful use of environmental resources and conservation. When people are on holiday they want to have a good time: anything from a different nightclub every evening to lectures on obscure beetles, according to preference. Few people are dedicated enough to put their ideals for a better world into practice – particularly when they have paid large sums of money for the experience. Many package tourists on adventure or cultural trips choose a different destination every year, and so have a short-term interest in insisting that they see and experience the 'best' the destination has to offer rather than a long-term interest in ensuring that the cultural or environmental integrity of their choice in any year is maintained.

The short time generally spent by tourists in any one location also mitigates against cultural integrity. The most obvious, colourful manifestations of culture are those most frequently seen, and if the performance or production of these do not fit in with tourist schedules they may be adapted to suit. Some of the dances of Bali are now performed in condensed productions made more accessible to foreigners and at set times. According to Picard (1993), the Balinese are now more preoccupied with artistic expressions which can be stage-managed for tourists than with the broader aspects of culture as it underpins their society and traditions. In the Philippines town of Banaue, a tourist resort has been set up where members of the local Ifugao tribe 'play the role of typical indigenous people and where the visitors play at being green explorers' (Gonseth, 1988, p. 34). But several benefits to this system are identified, in that both sides know that the set-up is artificial. The tourists lack the time and the language to have access to the 'real' culture, and for the Ifugao the resort protects them from a more pervasive expansion of tourism into their villages.

It is in any case rather patronizing to insist that tourism should have no cultural impact. As Gonseth says, 'industrial societies do not have the monopoly on the desire to change' (Gonseth, 1988, p. 41). A related weakness in assessing the sustainability of ecotourism focusing on indigenous, 'exotic' peoples is that it is naive to assume that tribal minorities can remain romantically primitive, divorced from the modern world. The influences for change are far wider than tourism alone: missionaries, government officials and television have all made far greater inroads into the sometimes fragile cultures of Indonesia's more isolated peoples than tourism, and it is very difficult to disaggregate the effects of tourism from these other influences.

Efforts by tour operators to provide their clients with reading lists and other pre-departure information on the culture of the area to be visited often meet with apathy or a move to read the information on the outbound flight (personal observation). Once at the destination, the ritual of photographing an aspect of local wildlife or culture before returning to the air-conditioned comfort of minibus or hotel, and to the 'safe' company of the group, replaces any true recognition of or interchange with the host culture. The requirement to see as much as possible in a brief time inevitably means that 'tourist contact favours caricature and inks in the pencil-lines of an inadequate reality' (Barley, 1990). Even young backpackers, who style themselves 'travellers' and roam the world in search of unspoiled destinations, are by no means free of the unconscious wish for an altered reality. Along the 'backpackers' route' through South-East Asia small hotels or 'homestays' offering banana pancakes have sprung up where backpackers can stay among their own kind – other young Westerners – swapping stories of avoiding rip-offs by locals and of the rigours of local transportation. At least the facilities in these cases are locally owned, but whether this kind of tourism is sustainable is highly debatable, given that developments of this type tend to be spontaneous and

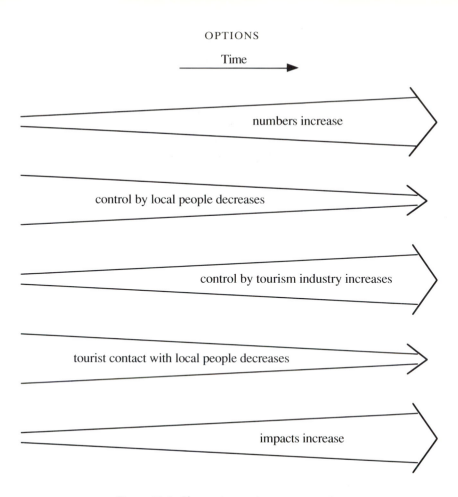

Figure 11.1 Change in tourist types over time
Source: R. Prosser in Cater and Lowman, 1994

uncontrolled, and that very often they are paving the way for investment in bigger hotels from outside the region or from abroad. This progression is shown clearly in the typology of tourists and the product cycle model shown in Figure 11.1 and Table 11.2.

ECOTOURISM IN INDONESIA

The question of scale is especially relevant in Indonesia, with its huge population of over 190 million people. Ecotourism is only ever likely to cater to a tiny proportion of total tourism arrivals, which reached 3.4 million in 1993 (Directorate General of Tourism) and grow at an annual average rate of 21.2 per cent (1988–1993). The vast majority of these tourists are accommodated in hotels at the increasing number of beach resorts and on cultural tours. The

Table 11.2 Typology of ecotourists

Category	Characteristics
Explorer	Individualist, adventurous, requires no special facilities. May be relatively well-off but prefers not to spend much money. Includes trekkers and some bird-watchers.
Backpacker	Travels on limited budget, uses local transport, cheap accommodation, often for months at a time. Enjoys trekking and other adventure activities, but often cannot visit remote areas because of expense. Requires low-cost facilities.
Special interest tourist	Dedicated to a particular hobby, adventurous up to a point, prepared to pay to indulge hobby and have others take care of logistics. Requires special facilities and services, e.g. bird-guides. Accepts discomfort and long travel where necessary to achieve aims. Includes research tourists. Prefers small groups.
General interest tourist	Often prefers security of group or tailor-made holidays, wealthy, interested in culture, keen on nature/wildlife where not too hard to see. Many enjoy 'safe adventure' activities such as trekking and white-water rafting. Dislikes travelling long distances without points of interest. Need good facilities although may accept primitive conditions for short periods.
Mass tourist	Prefers to travel in large groups, may be wealthy, likes superficial aspects of local culture, enjoys natural scenery and wildlife if easy to see, needs good facilities, will only travel far in comfort. Includes coach-tour groups and cruise ship passengers.

Source: Cochrane, 1993a

question is how to attract, in addition to these 'traditional' tourists, people whose vocational needs can be satisfied by the ecotourism product that Indonesia can offer in order to achieve some of its undoubted advantages.

Throughout the 1980s when Indonesian tourism moved into a new phase of deregulation and professionalism, high-spending package tourists and business travellers were targeted (Schwarz, 1989). Visitor numbers increased fourfold from 561,178 in 1980 to 2,177,566 in 1990 (Directorate General of Tourism, Jakarta). More recently, promotional efforts have spread out from the traditional emphasis on Bali and Central Java to include the more remote regions – and to cover a more diverse range of attractions.

Indonesia sits upon a natural ecotourism bonanza. If the country . . . can effectively articulate the diversity of its landforms, ecosystems and natural areas and demarcate the uniqueness of its biological endowments . . . then the rice fields of Bali, the coral reefs of Ambon, the orang-utan of Bohorok and the Javan rhino of Ujung Kulon can all realize their potential to enhance an appreciation of the natural world and the beauty of Indonesia.

(Ratnapala, 1994, p. 18)

Indonesia would certainly like to have more ecotourists. The under-resourced Directorate General of Forest Protection and Nature Conservation (PHPA), which manages protected areas, ascribes an important role to ecotourism in helping to conserve these areas: 'community-based eco-tourism ... is one example of how people in the area can become aware that the existence of a national park will help improve their standard of living through exploiting tourism opportunities' (Sumardja, 1995, p. 15). While this aim, which specifically excludes 'investors who would then be the only ones to profit' (ibid.) can only be welcomed, establishing community-based ecotourism is fraught with difficulties, as will be seen in the case studies later in this chapter.

Indonesia aims to establish forty national parks in total, and by the end of 1994 had declared thirty-one, covering 8 million hectares (Mulyana, 1995). All have been declared since 1980. There are a further 10.8 million hectares in other conservation areas such as strict nature reserves. The approach of Integrated Conservation Development Programmes (ICDP) or Integrated Protected Area Systems (IPAS) is considered an important strategy for management of the parks and reserves, and several programmes are being implemented throughout the country. Tourism and the sustainable use of forest resources are allowed in national parks but not in strict nature reserves.

The practical difficulties of developing ecotourism in Indonesia are threefold: physical, organizational, and cultural. Taking the physical ones first, Indonesia is a big country with a widely scattered sprinkling of national parks. Many of these are remote from large centres of population and can only be reached after arduous and time-consuming journeys, which many tourists are reluctant to endure. Once they arrive at a park, their aspirations to see the wealth of wildlife of which Indonesia is justifiably proud will probably be frustrated by the characteristics of the animals themselves: shy, solitary, arboreal and nocturnal. The problems of accessibility will in time be eased by continuing improvements in the infrastructure, particularly roads and flight links. Appreciation of the national parks could be fairly easily enhanced by improved interpretation and viewing facilities, such as visitors' centres, hides and walkways.

The organizational barriers are perhaps harder to surmount because they are deeply rooted in bureaucracy or cultural norms. The PHPA is in charge of national parks management, while the Directorate General of Tourism (DGT) is in charge of planning and promotion of tourism. In the past there was a lack of mechanisms whereby the two sides could co-operate, although in the 1990s the actions of the two Directorates have converged to a greater extent. Regulations drawn up by the PHPA in 1992 covered the payment of entrance fees to parks and reserves (previously a haphazard and unregulated affair), and in 1994 provided for the operation of ecotourism business in the parks. Meanwhile the DGT sponsored a series of conferences on Indonesian flora and fauna to coincide with Indonesia's Year of Environment in 1993, and produced attractive leaflets on Indonesia's natural attractions.

The cultural problems are rooted in the perception many Indonesians have of wilderness areas and their use. As previously noted by this author (Cochrane, 1993b), Indonesians tend to prefer a crowd to solitude, and, indeed, consider the desire to be alone in the wilderness a manifestation of odd behaviour. Observation of birds and other animals – highly developed and popular hobbies in the West – are almost unknown in Indonesia. There is therefore no pool of enthusiastic amateurs from which a potential corps of skilled nature guides could be drawn, and it is notable that there are extremely few tour guides working in the private sector who appreciate international tourists' expectations on a visit to a national park.

Even park wardens and managers still receive very little training in visitor handling. The lack of expertise in tourism in the national parks bureaucracy is one reason for the slow development of wildlife tourism in Indonesia, in that it has been impossible to guarantee basic standards of service, food and accommodation in national parks. Tour operators have therefore tended to base their clients in accommodation outside the parks and make day excursions to them, thus limiting substantially the visitor experience since most animal activity takes place in the early morning or at night. In any case, as the preceding discussion has shown, not all ecotourism takes place in protected areas or even necessarily in wild, natural areas. The indigenous cultural manifestations of an area are often just as much an attraction as the wildlife and plants, such as on the island of Nias off western Sumatra, where the preferred solution has been to use cruise ships where good accommodation and hygienic food can be provided.

CASE STUDIES

To look at some of the strengths and weaknesses of ecotourism in Indonesia, as well as at the application of sustainability, a number of case studies of tourism in national parks and to remote areas follow (see also Figure 11.2).

Siberut National Park

The Sumatran island of Siberut, part of which was recently declared a national park, has become a popular destination for budget travellers in the 1990s. Tourism doubled from 1,054 visitors in 1989 to 2,145 in 1991 (Cochrane, 1992), and certainly several thousands per year by now. The local inhabitants are Mentawaians, many of whom still find most of their needs in the forest and trade little with the outside world. Their costumes, rituals and

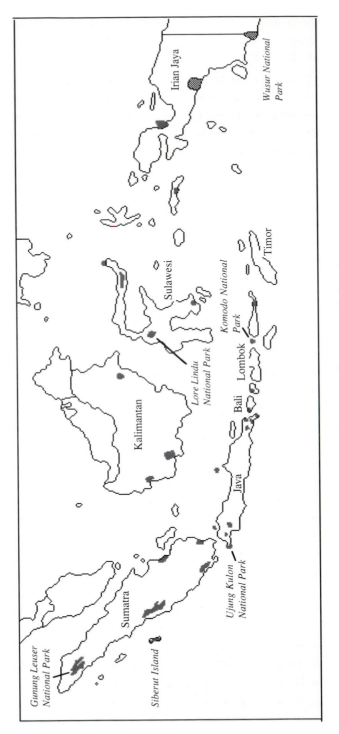

Figure 11.2 Ecotourism locations in Indonesia

hunter–gatherer practices are attractive to see and photograph. Visitors are lured to the island by the promise of living with primitive tribes for a few days. However, the Mentawaians have very little involvement with the organization of tourism, as the visitors are brought in by guides from the Sumatran mainland, and the tourism situation is a clear case of exploitation of indigenous peoples by more educated, entrepreneurial outsiders.

For tourism to be sustainable according to the criteria already discussed, the communities have to be involved at an organizational level, in which case there has to be an established community structure with valid decision-making capabilities, and the local people have to have something to sell, whether it is goods or services. But most of the islanders cannot speak Indonesian, let alone English or other tourist languages; there are no handicrafts to buy; and the accommodation (which the local people share with their pigs) is unsuitable for foreigners. If the islanders work at all it is as porters – for which they are often paid in tobacco rather than money. Thus there are no means by which the communities can earn much through tourism, and they have no control over how many tourists come or where they go – yet the tourists bring cultural dislocation in their wake, inadvertently causing the destruction of traditional barter systems and hospitality (ibid.).

The Mentawaians have self-expressed aspirations to take advantage of aspects of the modern world which can improve their standard of living, particularly better educational and medical facilities. In theory, tourism to the island could be a classic example of how tourism can contribute to fulfilling these aspirations. But under the current circumstances it seems unlikely that the Mentawaians can be more than 'stage props' for the tourists who arrive.

As far as tourism's contribution to conservation here is concerned, it is non-existent. Siberut is home to four indigenous species of primate, but tourists are not shown any of these and most are probably unaware of their existence; even for those who are, it is practically impossible to see any in the wild due to extensive hunting. The island's timber resources are far more valuable to the government than tourism, and so there is no incentive for either the government or the communities to conserve the natural habitat.

Gunung Leuser National Park

In the Gunung Leuser National Park, in northern Sumatra, ecotourism takes place in two locations: along the Alas River, which roughly bisects the park from north to south, and at the orang-utan rehabilitation centre of Bohorok, on the park's eastern edge.

The Alas river supports rafting trips. The first to be developed were up-market tours for which the tour operators bring all their needs – food, boats, rafting guides, tents – from the provincial capital, Medan. Money entering the local economy is minimal or non-existent, and the local community are not involved. Later, a second type of rafting trip started up, with local men acting as guides and providing simple rafts (inner tubes with a wooden platform) and basic food. These tours mainly attract young backpackers. In the first case, tourism provides no incentive for the local communities to cease illegal activities such as poaching and fishing by poisoning, and the presence of the tourists is resented, with problems of theft from the tour parties by villagers (Heckman, white-water rafting consultant, personal communication) In the second case, there are very few local people involved in the tours – not more than 20–25, and on a very irregular basis – so tourism's contribution to local incomes is currently too small to make it likely that people would give up their potentially destructive activities. The second type of tours also tend to cause more environmental degradation in terms of litter because the guides are untrained in environmentally sympathetic practices.

On a more positive note, the area is currently the subject of a European Union-funded ICDP, whose organizers are fully conscious of the need to make local people aware of the link between development programmes such as ecotourism and the continued environmental health of the park.

The second Gunung Leuser ecotourism site is the village of Bukit Lawang, near the orang-utan station of Bohorok, which receives an increasing number of visitors who mostly come to see formerly captive orang-utans being trained to live again in the wild. In the early 1970s the few tourists who went there were prepared for a mildly difficult trek into the forest and for basic facilities. Twenty years later, the path to the centre has been concreted over, and large coach parties arrive (Cochrane, 1993b). A considerable amount of building of accommodation is taking place. In 1989 there were no more than three guest-houses, while by mid-1993 there were fourteen, with more under construction. A large number of stalls selling food and cheap clothing have sprung up to cater both to the international visitors and to domestic tourists from Medan, for whom Bukit Lawang is a popular destination for day trips. In a clear example of the 'magnet effect' of economic development attracting outsiders, the local guides who take tourists on walks into the forest are mainly from outside the area rather than local men.

The tourists' desire to grasp the best available at each destination may lead them to disregard local regulations, with detrimental consequences. Wardens at Bohorok and at other rehabilitation centres in Indonesia and Malaysia frequently report that tourists pick up the animals in order to satisfy the desire to cuddle an orang-utan or have a photograph taken with one (personal communication), despite clear regulations prohibiting this and regardless of the impact of their behaviour on the animals. One of these impacts can be the transfer of disease from humans to orang-utans (or, indeed, from orang-utans to humans).

Although viewing orang-utans in the rainforest seems an obvious example of ecotourism, it is hard to see how it contributes much to conservation other than through the provision of entry fees. Biologists now agree that the individual orang-utans do not benefit from the attention, in that it takes longer to return them to the wild when they are provided with entertainment (and occasional handling) by the tourists. Plans for a new orang-utan rehabilitation centre in Kalimantan, based on fifteen years' evidence from existing centres, concluded that '[r]ehabilitation should never be conducted in combination with a visitor attraction [and that] some ten years after feralization the group can be subjected to well-controlled visitor attention' (Government of Indonesia, Directorate General of Forest Protection and Nature Conservation, 1991, p. 31).

The sustainability of the Bohorok attraction must also be in doubt. Attempts by park managers to limit the number of people at the feeding sessions by suspending ferry crossings across the river giving access to the centre are frustrated by people simply wading across. No attempt appears to have been made to limit the numbers of buildings or of tourists. It seems inevitable that at some stage word will get around that the centre has become crowded and down-market, Bohorok will decline in popularity, and explorers, backpackers and special interest tourists will start looking for new destinations to enjoy the tranquillity and wildlife observation opportunities they expect from a national park.

Gunung Leuser also provides a good example of the difficulties faced by the PHPA in implementing protected area management. Due to the pressure of land-hungry farmers all around the borders, the regional authorities were found to be encouraging poor families to clear land and plant crops within the boundaries of the national park as a way of alleviating poverty (Costello, 1990). Even within the PHPA itself there is a shortage of staff sufficiently well-motivated to enforce regulations, and these are therefore widely flouted. A field trip in October 1993 to Gunung Leuser revealed that the main interest of the ranger at one post was taking tourists into the forest to increase his income, while two days later a man carrying a gun was observed dragging a dead orang-utan out of a protected part of the forest near the guard-post and on to the roof-rack of a bus (personal communication).

Wasur National Park

Wasur National Park, at the south-eastern extremity of Irian Jaya, was only gazetted in 1990 and was managed from the start as an ICDP. The park is the homeland of several indigenous groups, and in addition there are several settlements of non-traditional peoples who moved there before 1990 (Worldwide Fund for Nature, 1995). Efforts to establish permissible uses of the park have included substantial consultation with members of the local communities. One of the options for reducing consumptive use of the park's resources was to encourage involvement in ecotourism. The chief attraction of the park is the ease of viewing wildlife, particularly birds, due to the open, savannah-like terrain or monsoon forest. To this end the villagers were helped to build simple guest-houses to be run for the benefit of the community, and the chief hunters from the nearest town (Merauke) were employed part-time as guides to take bird-watchers into the park (Cochrane, 1991).

In its early stages this project has all the signs of being very successful, in that the links between ecotourism and conservation were made explicit by the programme and seemed to be understood by the villagers. Factors which made the programme more likely to succeed were, first, that threats to the park involved a relatively small number of people, and second, that for the fore-seeable future at least, tourist numbers would be limited by the difficulty and expense of getting to the area: there are only four flights a week from the provincial capital of Jayapura, and overland travel is not possible. It remains a distinct possibility, though, that if tourism becomes a significant activity more educated and entrepreneurial outsiders will move in to reap the benefits.

Lore Lindu National Park

Lying to the south of Palu, the capital of Central Sulawesi province, the main focus for ecotourism in Lore Lindu is the Bada Valley, on the southern boundary of the park. The area is hard to get to, involving a 3–4 day journey from the nearest towns including 2–3 days on foot, or a journey by four-wheel-drive vehicle over a 64-kilometre track which can take up to two days to negotiate. Until recently it was thought that tourists came there to admire the megalithic statues and other stone remains which lie around the valley. A survey of visitors in 1994, however, showed that at least half came specif-ically because of the trek through the rainforest (Cochrane et al., 1994). This is one of the best forest treks in Indonesia, with excellent forest and good views of rivers and mountains, and a fairly good chance of spotting some of Sulawesi's endemic wildlife. Major threats to the park are the high level of illegal harvesting of the valuable rattan cane, and regional plans to construct a road into the Bada Valley. Part of the justification for this road was to open up the area for tourism, and it was depressing to find that tourism officials at all levels were openly sceptical of the desire of foreign tourists to trek rather than to drive to a destination.

The survey also included calculating the economic benefits to the area from developing ecotourism, concluding that around US$350,000 per year could be generated solely for local inhabitants through small-scale ecotourism activ-ities such as trekking, white-water rafting, horse-riding and bird-watching. This amount excluded expenditure on hotels etc. in the nearby main towns. When the findings of the survey were presented to the provincial government, the tourist department became very supportive of ecotourism. Lobbying of the Governor's office produced a signed agreement that the road would be re-routed in order not to affect the protected area (M. Fujita, Director of the Nature Conservancy, Indonesia, personal communication).

It is too early to say whether ecotourism in this case will prove sustainable, but the detailed planning process already completed and the support of local communities and government departments give the project a good chance of success.

Komodo National Park

In Komodo National Park, tourism was deliberately fostered to provide the regional government with an alternative source of income to developing agriculture or transmigration settlements on the islands (Hitchcock, 1993). Tourism arrivals have increased tremendously since 1983/84, when there were 742 foreign visitors, to 1992/93, when 13,656 arrivals were recorded (Durrell Institute of Conservation and Ecology, 1995). In 1994 one tour operator alone (P&O Spice Island Cruises) took around 4,900 visitors to Komodo, and are projecting an increase to 15,000 per year by 1997 (ibid.).

However, although tourism development has certainly stimulated aspects of the regional economy, the industry mainly benefits providers of tourist services on the neighbouring islands of Flores and Sumbawa, and on Bali where the tour operators who organize package tours to the national park are located. The village of Kampung Komodo on the main island of Komodo National Park has greatly increased in size since the park was gazetted in 1982, but the villagers' involvement in tourism is limited to selling occasional artefacts such as pearls or carved dragons to the tourists, and goats as bait for the Komodo lizards. Tourism has therefore presented the villagers with no alternative to their traditional livelihood strategy of fishing the rich waters around their islands, which the national park was partially designed to protect.

It is also highly debatable as to whether tourism has contributed to conservation of the Komodo lizards. A small number of lizards (around fifteen) are *habitués* of the feeding site, but they have become lethargic and somewhat tame as a result of their twice-weekly meal of goat as they no longer have to hunt or scavenge for their food. Feeding of the lizards was suspended by the head warden in May 1994 (Durrell Institute of Conservation and Ecology, 1995) and a study of the lizards' behaviour, population and the impact of tourism both on the lizards and on the local community is expected to produce results in 1997.

The lack of expertise of both private sector tour guides and park wardens is frequently demonstrated on trips to national parks, and a visit to Komodo by the author in 1992 gave an example of this. During the 2-kilometre walk to the lizard feeding site a considerable amount of wildlife (rusa deer, wild boar, scrub fowl, jungle fowl, sulphur-crested cockatoos and a sea-turtle) was spotted; however, all the PHPA wardens and the private tour guides with the group were oblivious to the animals, and tended to frighten them away by moving too fast and making too much noise.

The Asmat people

In a case of ecotourism focusing on indigenous people, Goodfellow (1993) describes the visit of a cruise ship to an Asmat village in Irian Jaya. Until twenty years ago the Asmat had little contact with the outside world and obtained all their needs from the swampy forests around them. Then their dramatic wood-carvings became known, and fetched high prices in the art galleries of industrialized countries. Conveniently located on the coast, their villages became a popular stop-over point for cruise ships – sometimes lauded by ecotourism commentators as one of the least harmful forms of tourism because all food and accommodation is provided on board and the passengers spend little time on land. On a visit to one of the villages one of the cruise ship passengers was led into a village house, where the family started to go through the traditional adoption procedure used to welcome strangers. All went well – until the family made it clear that they were performing the ceremony in the expectation of receiving money. Other passengers were subjected to aggressive selling techniques by the young men. Many of the passengers were doubtful about their effect on the Asmat, feeling that they

> had shown them things that they wanted and couldn't have continually, such as medical aid and easy food ... They had come to find common ground with another culture and perhaps to discover warmth and acceptance lacking in their own. In doing so they threaten to destroy the very values they held in high esteem.
>
> (Goodfellow, 1993, pp. 2–3)

This kind of tourism, which is well suited to the archipelagic geography of Indonesia, is becoming more and more popular. Yet is it sustainable? No special structures have been created for the tourists and there is a minimal impact on the environment, yet it is not small-scale because cruise ships commonly carry 600–1000 passengers, out-numbering villagers when they land; the host communities have no power to decide whether they are visited or not; and it certainly damages the host culture in altering traditional host–guest relationships.

TOWARDS THE FUTURE

The difficulties in achieving sustainable ecotourism, as in the examples given above, have too often been cast into the shadows by the 'green glow' cast by fine words about responsibility and respecting nature. Pointing out the drawbacks is not a particularly constructive contribution to the tourism-with-conservation debate, although it is essential that these should be considered in any realistic approach to the problem. And the problem will definitely not go away: with an expanding demand for new and exciting destinations and growing interest in the environment, added to the ever-increasing amount

Plate 11.1 Ecotourism in Kabupaten Maros, South Sulawesi: this area is famous for its butterflies, many of which are taken home in boxes as souvenirs

of leisure time and disposable income amongst industrialized countries and the newly-emerging economies of Asia, tourist numbers are destined to grow and grow. Under these circumstances, any contribution to mitigating the negative effects of tourism on sensitive ecosystems is likely to be beneficial in itself: the mere effort of striving to justify the labels 'ecotourism' and 'sustainable' may be better than ignoring the principles entirely – although there is a danger that using these labels without an accompanying change in behaviour or procedures may simply paper over the underlying conflicts and contradictions in sustainability as outlined above.

On a positive note, there are certainly productive aspects of the 'eco' trend in tourism, just as in other fields. Many people involved in adventure tourism care deeply about the places they send their clients to and try to minimize the damage caused by tourism, either by trying to inform their clients thoroughly about the environment and cultures they will be visiting, by limiting the size of groups, or by contributing to funds supporting development programmes in the target area (although their contributions are limited by the need to remain competitive in the market-place) (see Plate 11.1).

Emphasizing the gaps in knowledge and the need for studies may mean that these are undertaken and realistic plans made before developments are embarked upon. It is surely a constructive factor that conservation and development organizations are taking seriously the reality of tourism in national parks and are beginning to work with government departments in planning for it. One of the most important elements of future tourism development programmes in protected areas is to ensure that the local communities' perceptions of the area in question are ascertained, and that their needs, aspirations and capabilities are discussed – with them, not just with outsiders – before plans are finalized. This should mean that there is a significant level of involvement by local people in contributing to and benefiting from tourism, which in turn should help ensure that over-ambitious expectations are not raised of what can be achieved by the communities – in terms both of their participation in tourism and of their commitment to the protected area. Community participation in tourism has, of course, to be tempered with advice or guidance from outsiders who may be able to help avoid some of the pitfalls encountered in previous attempts to create genuinely sustainable ecotourism elsewhere.

This participation is, of course, an important element of the Integrated Conservation Development Projects discussed earlier. The failure of ICDPs to achieve the hoped-for results should not be seen too pessimistically: in Indonesia, at least, community participation in such projects has, so far, generally been more theoretical than actual. Several community-based tourism projects have recently been proposed or sponsored by conservation-oriented development agencies such as the World Wide Fund for Nature (in Wasur National Park) and The Nature Conservancy (in Lore Lindu National Park), and it is possible that if these projects are implemented as planned, local people will be empowered to take a part in tourism while it is still small, before the places become popular and the bigger tour operators from outside the area are tempted to move in.

The tourists themselves can also make a contribution towards conservation. Infringements of anti-poaching legislation and exploitation of the resources of a protected area become harder the more witnesses there are: the orang-utan on the roof-rack incident was reported by some Dutch tourists, and a letter campaign orchestrated by a tour operator complaining of illegal logging in the Alas river valley in North Sumatra and directed at the Indonesian Ministry of Forestry prompted concerned replies from the Ministry. The instinctive herd behaviour of tourists is also beneficial for conservation in that visitors can be 'managed' (or even manipulated) to limit their impact: it has often been noted that visitors tend to stray little from designated paths – or even from their cars! – and tourism's effect on wildlife is therefore not always great, particularly where the discouraging terrain of many Indonesian national parks (hot, humid, leech-ridden rainforest) is concerned.

Visitor numbers in most Indonesian parks are still too small to make a serious contribution to the economy. If widespread economic benefits are to be derived from wildlife tourism numbers will have to increase. This means that wildlife viewing facilities and tourist accommodation must be improved, through increased investment from development agencies, government departments and from the private entrepreneur who is allocated concessions to operate tourism facilities and services. At the same time, as mentioned earlier in this chapter, care must be taken not to over-emphasize the economic justification for preserving natural areas to the detriment of their conservation value.

On balance, although ecotourism in national parks in Indonesia is not currently making much of a contribution to conservation, and there are certainly few examples of it being sustainable according to accepted definitions of the concept, it seems likely that the situation could improve if past mistakes are taken into account and rectified in future planning and implementation of tourism projects. There is evidence to suggest that the political will to achieve this does now exist in the relevant ministries, but whether the mere expressions of good intentions will be backed up by actions remains to be seen.

12

THE BAJAU

Future marine park managers in Indonesia?

Rili Hawari Djohani

THE NEED FOR MARINE CONSERVATION

This chapter seeks to explore the potential role and contribution of local peoples to the management of the marine environment and resources. Focusing on the sea-faring Bajau people of the Indonesian archipelago, it asks how their innate skills, knowledge and experience might be utilized in the simultaneous conservation and controlled development of the country's coastal and marine environments, possibly through the establishment of marine parks. As we have seen in the previous chapter, local involvement in the formulation and management of conservation strategies is a fundamental prerequisite for sustainable development. A supplementary but significant benefit for the Bajau would be the opportunity that their stewardship of the marine environment might provide for them to buffer external pressures which are presently threatening their established way of life.

As the world's largest archipelagic state with over 17,000 islands, and approximately 580 million hectares of seas within its Exclusive Economic Zone (EEZ), Indonesia recognizes the need for sustainable development. The coastline is approximately 81,000 kilometres of which a large part is fringed by coral reefs and mangroves. Lying in the middle of the Indo-Pacific region, these ecosystems and their associated habitats such as reef flats, seagrass beds and lagoons, support possibly the richest marine fauna in the world. The nation's outstanding marine biodiversity is widely recognized (IBCSSC, 1992).

With a growing population, which may exceed 265 million in 2020, and a rapidly expanding economy there are pressing concerns for supporting coastal conservation within the context of the sustainable use of natural resources. Two central problems are habitat destruction of important coastal resources to local and national economies, and localized exploitation of fish and crustacea (Salim, 1988). Other unsustainable practices such as coral rock mining, blast fishing and land reclamation have long been recognized (Soegiarto and Polunin, 1981). In particular, the widespread use of dynamite has a devastating effect on coral reefs. Dynamited reefs may take a long time to recover even if left alone: about forty years are thought to be required

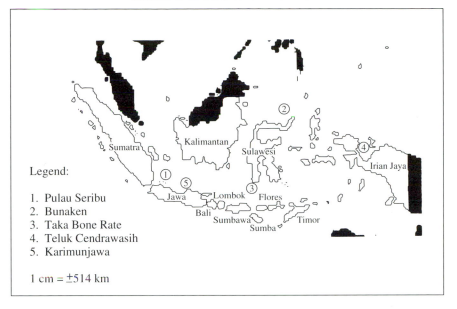

Legend:

1. Pulau Seribu
2. Bunaken
3. Taka Bone Rate
4. Teluk Cendrawasih
5. Karimunjawa

1 cm = ±514 km

Figure 12.1 Marine National Parks in Indonesia

for a reef to recover to 50 per cent hard coral cover. Dynamite fishing is completely non-selective. In addition to target species and target size, larger numbers of non-target species and very small juveniles are killed by each blast (Kenchington and Ch'ng, 1994).

Conservation is imperative as Indonesia will become increasingly dependent on marine and near-shore resources for subsistence, tourism development and export earnings. Environmentally sound and sustainable coastal resource development is both a national and primary goal of the development strategy which started with the Sixth 5-year Development Plan (Repelita VI, 1994–1999). Strong emphasis will be on eastern Indonesia with its many islands and extensive pelagic areas.

The need is to evolve from unsustainable single-purpose resource uses towards sustainable multiple resource uses supported by integrated planning and management, particularly at the regional (provincial) level (Sloan and Sugandhy, 1994). The government has responded to these pressures by declaring a policy to protect 30 million hectares of marine and coastal areas by the year 2000. Protected area planning is an important link between development and conservation (Djohani, 1989). At present, over 2.8 million hectares at twenty-four locations have been legally established as marine conservation areas (Hutomo *et al.*, 1993) including five marine national parks (see Figure 12.1). In addition, many existing terrestrial conservation areas have marine and coastal components, such Komodo National Park (see also the chapter by Cochrane in this volume).

261

Plate 12.1 Sea nomads as a tourist attraction, Phuket, Southern Thailand

A SEA-NOMADIC LIFESTYLE UNDER PRESSURE

There are more than 7,000 coastal villages in Indonesia, most of which rely on inshore fishing. The villages are characterized by poverty, underdevelopment, and low levels of education. The current national consensus is that developing coastal communities will involve their integration in sustainable coastal resource use and environmental management (Government of Indonesia, Ministry of State for Environment, 1992). Future economic growth will require employment diversification that depends on healthy environments. Most of Indonesia's marine fish landings originate from coastal waters in which small-scale fisheries dominate in terms of landings, numbers of vessels, and employment (Bailey *et al.*, 1987).

The Bajau are unique in that they are among the last of the once widespread, boat-dwelling people of South-East Asia. Their traditional subsistence technology was basically an adaptation of hunting and gathering techniques to a marine environment (Sopher, 1977). Today, most of these people have abandoned their boat-dwelling way of life and have settled on the land. Boat dwellers still remain in the Mergui Islands, Sulawesi and Sulu, but these people, too, are rapidly leaving their sea-borne homes and, if present trends continue, the sea nomad culture will eventually disappear from South-East Asia. There are now only a few enclaves of boat-dwellers found scattered

262

around insular South-East Asia, spending their entire life on small boats (*perahu*) and totally dependent on what the sea has to offer. The name *Bajau* is applied both to those people who now have a permanent dwelling on land and those who still live on boats.

More and more Bajau people have settled along the seashores of the Indonesian archipelago. The villages on stilts have become a common feature along the coasts of Sulawesi, Kalimantan Barat, Nusa Tenggara Timur, and Halmahera. This change in lifestyle has brought environmental issues to the forefront. Traditionally, the Bajau lived on boats. Gradually, temporary constructions were built on flat reefs where a fisher family would camp for several weeks. Ultimately, more permanent settlements in the form of houses on stilts were also built (Djohani, 1995b).

Coral rocks were used to build breakwaters and to strengthen the underwater base of the stilts which supported the wooden houses. Coral reefs are important habitats for fish and protect the coastline against wave action and storm surges. In the absence of alternative building materials affordable to the Bajau, coral-mining still occurs for construction purposes and this may lead to localized coastal erosion and depletion of fish stocks depending on the scale of coral mining. The settlements along the shore also face a sewage and garbage problem. The need for sanitation on the *perahu* was minimal. The little garbage which was produced was thrown overboard. Much larger quantities of sewage and litter are now being produced in the Bajau settlements ashore. This is thrown into the sea and on to the beach. The total lack of any sewage treatment has become both an environmental and a health hazard which is closely linked to the scarcity of fresh water. The poor hygienic situation results in dysentery, malaria and skin diseases. In this way, both the Bajau and their environment are affected by this change in lifestyle. Sewage problems were not an issue during their former sea-nomadic existence.

The Bajau population is growing rapidly. Their population in Sulawesi and Sumbawa is estimated to be approximately 250,000 people (Djohani *et al.*, 1993: based upon an eight-month field study (see Figure 12.2, locations 1–6) conducted under the auspices of WWF Indonesia and the Directorate General of Forest Protection and Nature Conservation (PHPA) of the Department of Forestry in Indonesia, in co-operation with the Foundation for Development of Indonesian Village Settlements (YPPI). The study aimed to obtain insight into the Bajau way of life through understanding of their traditional knowledge, ecological sensitivity, and marine wisdom.) Consequently, a growing need for construction materials, fuel wood, and food has led to the exploitation of coral reefs and mangroves in a non-sustainable way. Pressures from outside, including the wide-scale expansion of urban and rural pollution which affects the health of coastal resources, and the depletion of fish stocks by habitat destruction and over-exploitation, pose a severe threat to the livelihood of the Bajau. The competition over fishing grounds with other fishermen who use destructive fishing methods such as blasting

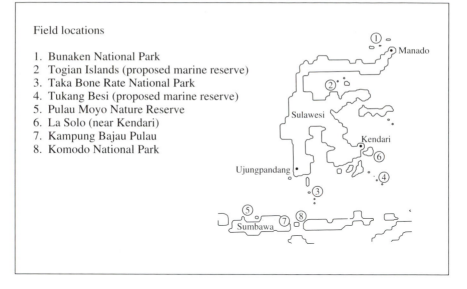

Field locations

1. Bunaken National Park
2 Togian Islands (proposed marine reserve)
3. Taka Bone Rate National Park
4. Tukang Besi (proposed marine reserve)
5. Pulau Moyo Nature Reserve
6. La Solo (near Kendari)
7. Kampung Bajau Pulau
8. Komodo National Park

Figure 12.2 Field locations

and cyanide, with commercial fishing interests, and with the tourism sector, do not leave much scope for the Bajau who rely centrally on their traditional fishing techniques and customs.

FISHING METHODS

Fishing is the basis of the Bajau's subsistence economy. Nowadays, the fishing techniques of the Bajau are mostly traditional, but with an increasing use of modern methods. The traditional methods include line-fishing (*pancing*), spears (*pano*, *sero*) and traps (*bubu*), which are common methods for fishing in shallow waters. Reef fish, crabs, lobster, sea cucumbers, and even sea turtles and dugongs are caught in this way. Spear-fishing is an effective and selective fishing technique. Bamboo traps are used in coral reef areas to catch fish alive. The modern methods including nets (*pukat*) and fishing with lamps (*bagang*), which attracts fish and squid at night, are used by Bajau fishermen with higher incomes. *Pukat* and *bagang* require relatively expensive gear which provides the Bajau with techniques for fishing in the deeper waters during both day and night. Pelagic species, such as tuna and sharks, are caught by line- and net-fishing.

Traders and middlemen of exporting companies such as pearl farms, aquarium exporters, live grouper and lobster business often approach Bajau communities for the collection of marine organisms such as in the case of the Bajau people at Kampung Bajau Pulau on Sumbawa (Djohani, 1995a: based on a two-month field study in 1995 in conjunction with the Nature

Conservancy (TNC) and PHPA. The aim of the study was to devise an outline strategy to decrease the use of destructive fishing methods in and around Komodo National Park: see Figure 12.2, locations 7–8.) Building upon their direct needs and affinity with the sea, the Bajau are provided with boats, fishing gear and food on credit. In the end, fishermen start using dynamite and cyanide to increase their fishing efforts over less time in order to pay their increasing debts. This process is often facilitated by the traders providing chemicals to catch live groupers and lobsters.

Accidents caused by the premature explosion of bombs in the air have occurred, killing or maiming the dynamiters. The presence of men with only one arm appears to be a reliable indicator of dynamite fishing in Indonesian coastal villages. Another cause of accidents, often leading to permanent disability or death, the bends or divers' disease, has afflicted compressed-air-breathing divers who exceed the safe limits of bottom time.

Traditionally, the Bajau people have three types of fishing grounds which can be distinguished on the basis of their distance from the village:

1 *Palilibu*: fishing close to the village, the fishermen returning home after a few hours.
2 *Pongka*: fishing takes place at a greater distance from the village, the fishermen returning home after a few days.
3 *Sakai*: fishing takes place at a very remote distance from the village, the fishermen returning home after some months or even up to one year.

Sakai can take Bajau fishermen as far as Australian waters. During such journeys, the Bajau are exposed to other fisher groups using modern techniques. After many months or even years, the Bajau fishermen return to their families providing them with goods and money.

The reason such fishing patterns have developed might be because of community-level controls on exploitation in the interests of conserving their limited and fragile resources, although this is disputable. It is evident that the small size of Bajau settlements, which can range from 500 to 2,000 people, and their consequently restricted needs with regard to the surrounding nature, are certainly important limiting factors in their own right (Persoon, 1994). The dispersed pattern of fishing grounds relieves the fishing pressure at certain sites. The result is that over-exploitation of fishing grounds by Bajau villages is as yet uncommon. However, pressure from outside fishermen using dynamite and cyanide poses a severe threat to the common fishing grounds in the Minahasa, North Sulawesi. A decline in the fish catch and an increase in fishing effort have recently been experienced by the Bajau people in Bunaken Park (see Figure 12.2). The reaction of the Bajau to such fishing pressure caused by outsiders varies according to location. On more than one occasion, the Bajau have relied on their traditional and cultural values regarding the sea.

MARINE COSMOLOGY

The sea has various levels of significance for the Bajau people. The sea is their home. The Bajau are born at sea and spend their entire life in the vicinity of vast oceans, but the dead are buried on land outside their home. The sea is a source of food for the Bajau. Fish, molluscs, and sea cucumbers form their main diet. The sea is also their friend, who helps them to navigate unknown waters. Over time, the Bajau have acquired the maritime wisdom necessary to forge their way through the Archipelago. Today, their knowledge of the sea is still very much alive. At the same time the Bajau fear the sea because of its waves and storms. All these meanings have led to an overall respect for the sea which has given rise to several taboos.

Taboos played (and still play) an important role in the Bajau communities in the Togians, Tukang Besi, and Sumbawa (see Figure 12.2). Sacred places at sea where people seem to disappear into the depths of oceans are still respected. Consequently, the Bajau did (and do) not fish in these areas. New conditions have created new issues which have led to the emergence of other values and taboos in Bajau villages with a local protected area or animal the result.

The growing extent of permanent Bajau settlements along the coast requires a clear allocation of fishing grounds among neighbouring families and villages to avoid conflicts about the fish catch. A boundary system for fishing grounds among the Bajau communities in the Togians (Figure 12.2) has been identified. Fishing grounds have been allocated for each village and the boundaries are respected by other villages (Djohani et al., 1993). Should outside fishermen want to fish in these waters, permission is asked from the village leaders. Community control over the fishing areas seems to be very efficient.

In other areas new taboos with a protective influence have arisen. For example, the Bajau in the island of Sumbawa (Figure 12.2) no longer fish the black coral which is found in deeper waters (circa 50 metres) because they have associated the coral collected with the onset of a sudden and permanent pain in their knees and arms. The misuse of scuba gear in many cases leads to such decompression symptoms among the Bajau. Often, they dive too long and too frequently using ill-maintained compressors. In the minds of the Bajau the black coral seems to be cursed and a taboo on the collection of this coral was placed by the community. In another case, in Tukang Besi (Figure 12.2), a local non-governmental organization (NGO) representing the Bajau people of Sulawesi Tenggara, established a replanting scheme for mangroves using the names of the Bajau people from the past. In this way, fear and respect for the deceased Bajau were anticipated. Perhaps because of awe, the cutting down of mangroves has been reduced since this taboo was placed on these areas.

266

STEWARDSHIP

In the past, traditional management may have worked well because of low human population densities (Johannes, 1989). Recently established community-based management reserves in the Philippines have been successful, partly because most of these are on small islands whose inhabitants' livelihoods depend on the surrounding coral reefs (White and Palaganas, 1991). The development of global trade and new technologies has altered many societies. The sea nomads are no exception. Settlement ashore has led to new issues, ideas and also to transformation of taboos, values and activities among the Bajau. Although their lifestyle, placed under pressure from outside, may not guarantee a continuation of sustainable exploitation, the Bajau may become important managers of the marine environment in Indonesia.

It is important to recognize that the local people can play an integral part in reef conservation, since it is they who are dealing with the day-to-day management of the marine environment and have a deep affinity and knowledge of the sea. The Bajau have survived for centuries on the open sea. The sea nomads relied on their maritime knowledge, values and ideas as their ancestors had done many generations ago. This marine cosmology guided the lifestyle of these seafaring people who lived in harmony with their environment.

Polunin (1990) questions the wider application of marine tenure in support of conservation since the reasons for the development of such systems by traditional communities are not clear. The establishment of new fishing boundaries by the Bajau in certain locations may provide an exciting entry-point to gain insight into which factors and processes are currently the basis for such community regulation. A similar process can then be used to involve the Bajau people living in and nearby (proposed) marine protected areas in the planning and implementation of zoning systems which is aimed at the regulation of compatible uses.

A number of Bajau people can be employed in the growing marine eco-tourism sector. They can be involved as guides, boat handlers, and in small businesses such as home-stays and *warungs*. A successful example is Nain Island in Bunaken Park (Djohani, 1995b) where the Bajau have found a new niche in the handicraft market. Ornaments are made from the small shells which are abundant in the area. Another home industry is the production of wooden beds with fine carvings. The families involved in these industries are able to benefit from the growing tourism development in Minahasa. They are better off than the Bajau families who depend entirely on fishing. The development of such home-industries may become an example to other Bajau communities in Indonesia with the potential for tourism development. Another opportunity would be to employ a number of Bajau people as park rangers who are responsible for the monitoring and enforcement of protected areas. Their knowledge of the marine environment such as the distribution of fish stocks, clams and other invertebrates would be useful skills.

267

The inventory of fishing grounds made with the help of the Bajau people would be an effective step towards designing and enforcing coral sanctuaries which would secure the replenishment of coral areas and fish stocks within the reserve. Their sense of belonging to the areas can be increased through their participation in such a process. Their current exploitation pattern, restricted for whatever reason, avoids over-exploitation of certain areas and the carrying capacity of coral reefs will not be exceeded by the Bajau. These patterns in combination with the traditional fishing techniques and the underlying processes for community decision-making are the instruments for discovering a sustainable way of exploiting the marine environment, if supported by the Bagai, the outside world.

The park management should help to allocate fishing grounds in the park and ensure the livelihood of Bajau people. The identification of sustainable, self-reliant alternative methods of fishing and livelihoods together with the Bajau people is the key to their participation in the planning and implementation of management in marine protected areas. In addition, the mobilization of initial funding and expertise for the development of a participatory training curriculum for local Bajau park managers is important. The Bajau, living in and nearby (proposed) marine reserves could play a significant role in park planning and management building upon their intimate knowledge and affinity with the sea which are unique and significant for an archipelagic nation such as Indonesia.

13

ENVIRONMENTAL DEGRADATION, NON-TIMBER FOREST PRODUCTS AND IBAN COMMUNITIES IN SARAWAK

Impact, response and future prospects

Michael J.G. Parnwell and David M. Taylor

INTRODUCTION

The gathering, use and trade of non-timber products, whether in the form of small-scale commercial enterprise or as a form of subsistence activity, play an important and generally understated role in supporting the livelihood and welfare of forest-dwelling communities (Peluso, 1991, p. 7). In addition to their established nutritional, medicinal, cultural and utilitarian functions, forest products may also yield a source of supplementary and emergency income and may help to provide income and subsistence stability in the face of fluctuations in farm production or market/price conditions for agricultural products. Their significance as an important and appropriate source of employment should also not be overlooked: 'More employment is generated by opportunities to collect, trade and process non-timber forest products than by tropical timber exploitation, at a far lower ecological cost to the forest' (Peluso, 1991, p. 7).

Notwithstanding their potential as a sustainable, renewable resource there is mounting evidence to suggest that the quantity and quality of non-timber forest products (NTFPs) are rapidly declining where the rainforest ecosystem has been transformed through the encroachment of commercial logging, agriculture and human settlement, and also by forest fires and other consequences of global climatic change, which might be seen as an indirect consequence of human agency. The commercial exploitation of NTFPs, often on a large scale, does little to facilitate their conservation, and may further be intensified as growing scarcity holds prospect of greater financial rewards, thereby intensifying a downward spiral of resource availability. Forest-dwelling

269

communities, whilst not entirely exonerated in contributing to this process, appear the most likely to endure hardship and economic vulnerability as a consequence, not least because of their typical economic, organizational and representational position *vis à vis* other groups in society.

This chapter explores the extent and implications of this situation in the context of forest-dwelling Iban communities in the Bintulu Division of Sarawak, based on research conducted in late 1992 and early 1993 (see also the chapter by King in this volume). The research has revealed some quite significant changes in the availability of certain non-timber forest products during the last fifteen years or so, coinciding with the most intensive phase of forest exploitation in the state. Wildlife in particular has become very scarce, and considerably more energy and money are now expended in pursuing previously plentiful and 'free' sources of animal protein. There is also evidence that traditional forms of social organization which served to regulate access to wildlife and the distribution of the spoils of hunting trips appear to be breaking down, and that these are becoming replaced by a greater degree of individualism and competition for dwindling forest resources.

THE IMPORTANCE OF NTFPS

The variety of NTFPs and the diversity of uses to which they are put are considerable. Mike Arnold, referring specifically to those products from which forest-dwelling peoples derive a source of livelihood (in the present chapter we extend this definition to include home consumption/use and barter/exchange) identifies the following:

> Gathered products include fuelwood, rattan, bamboo, fibres, medicines, gums and wild foods ... Forests also provide the raw materials for many small scale rural processing and manufacturing enterprises, such as wood for furniture and implement making, cane, reeds and vines for basket, mat and handicraft production, wood for charcoal production, nuts and seeds for oil processing, and bark for tanning processing. In addition fuelwood provides the main energy source for many other small scale processing enterprises.
>
> (Arnold, 1993, p. 1)

Beyond providing a regular source of livelihood for peoples who specialize in their collection and trade, NTFPs are also of particular importance in the way that they provide a means of supplementing insufficient or variable food supplies and income from agriculture, and in seeing people through occasional emergencies – a buffer which provides products and opportunities for income-earning at times when other options fail (Falconer and Arnold, 1991). Such activities can quite flexibly be intensified or scaled down according to need, demand, opportunity, and so on. The collection of NTFPs is also a particularly appropriate avenue for the rural poor, given the minimal skills and capital required, and the relatively unrestricted levels of access which are enjoyed (Arnold, 1993, pp. 8–9). Ease of entry and access, however, may be traded off against potential competition for what are essentially finite, if potentially renewable resources.

Competition may also take other forms. People who rely on trade in NTFPs may see their income dwindle as consumer tastes shift towards superior goods made from more durable or marketable substitute materials, or manufactured in formal, large-scale industrial enterprises. Conversely, where demand for NTFPs increases, traditional forest collectors may face growing competition from 'outsiders' who may be quick to exploit their lack of formal ownership and control over the resources and the land from which they are derived (ibid., p. 16).

Such changes may carry quite severe implications for the sustainable exploitation of NTFPs, and by continuation the sustainability of the livelihoods and welfare of those who depend upon them. An uncontrolled 'free for all', where all parties concerned pursue the intensive exploitation of NTFPs, either for commercial gain or economic and nutritional survival, may lead to a downward spiral from which the resource may never recover. Conversely, the collectors may themselves be vulnerable to other environmental changes which may be affecting the ecosystem upon which they depend for their livelihood. The commercial exploitation of timber resources may be seen as the most obvious cause, but more general global processes of environmental change, such as the increasingly erratic nature of the El Niño-Southern Oscillation (see also King in this volume), and the associated incidence of devastating forest fires where the climax vegetation has been transformed by human activity, may also have a severe, if somewhat localized, impact on forest-dwelling communities. Mike Arnold reiterates this point based on his survey of several forest locations in the Developing World, and also identifies other, related factors which may be contributing to change in this regard:

> Nearly everywhere users of forest products are faced with a decline in the size of the resource from which they obtain their supplies. . . . The principal causes of raw material shortage are usually connected with developments other than harvesting [by collectors of NTFPs] . . . Timber harvesting in the forest is likely to damage or destroy other components of the forest that provide products for small enterprises. . . . Clearance of land for agriculture . . . distances users from remaining supplies, and is likely to result in more intensive use of what remains. The opening up of forest areas is likely to precipitate other changes that make inroads on the resource.
>
> (Arnold, 1993, p. 15)

The link between environmental degradation and the depletion of NTFPs provides the principal focus of this chapter. Of central concern is the extent to which changes in the rainforest ecosystem resulting from commercial logging and its associated environmental effects have influenced the availability of NTFPs in parts of Sarawak, East Malaysia. Of further interest is the way in which the study communities have been affected by these changes, how they have responded, and how they might be expected to respond in future if environmental degradation and resource depletion continue to worsen. The following discussion will first present a brief assessment of the importance of NTFPs to Bornean societies, before moving on to look at this question within the specific context of the thirteen Iban communities which provide the focus for the present study. The chapter will then conclude by considering various policy options which emerge from the examination of the empirical material, including the proposed sustainable management and cultivation of certain NTFPs in support of small-scale forest-based enterprise.

THE CHANGING IMPORTANCE OF NTFPs IN BORNEO

A cursory examination of the literature reveals the considerable significance of NTFPs in the livelihood, welfare and social organization of forest-dwelling societies in Borneo (for a further discussion of the types and uses made of NTFPs for various Bornean societies, see e.g. Vadya *et al.*, 1980; Kartawinata *et al.*, 1982; Weinstock, 1983; Hoffman, 1984; Caldecott, 1988; Padoch, 1988; de Beer and McDermott, 1989; Cramb, 1989; Goody and Feaw, 1989; Brosius, 1991; Leaman *et al.*, 1991; Peluso, 1991; Dove, 1993; King, 1993e). A number of writers have also made passing reference to the extent to which the availability and use of these products are changing as a result of several, mostly externally induced changes, such as the impact of commercial logging, changing attitudes and preferences, and these communities' increasing penetration by and exposure to the outside world. However, to our knowledge no study supports these general observations with empirical evidence: it is hoped the present study may make a modest contribution in this regard.

Dr Chin See Chung (1985), in his study of a lowland Kenyah community in the upper reaches of the Baram River in Sarawak, has provided a comprehensive description of the kinds of NTFPs which are regularly utilized and the myriad uses to which they are put. The general impression obtained from Chin's work is that the Kenyah were still able to enjoy the relative abundance of the rainforest ecosystem on which they depended to a significant degree for their subsistence and livelihood: for instance, fish accounted for up to one-third of their food intake, and products obtained from collecting forest produce a further 10–25 per cent (Chin, 1985, p. 106; p. 115). Chin observed that the gathering of wild ferns, fungi and fruits involved little effort and time, and people seldom returned empty-handed from collecting activities. Fishing is similarly a quite reliable venture, and whilst hunting always carries a much greater element of risk of failure (Chin suggests a 25 per cent success rate), wild animals were still common in the area at the time of his study, particularly the bearded pig and wild deer. Because of the diversity and abundance of the rainforest ecosystem (and perhaps also because of relative isolation), the contribution of purchased foodstuffs to the diet of the Kenyah study community was negligible (Chin, 1985, p. 245).

However, Chin finished his discussion with an ominous warning:

> Of immediate concern to the Kenyah . . . is the potentially destructive activities of the loggers, to their rainforest environment. Logging when it reaches the Kenyah territory will not only destroy the primary, but also the old secondary forests. This will result in a scarcity of large trees for the Kenyah's timber requirements (particularly for boat-building) and a destruction of the wild illipe nut trees. It will also be detrimental to their other collecting activities and will similarly affect hunting and fishing through habitat destruction.

None the less, Chin suggests that the Kenyah are characteristically a quite adaptable people, and may thus be able to adjust to gradual changes in this regard. He highlights their preference for familiar foods, but also indicates their willingness to resort to less desirable food items at times of scarcity and need (ibid., p. 96). An illustration is provided by wild game: the Kenyah prefer pork, and may typically sell the barking deer and mousedeer that they derive from their hunting trips (although meat is typically sold only when there is an urgent need for cash), but will readily devour the latter when the bearded pig becomes scarce, as is occasionally the case.

Another illustration of the adaptability and innovativeness of the Kenyah is provided by their spontaneous response to the market potential of illipe nuts (Kenyah: *a'bang*; *Shorea macrophylla*). These edible, oil-bearing nuts have an important export market, the oil they produce being in demand as a substitute for cocoa butter in the confectionery industry, and also as a raw material in soap and cosmetics manufacture. They flower and fruit irregularly and infrequently: every 2–7 years there will be a good crop, which will yield a short-term income windfall. The commercial value of the nuts to the Kenyah (they are gathered entirely for sale), the need to harvest and process the illipe nuts within 2–3 days before the oil content falls, and the fact that the fruit-bearing trees are communal property, lead to frantic, and occasionally highly competitive activity between longhouse residents during the fruiting season.

Of significance to the discussion later in this chapter is Chin's observation (1985, pp. 229–31) that several Kenyah in his study community had attempted to domesticate the illipe nut-bearing tree. Eleven out of sixteen households had planted up to twenty-five trees each on the land to which they had customary rights of access. Wildlings had been taken from the rainforest, and seedlings had been propagated near the longhouse, all on their own initiative without any government support. Several had yielded nuts on at least one occasion, and Chin detected interest in expanding the domestication of *Shorea macrophylla*. Although this innovation had occurred spontaneously in response to prevailing market conditions, it might also be employed more widely as a model for coping with the exigencies of resource depletion resulting from over-exploitation and environmental degradation, as shall be discussed later.

Ilik Saccheri and Daniel Walker (1992), modelling their approach on that employed by Chin See Chung, have provided information relating to a highland Kenyah community in the Apo Kayan, a highland region between the Kayan and Maharam Rivers in the Indonesian province of East Kalimantan. Their study, conducted almost a decade after Chin's survey, suggested that the community had been much more significantly affected by growing resource scarcity. The division of the study longhouse, for instance, with some *bilek* families moving to clear new land further up-river, had been necessitated partly by increasing scarcities of forest products and wild game, especially the wild boar (Kenyah: *babui*; *Sus barbatus*) upon which the Kenyah depend for a significant proportion of their protein intake. An important contributory factor had been the parallel forces of population expansion and environmental degradation.

For those Kenyah who had remained in the old longhouse, it was noted that forest plants, which previously had constituted the majority of their dietary intake, had been greatly superseded by cultivated plants (Saccheri and Walker, 1992, p. 220). Superficially, this might be attributed to the effects of change in the local environment, but one must be cautious both about using circumstantial evidence to explain changes taking place in the make-up of the Kenyah diet, and in attributing change to a single source. Several factors may be exerting a simultaneous influence on the sources of Kenyah livelihood, and the various social and organizational processes which are associated with them, such as a growing awareness of and access to the 'outside world', and the penetration of the market economy. Indeed, it is interesting to note that some of the older respondents suggested that there is more pork to eat now than in the past, although this may be attributable to the fact that the present longhouse population is only about one-tenth its original size – in other words, population has readjusted to environmental and resource potential.

Saccheri and Walker obtained further evidence of resource depletion from the size of pigs, tusks and skulls which they observed in their study longhouse, with fewer wild boars reaching full maturity. In seeking an explanation, they concluded that

> of . . . importance in determining the size of the Bearded Pig populations is the massive habitat destruction which is now taking place in many parts of Borneo. Logging can alter the flowering and fruiting of rainforest trees, as well as removing a large proportion of dipterocarps.

The removal of fruiting refuges through logging and plantation agriculture is thought to have been the principal cause of the virtual extinction of the Bearded Pig in many areas of Borneo. At present, no commercial logging operations exist in the vicinity of the Apo Kayan as it is too remote and only swidden agriculture is practised. However, there are plans to build a road into this area and a more modern form of agriculture is being promoted by government. The pigs have managed to live with the swidden cultivators of the Apo Kayan for centuries, but it seems unlikely that they will be able to adapt to the more drastic changes to their environment which may now occur.

(ibid., p.227)

A final point which is of relevance to the later discussion in this chapter concerns the way in which the spoils from hunting trips are shared among other members of the community. The largest share is given to the owner(s) of the dogs used by the hunting party, and smaller shares are received, in descending order, by other members of the party, close family, longhouse families/individuals who are most in need (single woman-headed households, the aged), close neighbours and, if anything remains, other longhouse residents. The important point, as Saccheri and Walker note, is that even 'households which, for one reason or another, do not bring in much meat will have only slightly less meat to eat than more fortunate households.' (ibid., p. 232). However, in instances where communal longhouses had been replaced by individual houses, community life had been radically transformed, one manifestation being the severely curtailed practice of the free-sharing of meat from hunting expeditions. Parallel changes resulting from environmental degradation have been identified in the Iban case study which will be discussed shortly.

A further view of a Kenyah community is provided by Carol Pierce Colfer's (1983a; 1983b) study of Long Segar in East Kalimantan, Indonesia. A point of relevance here is that the longhouse has lain within a timber concession area since the late 1960s, although, because it possesses few merchantable species, it has been spared the full potential effects of logging. Accordingly, whilst the longhouse residents have been affected by logging activities, they have also retained more or less full access to the forest for the products they require for their gathering and cottage industry activities (Colfer, 1983a, p. 9), even though the supply of these has occasionally been affected by logging activities. As such, there are some useful contextual similarities between Colfer's study and the discussion of the Iban in Sarawak which follows.

Colfer's study also highlights the impact of the use of technology, in the form of the outboard motor and chainsaw, on rainforest exploitation (another theme which we shall develop later), drawing particular attention to the effects on gender roles and status. The outboard motor (ces) has given longhouse residents, men in particular, access to wider territories than hitherto, both of which have substantially increased the potential for the more intensive exploitation of non-timber products. The possession of a chainsaw enables men (mostly) to fell ten times as many trees in a day as is possible using traditional methods, which similarly represents a substantial shift in the balance of power between people and forests in East Kalimantan (ibid., p. 16). Not only can more territory be cleared for cultivation, but the fields can also now be more completely cleared. Thus larger trees which may previously have been left because of the energy needed to fell them may now also be felled, thereby reducing seed-producing potential for the regeneration of rainforest species. Colfer notes that, whilst the lifestyle of the Kenyah has been made easier through the use of modern technology, and standards of living have generally improved, this has been achieved at a significant cost to the forest ecosystem (ibid., p. 17).

Colfer highlights the effects of rainforest exploitation on the declining availability of some kinds of so-called minor forest products, but also emphasizes the role played by the availability of convenient and durable commercial substitutes which has resulted in a reduction of forest harvesting for personal use (Colfer, 1983b, p. 73). Plastic buckets have replaced bamboo products for carrying water, tin roofing is used in place of traditional forest-based materials such as *nanga* and *sip*, and the need to travel widely in search of *damar* has been obviated by the increasing availability of kerosene and electricity. In other instances, the expansion of commercial outlets for forest products has led to their intensified exploitation and utilization (Colfer, 1983b, p. 74; see also Guerreiro, 1988 for the Kayan in the Belaga region of Sarawak), and has also led to a readjustment in the allocation of time, with more time being devoted to the gathering of forest products (again, contributing to their more intensive exploitation) and proportionately less time and priority being afforded subsistence food producing activities: 'forest clearing is profitable, and profitability is a potent incentive' (Colfer, 1983b, p. 84).

Finally in this brief review of the literature, Katharine Pearce, Victor Luna Amen and Surik Jok's (1987) ethnobotanical study of a long-established Iban community in the Second Division of Sarawak brings us a little closer to the ethnic and locational context of the present discussion (see also the chapter by Terry King in this volume). Of significance to the association between forest disturbance and the availability of forest products is Pearce *et al.*'s suggestion that useful plants are not restricted to any particular type of forest, and are to be found in both primary and secondary forest types (ibid., p. 258). This is consistent with Saccheri and Walker's observation that pig populations had not been adversely affected by the practice of shifting cultivation (indeed, there is evidence that other forms of wildlife, for example squirrels and rodents, may have become more abundant as a result of the transformation of the primary forest). However, they warn that 'exploitation of forests due to logging, shifting cultivation and collection of jungle produce by settled communities . . .

can be expected to result in the loss of more and more useful plants' (ibid., p. 258; a point reiterated by Leaman *et al.*, 1991).

Pearce *et al.*'s study is also useful because the community under study is located in much closer proximity and with much better access to 'the outside world' than was the case with the Kenyah studies discussed above. In this connection, they question the continued importance of plants and materials gathered from the rainforest, in spite of the growing availability of alternative or substitute products from the market which in turn has resulted from improved communications access to nearby towns and bazaars, and increased media- or migration-induced awareness of the outside world and the products it offers (Pearce *et al.*, 1987, p. 258). The continuing availability and relative ease of access to natural products were cited as an important explanation for the continuing use of traditional products, as was the fact that they were in effect 'free'. A further explanation was the psychological importance that the Iban study community attaches to familiar things, and also to their lack of confidence in alternative or substitute products.

None the less, Pearce *et al.* also revealed growing evidence that some traditionally used plants had become redundant. Modern alternatives were replacing traditional items, especially materials for construction, furnishing and ornamentation. Barkcloth and washing agents were among a number of traditional items/products which were no longer being used, as was the case with materials used in pagan rites and rituals where longhouse dwellers had converted to Christianity. A trade-off was identified in relation to the amount of time which was usually expended in the collecting, processing and manufacture of 'free' traditional items set against the cost of purchasing convenient, fabricated alternatives. A further complication is added by the prestige value which may be attached to the possession and use of certain externally produced, purchased products. However, materials which were still abundant locally were often widely used in spite of the availability of modern alternatives.

Pearce *et al.* also discuss the potential for the further development of enterprises based on forest products, pointing particularly to that offered by the tourist, urban and wider international markets for handicraft products such as *parang* (knife/machete) hilts, spears, trap charms, baskets, hats and mats (ibid., p. 260). In addition to providing a supplementary source of income, they argue that the expansion of such forms of enterprise may also help to keep alive cultural traditions (but see Parnwell, 1993b; Graburn, 1976). They also recognize that increasing resource scarcity may constitute an important constraint in this regard, but suggest that the domestication of certain forest products which are widely used in the manufacture of handicrafts could be a possible policy option, building on the Iban's traditional knowledge of species and habitats, and their experience in 'domesticating' *bemban, senggang* and *pandan* (Pearce *et al.*, 1987, p. 260; see also, in this regard, Burgers, 1992, in respect of the Bidayuh of Sarawak; and Leaman *et al.*, 1991, in respect of the Kenyah of Kalimantan).

This brief examination of the few studies which discuss the dynamics of NTFP exploitation has identified a number of points of interest which are also relevant to the discussion which follows:

- Frequent reference to the actual or potential influence of commercial logging and habitat transformation on the availability and quality of NTFPs.
- The influence of population expansion and increasing land pressure on the exploitation and growing scarcity of NTFPs.
- The societies' responsiveness to external market potential, influenced quite strongly by accessibility, awareness and attitudes.
- The apparent erosion of traditional, communal means of regulating access to natural forest products, and the parallel increasing prevalence of individualism.
- The growing tendency for technology to play a role in the process of rainforest exploitation, with both direct and indirect effects on the quality and availability of NTFPs.
- The variable response of communities to resource scarcity, leading in some instances to the gradual pursuit of alternative, substitute products, and in others to the intensified exploitation of natural resources as prices adjust upwards.
- Changes in the make-up of diet and sources of livelihood, the substitution of cultivated crops for those previously collected from the rainforest (with the associated imperatives of investment, organization, time allocation), and also the erosion of some important social and cultural roles associated with communal or group activities associated with hunting, fishing (especially using *tu'ba* or fish poisons: Chin, 1985, pp. 113–15) and the gathering of forest products.
- The communities' cultural or psychological preference for the continued use of natural forest products on the one hand, and their readiness to switch to alternative/substitute products, for a variety of reasons, on the other.
- The growing importance of non-farm activities as a source of income.
- The potential offered by tourism (a rapidly expanding but volatile industry in Borneo) and the urban and export markets for handicraft products manufactured from NTFPs.
- Evidence of moves towards the domestication of certain natural forest products, both as a response to their declining availability and to increasing market demand.
- The spontaneous nature of sustainability mechanisms which have been developed in response to resource depletion, and the conspicuous absence of government initiative and involvement in this regard.

In the next section we will attempt to develop some of these themes, using empirical material derived from a study of thirteen longhouse communities

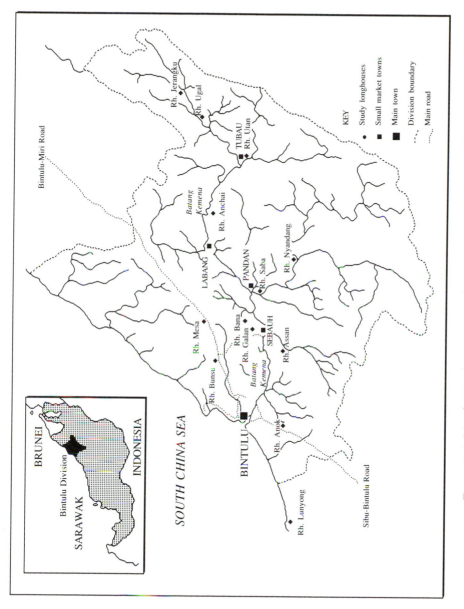

Figure 13.1 Bintulu Division, Sarawak, showing location of study longhouses

Table 13.1 Summary characteristics of the study communities

Longhouse (location/terrain)	Number of doors	Population density	Quality of environment	Access from longhouse	Proximity to a town	Proximity to Bintulu	Level of development	General profile longhouse of community
Rh. Anok (R, L*)	23	Medium	Poor	Good/Road	Near	Near	Fair	High level of male labour migration
Rh. Mesa (N–R, L)	7	High	V Poor	Good/Road	Near	Near	V Poor	Squatters
Rh. Ugal (R, U)	21	Low	Fair	Poor/River	Far	V Far	Poor	'Traditional' upland
Rh. Jerangku (R, U)	21	Low	Fair	Poor/River	V Far	V Far	Poor	'Traditional' remote
Rh. Utan (R, U)	46	Med/High	Fair	Fair/River	Near	Far	Fair/Good	Most men away at logging camps
Rh. Anchai (R, U)	12	Medium	Fair/Poor	P-F/Riv-Rd	Q Near	Q Far	Fair	Near a plantation where many find work
Rh. Bana (N–R, L)	30	High	Poor	Good/Road	V Near	Near	Good	'Model' agricultural longhouse
Rh. Nyandang (R, U)	34	Medium	Fair	Good/River	Q Far	Q Far	Fair	Reduced self-reliance due to migration
Rh. Saba (R, L)	9	Medium	Fair	Good/River	Q Near	Q Near	Poor	Landless
Rh. Assan (R, U)	24	Medium	Fair	Good/River	Near	Q Far	Fair	High involvement in external work
Rh. Galan (N–R, L)	10	High	Fair	Good/Road	Near	Q Near	Fair	Non-riverine
Rh. Lunyong (RC,L)	43	Low	Poor	V Poor/Sea	Far	V Far	Poor	Coastal swamp
Rh. Bunsu (N–R, L)	14	High	Poor	Good/Road	Near	Near	Fair	Many were formerly plantation workers

Note: * R = Riverine   N–R = Non-riverine   L = Lowland   U = Upland   C = Coastal

Plate 13.1 Iban longhouse, Ulu Jelalong, Kemena Catchment, Sarawak

in the Bintulu Division of Sarawak. Whilst the survey utilized both quantitative and qualitative methodologies drawn from a broad cross-section of study communities, the following discussion is based mainly on an aggregate examination of the quantitative data. The main objective is to identify the nature and extent of change in the availability and use of NTFPs, and to assess the relative influence of some of the above-mentioned factors in instigating change in this regard. In assessing how the study communities have responded to the exigencies of change, we will also be able to judge the desirability and feasibility of alternative models of development which may both help them to adjust to the continued degradation of the forest ecosystem and build diversified economic systems based upon the continued, sustainable exploitation of non-timber forest products.

THE IBAN STUDY COMMUNITIES

Figure 13.1 depicts the location of the thirteen Iban longhouses under study, and Table 13.1 summarizes some of their main general features. Because of constraints of space these longhouses will mainly be referred to collectively, although Table 13.1 indicates quite clearly that the data are drawn from a broad cross-section of communities in terms of size, population density, location, accessibility, environmental conditions, level of development, and so on (a disaggregated analysis of the data is presently being undertaken: see King, Taylor and Parnwell, forthcoming). Thus whilst we should be cautious in interpreting some of the generalized findings which are presented below, it should also be borne in mind that the data offer a more broadly representative view than can be obtained from the studies of individual communities, such as we have examined in the previous section. However, the fact that the data relate to just one ethnic group means that the findings can not be considered representative of the Bintulu catchment as a whole. Furthermore, as there are some significant differences between this fairly recent locus of Iban settlement and other parts of Sarawak (notably the First and Second Divisions) where the Iban are much longer-established, the study cannot be considered to be representative of the Iban as a whole, as King emphasizes in his chapter in this volume (see also Padoch, 1982).

We commence our enquiry by assessing ecological conditions and the nature of environmental change in the study area. Respondents were asked to describe their environmental surroundings according to thirteen key words or phrases (see Table 13.2). Whilst obviously liable to a degree of subjectivity, it can be argued that longhouse dwellers' perceptions are very important in the context of economic decision-making. On this basis, it is quite clear that a fairly significant number of respondents have both a negative and pessimistic view of their physical surroundings. Two-thirds considered the environment to be 'unpredictable' (*tusah ka nemu* in Iban), almost half thought it was 'changing' (*berubah*), and one-third felt that the terms 'unstable' (*enda tetap*) and 'degraded' (*kusi*) accurately described local environmental conditions. If we disaggregate the data, we find that a small number of longhouses scored highly in each of these aspects, including Rh. Nyandang, Rh. Bana, Rh. Jerangku, Rh. Utan and Rh. Anchai. It is interesting to note that, with the exception of Rh. Bana, all of these longhouses are upland riverine communities which have been very directly affected by changes in the forest ecosystem and by associated environmental problems such as the sedimentation of water courses. All four communities have also been affected by commercial logging which has largely taken place in the hill forest areas over the last fifteen years, Rh. Utan and Rh. Anchai particularly so. In many areas the forest has been 'selectively' logged several times over the last 30–40 years, and in some instances, especially in the vicinity of Rh. Anchai has, effectively, been clear-felled. As if to deflect attention from this blatant transgression of forestry regulations, an innovative form of land use – the establishment of a rattan plantation – has been introduced (this will be discussed a little later in this chapter).

Logging in many respects represents the crucial agent of environmental change in the study area. The commercial exploitation of timber resources commenced in the area in the early 1960s, several years before some of the longhouses (e.g. Rumah Anok and Rumah Bunsu) had been established. Logging in Ulu Jelalong, where both Rumah Ugal and Rumah Jerangku are

Table 13.2 Perceptions of environmental conditions in the vicinity of the longhouse

	Rh. Anok	RH. Mesa	Rh. Ugal	Rh. Jerangku	Rh. Utah	Rh. Anchai	Rh. Bana	Rh. Nyandang	Rh. Saba	Rh. Assan	Rh. Galan	Rh. Lunyong	Rh. Bunsu	Total
Fertile (subur)	4	0	1	0	2	1	1	1	2	0	0	2	0	14
Marginal (enda tentu manah)	6	4	9	10	6	4	18	13	3	15	5	3	2	98
Unpredictable (tusah ka nemu)	5	1	17	18	13	12	25	15	4	10	8	1	9	138
Poor (jai endar)	2	0	0	1	0	1	1	0	0	0	5	0	0	10
Unstable (enda tetap)	11	1	6	11	6	2	10	9	1	8	7	2	6	80
Degraded (kusi)	7	1	6	7	6	4	10	3	1	5	3	8	8	69
Constant (tetap)	1	1	0	1	1	0	2	4	1	2	0	13	1	27
Changing (berubah)	5	0	10	12	12	10	17	14	0	13	1	1	6	101
Attractive (manah)	6	5	14	9	9	9	18	13	2	12	1	27	7	132
Bountiful (terlalu subur)	6	1	4	0	0	3	0	1	0	0	0	18	3	36
Rich (ami)	4	4	3	0	0	0	1	2	2	8	0	17	3	44
Much potential (maioh potensi)	6	3	2	2	0	2	9	6	4	7	2	29	7	79
Number of Bilek	21	6	21	20	15	12	28	21	6	20	8	31	13	222

Plate 13.2 Logging operations in the Kemena Catchment, Bintulu, Sarawak: the use of technology has significantly tipped the balance in relations between humans and the natural environment

located, commenced fairly recently, but it has already been responsible for removing some of the timber species, such as *belian*, which these communities rely upon, *inter alia*, for house-building. A little further down-river, the logging camps around Rumah Utan, where the forest has been intensively logged, are already closing down because of the scarcity of commercially exploitable timber species. Rumah Mesa and Rumah Saba both have the benefit of location in close proximity to a nature reserve, although in the former case, the insecurity and vulnerability of their illegal presence in the area (the government to date has refused to recognize their claim to the land they occupy), together with their prohibition from exploiting the resources of the Similajau Forest Reserve, detract from the potential advantages of its location.

Respondents were asked to assess the impact of logging in the study area over the last fifteen years, as it was felt important to obtain a 'view from within', however subjective this might have been (see Plate 13.2). This matter was addressed at the level of the household and the community. At the community level, respondents were universal in bemoaning the impact of commercial logging on water quality, the availability of fish and prawns, wild animals and natural forest products, the availability of primary forest to be cleared for agriculture (mostly shifting cultivation), and also the market price

Plate 13.3 Raft of tropical hardwood, Batang Rejang, Sarawak

for ironwood and other important woods for construction, boat-building, and so on. In many cases the make-up of their diet had changed significantly, and health problems had sometimes increased. Government restrictions on access to forest, and the conversion of forest to plantations, were a further source of grievance. The logging companies were not entirely to blame, however: the over-collection of forest products because of population increase, more restricted access to forests and, to a lesser extent, the breakdown of communal responsibility and sanctions, were also cited as having contributed to this situation. At the household level, some of the effects of logging activities were listed, in order of declining frequency, as a general deterioration in the quality of the environment, increasing water pollution, declining soil fertility, soil erosion, a decline in the availability of forest/river products, and a loss of virgin forest for shifting cultivation.

There is further evidence regarding the nature and impact of environmental change in the study area. Shifting cultivation, still practised by more than one-third of survey households, clearly involves a very high degree of reliance upon the forest ecosystem – more so than permanent forms of farming, which involve a greater element of control over the environment, and a higher level of use of modern farming inputs. Where still practised, the study households generally used the shifting (swidden) cultivation system to produce

Plate 13.4 Wood mill near Bintulu town, Sarawak

rice. An indication of the constraints imposed by physical conditions in this regard is provided by the fact that households on average produced sufficient rice for only 9.2 months of the year, and slightly less than half of farming households were self-sufficient in rice production (this none the less compares favourably with the 32 per cent self-sufficiency level for the Iban which was reported in a study by Freeman in 1970: cited in Dove, 1993, p. 146). Almost two-thirds of *padi*-producing households considered that levels of rice self-sufficiency had fallen over the last fifteen years, with the over-exploitation of land resources (38 per cent) being given as an important explanation. For many, the fallow period in the shifting cultivation cycle was less than the minimum required for restoring soil fertility. Many had given up growing rice altogether. Some 130 households (58.6 per cent) indicated that they have to purchase rice on a regular basis; 217 (97.7 per cent) regularly purchase other foodstuffs.

Almost two-thirds of respondents (65 per cent) indicated that the produce they obtained from non-timber forest products (including wild game) and farming was insufficient to meet their general consumption needs. Furthermore, all bar two respondents claimed that the situation had deteriorated during the last fifteen years, with commercial logging (see Plates 13.3 and 13.4) again being given as the principal explanation (n = 196, 88 per cent), along with population increase (n = 75), and over-exploitation of the forest ecosystem by the residents themselves (n = 13: several gave multiple answers). Thus already the image of self-sufficient communities enjoying the abundance of the rainforest ecosystem is beginning to evaporate.

In spite of this deterioration in the availability and sufficiency of forest products, the study households still rely quite heavily on the forest ecosystem as a source of food (especially protein-rich meats), medicines, building materials, raw materials for making handicrafts, and so on. Table 13.3 contains a list of some of the forest products which are more commonly used by the study communities. In terms of the collection of forest products for sale, it is interesting to note (a) that only a quite small proportion of longhouse families are engaged in such a form of enterprise (n = 15, 6.8 per cent), and (b) that a much larger proportion (n = 40, 18 per cent) derive a significant proportion of their cash income from selling planks cut from trees in the rainforest. Some thirty-five respondents listed the selling of planks as their principal non-farm activity.

A large proportion of respondents utilized natural forest products in the production of handicrafts. Altogether some 71 per cent of *bilek*-families listed handicrafts production as one of their non-farm activities, although it is interesting to note in relation to the later discussion that only four households manufactured craft products (mats, baskets, *pua*, purses, clothes and boats) commercially: the vast majority retained their products (mostly mats and baskets) for home use. Among the more commonly used materials in the production of handicrafts were rattan (146 out of 158 producing families), *bemban* (132), *senggang* (94), *keropok* (36), *daun biruk* (26) and bamboo (19). In relation to the later discussion, it is alarming to note that 104 households indicated that such natural products had become 'more scarce' over the last fifteen years (the reference period used in the study, and a further 53 considered these NTFPS to be 'much more scarce'. A large number of households also indicated that they had ceased producing handicrafts, or had severely curtailed such activity, because of difficulties they had encountered in obtaining natural raw materials from the forest. This is an especially alarming situation because, as we shall see shortly, the commercial production of handicrafts represents a potentially valuable *in situ* means of overcoming declining levels of livelihood derived from on-farm activities.

Another stereotype of the Iban, as rapacious hunters and skilful collectors of other forest produce is challenged by data obtained from the study communities (Table 13.4). Altogether, slightly less than one-third of all respondent households continue to practise hunting (30 per cent) and fishing (32 per cent), whereas two-thirds engage in the collecting of forest products. Furthermore, of those who still practise these traditional pursuits (see Sutlive, 1992, p. 87), very few do so on a regular basis. Only 10 per cent, 18 per cent and 8 per cent respectively engage in such activities more frequently than once a month. This in itself is a quite remarkable finding. Such activities have traditionally formed a cornerstone of the Iban way of life with

Table 13.3 Summary of the main forest products used by the study longhouse communites

Tree products	Handicrafts	Fruits	Fishing
For house-	For mats and baskets:	*mawang*	Fish
building:	*rotan sega*	*raba*	Prawns
keladon	*rotan lepoh*	*sibau dara*	Turtles
serangan batu	*rotan semut*	*jeruit*	
bintangor	*rotan tut*	*puak*	
kelakup	*rotan lia*	*kemayau*	
kapur	For baskets, hats,	*kubar*	
belian	fish traps, mats:	*asam paya*	
tekam	*tunggal*	*binjai*	
merebau	For baskets:		
perawan	*seru*		
tekam	For baskets, hats,		
meranti	basket rims:		
For boatbuilding:	*cit*		
meranti	For animal cages:		
tekam	*danan*		
For fuelwood:	For baskets,		
mengilas	basket rims:		
mengasing	*batu*		
merapinang	For mats and baskets:		
entegaram	*senggang*		
ru	*bemban*		
buan	*akas*		
kepapar	*keropok*		
kemirit	Also:		
kemunting	Bamboo		
kelampai	*daun biruk*		
beletik			
kumpang			
meregalang			
beron/baron			
bantas			
Support for pepper bushes:			
tekam			
String:			
tekalong			
Knife handles:			
kempait			
For sale:			
terentang			
gerungang			

Medicines	Hunting	Also
For fever:	*babi* (wild boar)	*ranggil*
kulit peredu	*rusa*	(soap)
For relieving	*kijang*	*kulat*
snakebite; eye	*pelanduk*	*kering*
infection;	*batih*	(mosquito
drunkenness:	monkey	coil)
randau	*wak-wak* (gibbon)	
penawas	*nyumbuh*	
To make a ghost weak:	(pig-tailed macaque)	
kupak ranggil	*kera* (long-tailed macaque)	
ubang buah empit	*landak* (porcupine)	
kayu ujan panas	*baya* (crocodile)	
(*selukai*)	bear	
	tupai (squirrel)	
	tortoise	
	Birds:	
	puna (pigeon)	
	empulu (*microscelis* spp.)	
	rangak (black hornbill))	
	burong ruai	
	(argus pheasant)	
	kenyalang (hornbill)	
	kekalau	
	sempidan	
	emperagam	
	sengayan	
	bedidik	

hunting in particular playing an important cultural as well as nutritional and occasionally economic role. Less than one-fifth of respondent households considered that the products obtained from hunting made either a 'vital' or 'very important' contribution to their livelihood, whereas the figures were slightly higher for fishing (24 per cent) and collecting (38 per cent).

Impact of change

The decline in the importance of such activities can be attributed, *inter alia*, to a growing incidence of male migration (which has reduced the labour available for these traditionally male pursuits), the availability of alternative products from commercial sources, and the more general modernization of the study communities. Equally, it is clear that the degradation of the rain-forest ecosystem through commercial logging activities and other forms of resource exploitation has played a major role in this situation. Information obtained from surveys conducted at the community and household levels indicates very clearly that the activities of logging companies were considered to have been the principal reason for the dramatic decline in hunting, fishing and collecting activities in the study area. Community-level interviews in each of the thirteen study longhouses yielded a consistently bleak saga of forest products having become much less readily available as virgin forest, and the habitat it provided for wild animals, had been felled by logging companies and, rather less frequently, as a result of over-exploitation by the longhouse residents themselves. Although a small number of our respondents were still able to obtain at least some of the products they needed, every longhouse visited mentioned that their previous abundance had been replaced by an extreme scarcity which had often rendered their pursuit unworthwhile, had significantly changed consumption patterns, had hurried the adoption of alternatives (often purchased at considerable cost from local markets), and had sometimes led to growing competition for these limited resources. The residents of Rumah Saba, Rumah Utan, Rumah Jerangku, Rumah Mesa and Rumah Bunsu all indicated that hunting and collecting activities had all but ceased because of logging activities. Indeed, several doubted that the products they formerly depended upon still existed in the rainforests adjoining their longhouses.

An equally sad situation appertained in connection with hunting and fishing. Fishing activities had been severely curtailed, according to respondents, principally on account of the severe pollution/sedimentation of water courses which had resulted from logging activities. Similarly, whereas previously individual or groups of hunters could reasonably have expected to return from a hunting trip with sufficient meat to feed not only the hunting party but also several other *bileks* in the longhouse (a past success rate of 6–8 out of every 10 hunting expeditions was often cited, as were lavish tales of meat being discarded because it was so plentiful), today this is far from

Table 13.4 Frequency of hunting, fishing and collecting activities in the study longhouses (n = 222)

Frequency	Hunting	Fishing	Collecting
'Daily'	8	12	8
'Weekly'	14	27	10
'Monthly'	23	26	11
'Twice Yearly'	14	6	84
'Annually'	8	1	35
Not Involved	155	150	74

Source: Survey data, 1992/3

being the case. Several respondents reported a zero success rate for ten trips, and many had given up hunting altogether, except during peak fruiting seasons when the population of wild boar (Iban: *babi*) proliferated.

A rather sorrowful illustration is provided by Rumah Saba where, because of being almost completely surrounded by logging camps, residents now have to embark on trips up-river to Binjo in order to find wild game and other forest products, at a cost of around 17–18 gallons of fuel per trip. Accordingly, some *bileks* now engage in hunting only once or twice a year, whereas previously it may have been a weekly activity. Even a hunting expedition to Binjo may yield nothing these days. Logging is not the only factor responsible, however: some respondents also pointed to the over-exploitation of wildlife by individual Iban hunters. Residents in Rumah Lunyong, for example, pointed to people returning with up to 100 birds from a single hunting trip – many more than they could reasonably expect to consume – and of people using electricity generators and chemical poisons to catch fish and prawns, so scarce had these become (a point we shall return to shortly).

Response to change

Residents were asked to specify how they had responded to the changed availability of NTFPs. For products to which the Iban could find no convenient substitute, the vast majority (78 per cent) indicated that they had either curtailed, or had considerably reduced the frequency of collecting activities. Others indicated that they simply obtained less from their collecting trips (30 per cent – several gave multiple answers), travelled further afield (3 per cent), had stopped making handicrafts which depended on these products (2 per cent), or had attempted to cultivate/rear these products in their home gardens (2 per cent). In the case of products for which the Iban could find substitutes, the overwhelming majority (87 per cent) indicated that they were now obliged to purchase them, others had turned to substitutes which they could cultivate or rear themselves (17 per cent), such as by digging fish ponds, and some indicated that they had changed the make-up of their diet (4 per cent).

The study identified further effects of the rapidly declining availability of natural products, and other rather more insidious responses. Perhaps the most alarming response of an admittedly small number of residents had been to turn to technology as a means of countering the dwindling supply of natural products. Fishing by electrocution (using power generators) and by using chemical poisons (replacing the organic poisons, *tubai*, which have traditionally been employed in this activity), were a couple of the techniques which residents would admit to using in a vain attempt to maintain fish and prawn supplies. The long-term effects of using technology to exploit already rapidly diminishing natural resources should be obvious. Chainsaws had also significantly increased the power of the Iban to intensify the exploitation of timber resources, although the majority of owners simply used this form of technology to ease the work burden involved in clearing land for swidden cultivation.

The misuse of technology was a method employed by only a small number of individuals – generally from the more wealthy households which could afford this particular technological 'fix'. Although limited in its extent, it was none the less one of several signs that traditional forms of communality were beginning to break down, and that a greater degree of individualism was becoming apparent. Local *adat* (customary law) was found in most longhouses to be the principal mechanism for mitigating the over-exploitation of forest resources by longhouse residents (and although fines were also imposed on outsiders, this has clearly acted only as a weak deterrent). However, it was noted that hunting *adat* had almost completely disappeared in many cases. Previously, when stocks of game and wildlife had been fairly abundant, hunters and hunting parties had been obliged by *adat* to share their catch with other *bilek* (a longhouse unit, usually a household), usually at a ratio of two parts each for the hunters to one part for other *bilek* and those which had provided hunting dogs but which had not participated in the hunt. Today, such a practice has become extremely rare, with longhouse residents tending either to consume all of the meat obtained from their hunting expeditions (in any case the meat was generally insufficient to satisfy their own needs), or to sell it to neighbouring *bilek*. Commonly heard statements included 'if we want to eat meat we must pay' and 'if I have no money I just have to watch others eat [meat]'. Such a situation would have been unthinkable a few years ago. However, it is difficult to say whether this is due to a breakdown in socio-cultural practices, or is simply a natural response to the scarcity of wildlife and the consequent increased value of meat obtained from the wild. Further research is required to provide a definitive answer to this question. What is certain, however, is that the dramatically reduced incidence of hunting, especially in groups, and *tubai* fishing has removed an important activity around which social systems and Iban cultural identity revolved in the past. The same can be said of the decline in the incidence and importance of handicrafts production, around which a great deal of social interaction (especially of women) usually takes place, and also the rapidly increased occurrence of labour migration, which is significantly changing the age and gender make-up of longhouse communities for extended periods of time (see also Parnwell and King, 1995).

In order to add some empirical substance to the social effects of the changes discussed above, respondents were asked to indicate whether they agreed with certain statements which related to their relationship with other *bilek*-families in their respective longhouses in connection with access to natural forest products (see Table 13.5). Whilst we should be cautious about interpreting these aggregated data, they do tend to support some of the findings which have been discussed in this section. Thus, one-fifth of respondents believe that their neighbours are exploiting local resources more than might ordinarily be deemed acceptable by other members of the community, although interestingly only five per cent indicated resentment of this practice. One-quarter viewed access to forest products as a competitive venture, and more than three-quarters were quite willing to sell to other households forest products that they had collected/obtained. Rather alarmingly, more than one half thought that the use of technology was the best means of maintaining the flow of NTFPs. It is a little surprising that only around a half of respondents believed in equality of access to natural forest products, and that more than one-third failed to indicate that they were willing to share the products derived from the forest with others in their community.

In spite of these findings, only a very small proportion (5 per cent) of longhouse residents considered that their relationship with other *bilek*-families in respect of access to NTFPs had deteriorated over the last fifteen years. The remainder felt that such relationships had remained consistently good during this period (93 per cent), or had in fact improved (2 per cent). More generally, some 83 per cent of respondents considered that levels of co-operation within their respective longhouses remained 'very strong', and only one person thought it 'very weak'. In terms of change, 5 per cent 'strongly agreed' that the level of co-operation between households had deteriorated over the last fifteen years, and a further 11 per cent 'agreed' with this view. Thus, whilst there may be signs that changes are taking place both in people's attitudes and behaviour regarding access to NTFPs, which in turn may threaten their longer-term availability and quality, there is not much compelling evidence to suggest that this is having a profoundly disruptive impact socially. Inasmuch as *adat* any longer has the power to control resource exploitation from within (which is doubtful, bearing in mind (a) the (often realized) potential for powerful élites to subvert the institution via its designated custodian, the *tuai rumah* (longhouse headman), (b) the lack of sharpness in the geographical, legislative and cultural boundaries of its jurisdiction, (c) its limited powers of punishment and thus deterrence, and (d) the broad range of activities that it encompasses: see, for example, Freeman, 1970; Hooker, 1976; Sandin, 1980), this must be seen as an encouraging sign. However, it is

Table 13.5 Perceptions of change in relationship with other *Bilek*-families in connection with access to natural forest products (n = 222)

Statement:	Agree	%
'I believe that we should all have equal access to these natural products'	112	51
'Some people are taking more than their fair share of these products'	41	19
'I believe that we are all in competition with each other for these products'	56	25
'I am willing to share with others the products that I obtain'	135	61
'I am willing to sell to others the products that I obtain'	174	78
'It is necessary to invest in technology to maintain a supply of these products'	130	59
'I resent others when they over-exploit these natural products'	12	5
'I am happy when one of my neighbours is lucky (e.g. in catching a *babi*)'	97	44

Source: Survey data, 1992/93

equally clear that the study communities are more or less powerless to influence pressures emanating from without.

Before turning to assess some of the implications of these changes, there are three further effects of the changed availability of natural products which are worthy of brief comment. The first, which has already been alluded to, is the rapidly growing imperative of having a source of cash income. The only realistic alternative to the collecting of forest products has been shown to be the purchase of substitutes, which is clearly difficult with the modest cash incomes which the study communities derive from their farms (see Parnwell and Morrison, 1994). One problem associated with this situation is that some households have found it easier to obtain a supplementary source of income than others, principally by engaging in labour migration, which has clearly had a divisive effect within the longhouses. Certainly, those without a cash income had experienced considerable difficulties as a result of the growing scarcity of natural forest products.

A second, related effect has been the change which has taken place in the make-up of the diet of several respondent households, and the nutritional and health consequences of this (see also the chapter by King in this volume). Not only have many of the households lost some important sources of protein, fat, vitamins and calories from the wild, but it has often been the case that the reared/cultivated substitutes that they purchase from markets in town or from *towkay* (traders, entrepreneurs: usually Chinese) who periodically visit their longhouses are either much more deficient in these properties or are laced with additives (such as mono-sodium glutamate in processed foods, and pesticides in fresh vegetables) with which the internal constitutions of many rural Iban have, and wish to have, no great familiarity. Although the evidence is largely circumstantial, several respondents associated such changes in the make-up of their diet with a large increase in the incidence of health problems such as high blood pressure, stomach aches and other gastric problems (although water quality problems must also have had an important impact in the latter two respects).

A third response to change was that people had begun to show an inclination to turn 'from capture to culture' – that is, to attempt to cultivate in their *temuda* (land under secondary forest) and home gardens some of the forest products which had become increasingly scarce, and also to rear or cultivate livestock and plants which could provide alternative sources of food and meat. Production of the former included *bemban, keropok, rotan sega, senggang,* bamboo and wild ferns, whereas the latter included rearing pigs and chickens and the cultivation of various vegetables and fruits, with assistance from the government Home Economics Programme. Such a response may be considered a very important 'mechanism of sustainability', and will be explored further in the final section of this chapter.

FUTURE PROSPECTS

This chapter has highlighted a number of changes which have taken place in the availability and quality of non-timber forest products in a part of Sarawak which has experienced quite significant environmental degradation as a result of intensive commercial logging over the last 10–15 years. Whilst not all of the changes described in this chapter can be attributed directly to environment degradation, it has clearly played a significant part. To a large extent the findings are consistent with the earlier discussion of NTFP dynamics elsewhere in Borneo.

If we see the availability and quality of NTFPs as a 'barometer of change' taking place in the tropical rainforest, the findings of this study should give some cause for concern. However, it is not only the forests which are threatened by such processes of change, but also the people who depend upon them for a significant part of their livelihood and welfare such as, in the Sarawak context, the Iban, Punan, Kenyah, Kayan, Bidayuh and several other ethnic groups. For many, as we have seen, the response has been to adjust, quite radically in some cases, the orientation of their economic and social activities. Out-migration, for the spatially mobile at least, offers the prospect of income continuity or even enhancement, but at some social and psychological cost both to the migrants and those who remain in the longhouse. For those who choose, or are obliged, to stay, a significant change is taking place in the sources from which they satisfy their subsistence needs and derive their livelihood, with the market economy playing an increasingly important role. This, in turn, creates further imperatives for out-migration, and thus a cumulative process of external dependence may be set in train.

We have also seen, however, that the Iban study communities have demonstrated a characteristically high degree of resilience in responding to the quite fundamental changes which have taken place in the forest ecosystem. In the main, these responses have taken the form of 'coping strategies' – means of sustaining their existence, but often at a cost to their self-reliance, self-identity and social cohesiveness.

> Within the past quarter-century, dramatic changes have occurred in Iban residence patterns, occupations, roles and relations. By implication these changes have affected and will affect the ethnic identity of the Iban. . . . Beyond question, increased dependence on the market and a monetised economy has psychologically, sociologically and culturally affected the values of many Iban.
>
> (Sutlive, 1989, pp. 41 and 43)

In responding to the exigencies of change, some have trodden a technological path which seems likely to speed up rather than reduce the rate of resource depletion. Others have shifted the balance of their activities either to allow more time to be spent in pursuing dwindling stocks of game and forest products, or in reducing their dependence on these natural products.

What, then, might be done about this situation? Fundamentally, the answer to this question depends upon how the information presented in this chapter is interpreted. A leading academic in the University of Malaya expressed to one of the authors the sentiment that the modern Malaysian state is no place for people still to be residing in and deriving their livelihood from the rainforest. Thus any economic and physical process which encourages people to move towards the urban centres and into the burgeoning industrial sector should be interpreted positively. An alternative perspective rests on the extent to which people retain the option to choose how they would prefer to derive their livelihood and how and where they would prefer to live (see Parnwell, 1993a, p. 24).

Ultimately, the answer to the question rests partly in the hands of those who have the power to make 'decisions against nature', and partly with the communities themselves. Just as we have attempted in this chapter to use the respondents' own views to judge the nature and impact of change in the availability of NTFPs, so it is important to base judgements of possible policy options upon their own needs, aspirations and preferences. Migration, for example, offers the prospect of better and, with the exception of logging, more reliable long-term income earning opportunities, and thus should not be overlooked or discouraged because 'keeping people down on the farm' is seen as the best way of 'doing development' in rural areas. Alternatively, if people are seen to migrate through necessity rather than choice, then there may be some validity in pursuing *in situ* forms of rural development which will at least give people the option of improving their welfare and prospects without having to move elsewhere to achieve this. What forms of *in situ* development might thus be deemed appropriate? We shall finish this discussion by considering just one policy option.

Plate 13.5 Iban man making rattan chairs, Kemena Catchment, Sarawak

Mike Arnold (1993) has assessed the potential for small enterprise development based on the utilization of forest products. In his review of the literature he has identified the important role of small-scale forest-based enterprise in terms of its employment and income-earning potential, especially for women, the poor and for people whose income from other sources is insufficient or not dependable. Flexibility, ease of entry and low skills and capital requirements were seen as further virtuous features of small enterprises, although these in turn contributed to very high levels of competition in economically backward areas where there are large numbers of poor and unskilled. Arnold decries the decline of such forms of enterprise as a result of competition from large-scale modern enterprises and increasing resource scarcity: 'The decline of labour-intensive one-person activities . . . is of concern for the erosion of income-earning opportunities that it implies for the very poor, and in particular for women' (Arnold, 1993, p. 12).

In the area under study, we need look no further than handicrafts production to identify a form of small-scale enterprise which presently fulfils a useful social and utilitarian function, and which could play a much more significant economic role than has been the case hitherto. We have seen that in at least three-quarters of the study households there is at least one person who has skills in making handicrafts of various kinds, although for the vast

Table 13.6 Sarawak: unprocessed rattan exports, by value (M$) 1979-89, and rattan exports as a percentage of total exports of NTFPs, 1985-89

Year	Exports	Year	Exports	Year	Rattan as % of NTFP Exports
1979	83,997	1985	361,105	1985	34%
1980	138,722	1986	54,849	1986	16%
1981	93,666	1987	5,969,825	1987	14%
1982	265,511	1988	6,965,083	1988	79%
1983	579,743	1989	7,760,302	1989	92%

Source: State of Sarawak Forest Department, 1986 *et seq.*

majority crafts production is used as a means of existence, and not a means of future investment (cf. Arnold, 1993, p. 10). The majority of handicraft items are produced by women, who are provided with an important source of both work and social interaction through their manufacture.

Whilst only a very small fraction of households have previously produced such items for sale, the potential market for the handicrafts of Sarawak's native peoples is very rapidly expanding, both for sale to tourists and for export. The Sarawak government is presently playing a very active role in the promotion of the state's handicraft products (Berma, forthcoming). It is thus entirely feasible for greater efforts to be made to support the commercial production of handicraft products within Iban longhouses as a means of supplementing earnings from agriculture and, possibly, of reducing imperatives for out-migration. A number of barriers and constraints must first be overcome, but there are many appealing aspects of this potential model of *in situ* development. Given that only four *bilek*-families had previously been engaged in the commercial production of handicrafts, it was encouraging that 57 households (26 per cent) indicated that they would be willing to consider this option as a supplementary source of livelihood in the future, provided a reliable market were made available. Of those who were rather more conservative, the main reasons given for their reluctance in this regard included a lack of appropriate knowledge and skills, especially for making saleable, high quality handicrafts (50 responses), a lack of time (18), a lack of labour/women (9) and, most importantly, a scarcity of the requisite raw materials (91).

One of the most serious constraints on the future development of small-scale handicrafts production within the longhouse, of course, is the rapidly declining availability and quality of the constituent raw materials – the non-timber products derived from the rainforest. It is not only the availability of these resources that is a problem: 'Shortages of raw materials are exacerbated for small enterprises by their lack of working capital, which prevents them holding stocks. Nor are they able to invest in the resource itself' (Arnold, 1993, p. 15). Thus, for the development and commercialization of handicrafts production to become a realistic proposition, a parallel effort

Table 13.7 Value of rattan furniture exported from Malaysia, 1980-90 (M$)

Year	Exports	Year	Exports
1980	2,775,389	1986	7,044,553
1981	2,853,954	1987	11,104,218
1982	2,160,829	1988	35,500,000
1983	1,747,418	1989	45,400,000
1984	4,229,657	1990	54,800,000
1985	5,498,521		

Source: Government of Malaysia, Ministry of Primary Industries, 1991

must be made to ameliorate the effects of some of the changes which have provided the main focus of this chapter. The one sign of optimism in this regard have been the spontaneous moves by some respondents to shift from 'capture to culture' – i.e. domesticating certain rainforest species as a way of overcoming their dwindling supply, regulating their availability and reducing the amount of effort which is expended in obtaining them (see also Chin, 1985; Pearce *et al.*, 1987; Peluso, 1991).

> farmers everywhere have shifted the production of particular tree products of value to them on to their own land by protecting, planting and managing trees of selected species. . . . The main impetus for this domestication is to meet household needs, but can also link into non-farm enterprise activities.
>
> (Arnold, 1993, p. 17)

As we have seen, attempts have been made to cultivate forest products used in construction and the production of handicrafts, and for food. More scientific research is needed to assess the prospects and means of cultivating other natural forest products, but it does represent an important potential means whereby forest-dwelling communities can come to terms with the exigencies of environmental degradation whilst also building future economic opportunities.

Rattan represents one forest product which the Indonesian case has shown to have considerable potential for domestication in support of trade and the commercial production of handicrafts (see Plate 13.5). Rattan plantations in East Kalimantan have been established since at least the late-nineteenth century (Peluso, 1991, p. 14), and more recently smaller clumps of rattan have been established on a smallholder basis (often inter-cropped with other tree crops), where they can be sustainably harvested every 7–10 years. Domesticated rattan production in Kalimantan supports thriving furniture and mat-making industries, and also supports a considerable proportion of the 83,000–100,000 jobs in collection, trade and rattan processing activities in Indonesia as a whole. It has been estimated that the rattan mat- and carpet-making industry in South Kalimantan could employ 15 per cent of the industrial labour force, thereby rivalling the employment capacity of the timber industry in the province (Peluso, 1991, p. 15, citing Liang, 1979).

Malaysia has lagged far behind Indonesia in exploring the potential for the domestication of rattan species, and, indeed, still exports considerable volumes of unprocessed rattan to countries such as Singapore and Hong Kong, where the value added through manufacture may be up to 100 times that in its raw state (see Table 13.6). However, there are signs that attitudes are changing. The Malaysian Forestry Department has been engaged in experimentation with the domestication of rattan species for almost two decades. More recently the Perkasa Rattan Estate, established close to one of our study communities, Rumah Anchai (principally, it should be said, to deflect attention from the extent to which the environment in the area had been so badly degraded by logging activities), provides an excellent example of what might be achievable if sufficient resources are invested in 'replicating' the conditions found in the forest ecosystem, and which produce natural products which are in increasingly short supply in their natural habitat. However, we should be mindful that such initiatives serve also to deflect attention from the root causes of the problems they are in essence being established to overcome.

Within the next 5–10 years the Estate expects to supply unprocessed rattan to downstream enterprises such as the Chinese-owned rattan furniture factory in Bintulu town. Such enterprises have contributed to the exponential increase in the value of rattan furniture exports from Malaysia over the last ten years or so (Table 13.7). In relation to the discussion above, the next important step would be to encourage the cultivation, processing and manufacture of rattan *in situ* within longhouses in the study area. Most encouragingly in this regard, a very high proportion of respondents (91 per cent) indicated that they would be willing to contemplate this 'capture-culture' option, considering it to be an easier way to obtain the products they need (n = 166); that it would obviate the need for them to purchase these materials (19); and that it would even give them the opportunity to sell some of the produce they cultivated (92). Labour, land, money and expertise were cited as reasons for people's reluctance to consider this particular 'sustainability mechanism'.

The capture-culture model could be extended to include other forest products of which forest-dwelling communities themselves presently make limited use, but for which there is a rapidly increasing world demand. For example medicinal plants, which the Iban study communities appear rarely to use, could, with external assistance, provide them with a dependable source of income and livelihood. In addition, such a strategy could help to add value to Sarawak's rapidly expanding areas of degraded or regenerating forest.

14

ENVIRONMENTAL CHARACTERISTICS OF BRUNEI'S TEMBURONG RAINFORESTS AND IMPLICATIONS FOR CONSERVATION AND DEVELOPMENT

Alan P. Dykes

Environmental change, sustainable development and conservation in South-East Asia are three major and inter-related issues of great social and scientific concern. In most of the countries in the region, 'environmental change' is often assumed to be synonymous with 'deforestation'. In many instances this is justifiably so as the loss of primary tropical rainforest, and the attendant biological diversity, are one of the greatest concerns of biologists throughout the world. However, deforestation is but one process by which environmental change may occur. The end products of forest exploitation range from thinned and damaged primary forest right through to abandoned wastelands. Other causes of forest destruction and localized environmental change are the reservoirs created for hydro-electric or water management purposes, such as the Pergau and Bakun Dams in Malaysia or, on a smaller scale, the Binutan reservoir in Brunei.

The case of forestry management in Brunei provides an interesting contrast to developments elsewhere in the region. Partly as a result of the rich natural resource base of the country (especially oil and gas) and a low population density, forest management as practised by the Forestry Department is by and large sustainable. This chapter thus considers how state control of the forests has enabled the substantive protection of the natural rainforest. If internal policy conflicts occur over development questions, it is nevertheless the case that Brunei's rainforests have been largely protected. With reference to Temburong District, this study examines the social and scientific bases of sustainable development and environmental change in Brunei.

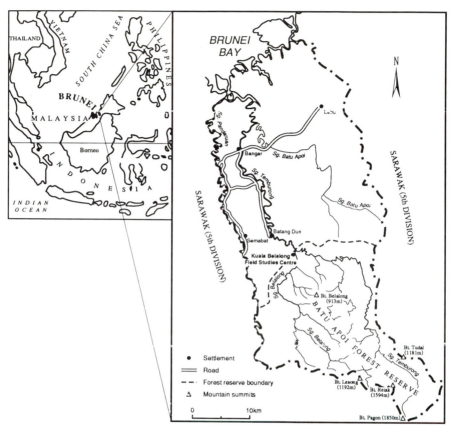

Figure 14.1 Brunei: Temburong district

Negara Brunei Darussalam is an oil-rich state of 5,765 km² situated on the northwest coast of Borneo, its inland borders entirely surrounded by the Malaysian state of Sarawak (see Figure 14.1). It has a population today of a little under 261,000, mostly confined to the coastal strip (Goh, 1992). Around half of the population lives in and around the capital city, Bandar Seri Begawan, with much of the remainder concentrated around the oil and gas industry headquarters at the western end of the country. Some inland settlements exist along the main rivers, but many of the up-river and head-water regions remain unpopulated. Revenue from its hydrocarbon resources and overseas investments result in a per capita Gross National Product which is among the highest in the world. As a result of this economic situation, for reasons which will be briefly explained later, 59 per cent of Brunei's land area retains an intact primary rainforest cover into the 1990s, and a further 22 per cent is also still covered with secondary and other forest (Collins *et al.*, 1991).

Much of Brunei's remaining primary rainforest is contained within designated Forest Reserves, within which public access is limited by ordinance and the destruction or removal of plants or animals is forbidden. Some of these reserves provide recreational facilities, but many occupy inaccessible upland regions. One is the Batu Apoi Forest Reserve in the Temburong District of Brunei. This district comprises the eastern part of Brunei, separated from the remainder by a narrow strip of Sarawak, and has a land area of 1,288 km² and a population of little more than 7,000. The Batu Apoi Reserve occupies the entire southern part of Temburong, 488 km² of some of the finest remaining generally undisturbed Mixed Dipterocarp rainforest in South-East Asia. Until 1990, human activity within this area was limited to occasional army training exercises, and hunting parties from the nearest villages to the Reserve. These latter activities have persisted for several decades and, although the impact on terrestrial animal populations appears to have been minimal, some local people do claim that fish catches have been declining in recent years. Small-scale research expeditions have occasionally entered the reserve, but prior to 1991 these have been few and far between (e.g. Allen, 1979).

FORESTRY IN BRUNEI DARUSSALAM

A recent history of Brunei's Forestry Department outlines the philosophy behind the policies adopted by the department since 1933 (Government of Brunei: Forestry Department, 1993). Forestry activity during the 1930s was minimal, and was primarily concerned with providing timber for the Brunei Oilfield. However, Forest Reserves were designated as early as 1934, with the first notification of the Batu Apoi Forest Reserve in Temburong District being issued in 1936. In 1937 the Forestry Department set about developing the timber industry, but timber was still being imported from Sarawak even for government building works in 1939, by which time 1,540 km² of inland forests were included within Forest Reserves. This area had increased to 2,430 km² by 1954, but forest destruction by shifting cultivation in up-river interior regions was by then regarded by the state as a problem. The export of timber from Brunei was banned in 1957 because the domestic demand for timber had until then continued to exceed the available domestic supply, requiring imports from Sarawak. This ban was relaxed in response to the surplus of timber which had accumulated by the late 1960s, but was reinstated in 1970. Further development of the timber industry saw some, strictly restricted, exports of sawn timber resume by the early 1980s. Department forestry policies were established through a ten-year plan in 1955 which specified that management of reserved forests should be based on 'sustainable yield by working circles' (ibid., p. 42). At the end of 1983, on the eve of full independence, Brunei had eleven Forest Reserves covering 2,119 km², and timber production totalled about 100,000 tons, of which 78,000 were from Forest Reserves.

As early as 1955 it was recognized by the state that Brunei's forest resources were in a unique economic position. First, there was little pressure on the land, with only around 50,000 people inhabiting 5,765 km^2, so that 75 per cent of this was under high forests. Second, there was also very little pressure on the forest, since 'the rate of cut was far below the potential cut on a sustained yield [and] the State's revenues from the oil industry were sufficient for expenditure and for development' (Government of Brunei: Forestry Department, 1993, p. 41): the government was thus largely indifferent to revenue from the forest. These factors together held broadly true until independence in 1984, although the population had been growing increasingly rapidly during this time. There continued to be little pressure on the forests throughout the post-independence changes. The Forestry Department was placed under the Ministry of Development following independence, and it expanded its activities 'in line with the new national economic policy of industrialization and economic growth' (ibid., p. 58). A comprehensive survey and analysis of the forest resources and potential economic value of those resources were therefore commissioned by the Forestry Department, part of which detailed the full commercial value of the timber standing within the Batu Apoi Forest Reserve. Despite these commercial findings, the consultants' final report identified this Reserve as being appropriate for development as a National Park (Anderson and Marsden, 1984).

In January 1989 the Department was transferred to the new Ministry of Industry and Primary Resources. 'Sustainable utilization of the forest resources and upgrading of the environmental quality' (Government of Brunei: Forestry Department, 1993, p. 60) formed the basis of the Department's subsequent administrative restructuring, which included a new section responsible for recreation, conservation and associated developments. In September 1989 a new National Forest Policy was issued, 'emphasizing the needs of industry and of sound environment, but at the same time strives for excellence in tropical forestry' (ibid., p. 61). The Reduced Cut Policy, which originated in the National Forest Policy and restricts total annual log production in Brunei to 100,000 m^3/year, was enforced in January 1990, leading to increasing timber imports throughout the next two years and increasing emphasis by the Forestry Department on plantation development for timber production. This was not a response to over-production of timber in the past, but instead a deliberate move to reinforce the sustainability of forestry activities. By 1993 the Forestry Department was turning more towards policies and activities centred on the themes of plantation and product development, environmental conservation and wildlife conservation.

Thus, to a degree unmatched elsewhere in the ASEAN region, the Brunei government has maintained strict control over the fate of the forests within its borders. Environmental change in Brunei has been limited by government restrictions on logging as well as because of the concentration of much of the country's population in the capital. As the following discussion of recent social and scientific developments in Temburong District illustrates, however, the nature of management and conservation may now be changing in the country.

CONSERVATION THROUGH RESEARCH?

For several years before 1990 there existed the possibility that the Brunei Government's Ministry of Development might begin to investigate the feasibility of building a hydro-electric plant in the Temburong valley, which would create a large reservoir within the Batu Apoi Reserve. During this same period, the Biology Department of the Universiti Brunei Darussalam (UBD) was developing plans to establish a field research centre within an area of pristine rainforest, with the intention that such a centre would, by its nature, be a unique facility for education and research within South-East Asia as a whole.

The intention by UBD to establish such a centre matched the aim of the Royal Geographical Society (RGS) in the UK to mount a tropical rainforest expedition to the region, and in 1989, around the time of the restructuring of the Forestry Department mentioned earlier, it was agreed that the RGS would provide a baseline environmental data set for the centre, on which future activities could be based (Cranbrook, 1993). A site near Kuala Belalong at the northern edge of the Batu Apoi Reserve was eventually selected in late 1989, and construction of the Kuala Belalong Field Studies Centre (KBFSC) commenced in 1990. Whilst sufficiently remote to be set within pristine rainforest and deter casual visitors from easily reaching it, the KBFSC could be reached from Bandar Seri Begawan in as little as two hours at certain times of year.

The UBD/RGS Brunei Rainforest Project 1991–92 was launched in London in October 1990 with the approval of His Majesty The Sultan and the support of the entire Brunei government. Most of the direct governmental involvement in the Project was channelled through the Education Ministry, which is directly responsible for UBD, but with the full co-operation of the Ministry of Industry and Primary Resources, and in particular the Forestry Department. The expedition phase therefore commenced in January 1991 under the leadership of the Earl of Cranbrook, an expert tropical biologist. The Project was overseen on site by the Deputy Leader, Dr David Edwards, then head of UBD's Department of Biology, and the screening of research proposals for the Project was delegated to a UBD scientific committee under the chairmanship of the Director of Forestry.

The scientific framework of the Project was tailored to provide the baseline environmental data required for UBD's long-term objectives involving the KBFSC which are first, to provide a focus for field-teaching, training and research in Brunei's lowland tropical rainforest environment, and second, to provide facilities for field-work at university and secondary school level

(Cranbrook, 1993). These objectives are in line with the Forestry Department's generally conservation-minded policies, and sit comfortably with the proposal to create a National Park in the area. Central to the scientific programme were five long-term environmental and ecological studies. The hydrology/ geomorphology study aimed to explain the dominant processes which shape the land surface on which the rainforest itself stands (Dykes, 1994). Two of the core studies serve to link the first one with the ecology of the forest, investigating botanical variations with altitude and recolonization of forest gaps due to seed dispersal by birds, and the other two studies were concerned with the biodiversity and ecology of beetle and ant populations respectively. In addition to these core studies, a wide range of smaller, dominantly biological investigations was also approved by the UBD scientific committee, forming a comprehensive programme of botanical, zoological and ecological surveys (e.g. Cranbrook, 1992). The combined results of these studies provided a sound knowledge-base from which the environmental impact of future human activity in the area could be assessed, and on which future environmental and conservation-directed research could be built. In other words, UBD could, through the RGS, support the Forestry Department's conservation policy with a considerable quantity of scientific findings relating to the potentially threatened area.

THE RAINFOREST ENVIRONMENT

Before considering the governmental policies that were announced during and following the Brunei Rainforest Project, it is necessary briefly to illustrate and explain the physical and environmental characteristics of the Batu Apoi Forest Reserve.

The Batu Apoi Reserve comprises a deeply dissected landscape which, although mountainous in appearance, reaches altitudes of 1,000 metres or more only around its extreme southeastern borders. With the exception of the extreme south-east of the region, and a few ridgetops further north, the entire region is underlain by a uniform shale lithology of 16–30 million years in age (Wilford, 1961), known as the Setap Shale or Temburong Formation (Brondijk, 1963). River channels occupy the entire floors of the steep-sided V-shaped valleys formed by incision of the channels into the continually – but slowly – up-lifting land mass. Many valley slopes are potentially susceptible to landsliding, although the actual frequency of even small landslides is extremely low (Dykes, 1993; 1994; 1995). On top of this landscape stands magnificent Mixed Dipterocarp rainforest (Ashton, 1964a; 1964b; Anderson and Marsden, 1984; Cranbrook and Edwards, 1994) which in places displays structural variations determined by the landscape itself, such as the general absence of very large trees on the steepest slopes. It is thought that depths of soil exceeding 1–2 metres are rare throughout the region.

The climate of Brunei is characterized by largely constant daily mean temperatures and relative humidities, and low wind speeds except during storms (Grant, 1984). The most important climatic descriptor is the rainfall regime. The mean annual rainfall at Semabat is a little over 4,100 mm, and the Batu Apoi Reserve can expect at least as much rain annually. Although noted for its inconsistency and variability, two distinctly wet seasons can usually be identified (400–600+ mm/month), coinciding with the overhead passage of the Inter Tropical Convergence Zone in May and November. January/February and July/August tend to be relatively dry (100–300 mm/month). Up to 1,900 mm/year of this water may be lost as evapotranspiration from the forest vegetation (Dykes, 1995), giving a mean annual run-off of over 2,200 mm, equivalent to an annual discharge of over 950 million cubic metres of water downstream of Kuala Belalong. The shallow but highly permeable clay soils covering the steep slopes allow rainfall during storms to be transmitted to the drainage network extremely rapidly, with peak storm-flows often occurring within an hour of the rainfall ceasing, and it is the extreme hydrological conditions experienced by the hill slope soils during these intense storms which provide the trigger mechanism for the smaller, shallow landslides mentioned earlier (ibid.).

The rainforest which covers this hydrologically dynamic landscape is characterized by an uneven canopy around 30–45 metres above the ground, with many individual trees standing as 'canopy emergents' 60 metres or more in height; some of these large, over-mature trees may be more than 500 years old (Whitmore, 1975). A third layer of shade-dwelling trees is sometimes identified beneath the main canopy, with a sparse ground vegetation under that. The emergent layer is dominated by members of the Dipterocarpaceae family, of which there are ten genera and around 270 species in Borneo alone, and there may be over 2,000 species of trees with stems at least 10 cm in diameter in the primary forests of Sarawak and Brunei (ibid.). The forest structure is spatially extremely heterogeneous, with topographic influences limiting the occurrence of large trees on the steepest slopes, and smaller-scale disturbances such as tree falls and occasionally landslides creating gaps which are available for recolonization by pioneer species, the overall result being a patchwork of mature, disturbed and developing forest. Such processes contribute to the maintenance of the high biodiversity of the forest.

The scope for biological, ecological, and many fields of geographical research in the Batu Apoi Reserve can thus be seen to be extremely large (see also Cranbrook and Edwards, 1994). Knowledge of the geomorphological and hydrological processes in undisturbed forest will improve the ability to predict direct and indirect physical effects of human activities in the forest, including the extremes of logging and flooding, whilst biologists may be able to improve the present best estimates of total global biodiversity on the basis of results from this area. Rainforests are also being increasingly regarded as a potential rich source of medicinal products (United Nations Environment

Programme, 1993), and investigations of traditional usage of forest products in this way have begun. Thus the combined potential for investigating the natural physical and biological processes of primary tropical rainforests is considerable. However, the very high annual run-off from the entire Batu Apoi Reserve, together with the quite hard, impermeable bedrock with just a thin soil cover which defines the narrow and steep-sided valley forms, provide some of the best environmental attributes for a hydro-electric scheme, and certainly the best available in Brunei. The very nature of the region therefore maximizes the potential conflict between two medium-term land-use possibilities; that is, between conservation for research, education and tourism in the form of a National Park, and significant environmental destruction by the construction of a dam and flooding of one or more major valleys.

RESEARCH, EDUCATION AND TOURISM

The expedition phase of the Brunei Rainforest Project ended in March 1992, with the Project finally culminating in the International Conference on Tropical Rainforest Research: Current Issues, organized and hosted by UBD in April 1993, and the publication of a book detailing the results of the scientific studies and surveys conducted during the expedition (Cranbrook and Edwards, 1994). The expedition, conference and launch of the book all received high-profile coverage in the local media in Brunei (e.g. Edwards, 1992), and the declaration in 1991 by the Minister of Industry and Primary Resources that the Batu Apoi Forest Reserve would become Brunei's first National Park received maximum coverage. The expedition and conference were regarded as highly successful by UBD and the Education Ministry in Brunei, in providing a preliminary database for the KBFSC and a detailed rainforest textbook suitable for high school students, whilst at the RGS, the scientific work undertaken during the expedition and the potential for publications derived from that work demonstrated the great success of the Project.

The 1993 conference identified in detail the research possibilities and, more importantly, priorities, which should be addressed by scientists intending to work at the KBFSC in the future, especially basic but essential long-term environmental monitoring programmes which should be implemented and continued indefinitely in order to build on the successful initial database. This was especially important given that prior to 1991 even the Forestry Department had only occasionally even visited the Temburong forests, and few surveys of any kind had ever been conducted (e.g. Ashton, 1964b; Anderson and Marsden, 1984). Professor Datuk Haji Sharom Ahmat of UBD said of the KBFSC in 1991: 'What we want to do is to create a major international centre of excellence for rainforest research and education. The Sultanate of Brunei will ensure that there is no human encroachment in the area.' Following the conference in 1993, UBD drew up a ten-year research plan for the centre in line with this statement and based on specific issues raised during the conference.

The declared intent that the Batu Apoi Forest Reserve should become a National Park has been followed up by significant progress in the establishment of the administrative and physical infrastructures needed. Although there was (and is) no provision in Brunei Law for National Parks, the reorganization of the Forestry Department has created a section dedicated to National Parks and Recreation (Government of Brunei: Forestry Department, 1993), which not only enables but also considerably strengthens the Department's plans for the area; accordingly, the implementation of the National Park is proceeding rapidly. Construction of visitor facilities and a Park Headquarters was well under way near Kuala Belalong by the middle of 1994, although well away from the existing KBFSC which, along with the Belalong valley, is to be reserved for educational and research activities only. Access for so-called 'ecotourists' will be strictly limited by the facilities available, but their presence and the desire by the Forestry Department to cater for them should ensure the long-term prevalence of conservation-minded policies for the region. It is possible that the limited access policy is intended in the interests of conservation and management of the protected forest. However, unlike in other countries in South-East Asia (see the chapter by Cochrane in this volume), it is worth noting that the economic attraction of ecotourism is perhaps less in Brunei at present for much the same reason that its forests are not systematically exploited for timber.

PRIORITIES: CONSERVATION OR ECONOMIC DEVELOPMENT?

As noted earlier, between 1984 and 1989, the Forestry Department had been part of the Ministry of Development in Brunei, which was responsible for all economic, industrial and infrastructure development as well as the management of all natural resources in Brunei. The latter included management of the country's hydrocarbon reserves and forests. Following the creation of a separate Ministry of Industry and Primary Resources in 1989, the Ministry of Development has focused on establishing and implementing a national policy of economic diversification, in order both to plan for the exhaustion of oil supplies during the next thirty years and to improve the national infrastructure. At present, Brunei's electricity is generated by oil-fired power stations, and in conjunction with the present programme of rural electrification, plans to generate power by alternative means are predictable. From an environmental conservationist perspective it is unfortunate that power supplies in Temburong District are of particular concern to the Ministry, and that the upper Temburong valley is ideal for a hydro-electric scheme. It is this situation which gives rise to the conflict between conservation and development policies in this forested region of Brunei.

The Batu Apoi Forest Reserve was designated a National Park during 1991, part-way through the Brunei Rainforest Project, and regular educational visits to the KBFSC commenced in May 1992. As stated, by April 1994, construction of headquarters and visitor facilities for the new National Park was under way near Kuala Belalong under the umbrella of the Ministry of Industry and Primary Resources. However, the Ministry of Development commissioned a Canadian engineering company to conduct a feasibility study into the proposed hydro-electric scheme, which was completed by the middle of 1994. Whilst hydro-electricity to some extent represents a sustainable energy resource, being powered ultimately by rainfall, the start-up costs are environmentally huge, and the feasibility report may have taken this into consideration in its final recommendations. It is unlikely that the Ministry of Development would implement any recommended hydro-electric scheme in the short term, given the considerable and highly publicized progress already made to establish and develop the National Park. The potential threat to the valleys within the National Park remains, however, and in the longer term the drive for economic development may yet see conservation goals sidelined in favour of practices that may lead to environmental destruction.

CONCLUSION

With a long history of conservation-minded forestry policies, recently reinforced by the imposition of the Reduced Cut Policy on logging concessions and the establishment of the National Park in Temburong District, the Forestry Department in Brunei presents itself as an environmentally friendly governmental organization. There are no non-governmental environmental organizations in Brunei, and in general little need for any as far as issues of environmental change are concerned, although the establishment of the new KBFSC by the Biology Department of UBD and the associated RGS expedition provided significant external support for, and promotion of, the conservation-minded policies of the Forestry Department. The greatest environmental threat in Brunei comes from the proposal by the Ministry of Development to flood a large area of the new National Park by constructing a dam and power station, although this plan is unlikely to proceed at least for the time being.

The importance of this study lies in the fact that Brunei is in a unique economic position compared with other countries in South-East Asia in that it does not need revenue from timber production. However, like its neighbouring countries it is keen to attain an even higher level of economic development. As a result of Brunei's valuable oil and gas resources, the government has so far been able to promote national economic development without significant environmental degradation. However, as oil and gas reserves are depleted, maintaining a balance between conservation and development will prove more difficult. Competing plans for the southern part of the country's Temburong District thus perhaps serve as a warning of the difficult political and economic choices that may lie ahead.

15

PHILIPPINE COMMUNITY-BASED FOREST MANAGEMENT

Options for sustainable development

Gilbert C. Braganza

Land is a loving gift from the MAGBABAYA for all men
Because land produces life's nourishment
Land itself is life
It cannot be sold or taken away by anyone
It has been kept under the stewardship of the great, great ancestors
whose spirits continue to live and keep watch
so the land could be of incessant use for generations
and generations
and their descendants

Higao-non saying (Vitug, 1993)

INTRODUCTION

It has been observed that many of the development approaches relating to environmental issues adopted by developing countries, particularly those by the least developed ones, are colonial in origin. Such an approach, or model, understands and analyses environmental issues as purely environmental in character rather than a complex social, political, economic, and ecological set of inter-related concerns. It also lays the blame for environmental destruction on the incapability of land users, links the issue to over-population, and establishes policy directions that insist on active participation in the market economy (Blaikie, 1985). Recent observations and analysis of local-specific experiences have indicated the failure of such an approach in that it has instead contributed to environmental damage. Faced with the prospect of uncontrolled environmental exploitation as a necessary consequence of economic growth and industrialization, it became essential that appropriate development paradigms be sought.

311

The emergence and recent popularity of the sustainable development concept are considered by some to meet this need. Sustainable development suggests a form of economic growth and productivity that is ecologically sound and takes into account the concerns of future generations. Although already popular as a concept among a few academics, environmental activists, and development strategists since the 1960s (Elliot, 1994) and further articulated during the 1980s (Lewis, 1992) in such publications as the Brundtland Report, it was only during the 1992 Conference of the United Nations Commission on Environment and Development (UNCED) at Rio de Janeiro that the concept of sustainable development was globally recognized and accepted. Some of the highlights of the principles embodied in the Rio declaration acknowledge sustainable development as that which is wholly concerned with the welfare and state of human beings, that which integrates environmental protection, promotes a higher quality of living, requires the participation of all people and states, recognizes the role of indigenous knowledge and traditional practices, encourages inter-state co-operation, provides effective and contextual legislation, fosters peace, and is scientific. It may be safe also to assume that a sustainable form of development is that which puts high regard on social justice, economic equity, and active ecological response.

Since then, sustainable development has become a sort of a battlecry in the movement towards economic growth that is socially and environmentally sound as much as it is also contextually relevant. Many countries, both developed and developing, have eagerly taken the opportunity to conceptualize and methodologically define its meaning and application. It is therefore not surprising to discover the variety of interpretations and strategies that have emerged, mainly brought about by the diverse historical and political economic experiences. There is a developing literature documenting local-specific cases where sustainable development, as an ideology or as a strategy or both, has been utilized as a prescription for relieving countries from the dilemma of economic growth and ecological stability. However, not many have confronted a careful analysis of sustainable development's contradictions, particularly in the Third World where there is an intense desire for economic prosperity, and yet where there is enormous pressure to exploit remaining natural resources. This is probably due to the fact that the dynamic relating development and environment is a very complex one and which struggles to integrate almost two opposing concepts: the preservation of the natural environment that is perceived to have global impact, and the pursuit of economic growth as a requisite for increasing human 'wealth and well-being' (Redclift, 1987; Lewis, 1992). Caught in the middle of this is the issue of social justice, for example 'environmental racism' (Bullard, 1993).

Yet, it has been suggested that, in addressing environmental issues, one must consider an analysis of structural and economic concerns in terms of a political economic approach (Redclift, 1984). Recent studies on state-determined development programmes *vis à vis* environmental protection in Burma

(Bryant, 1995), Malaysia (McDowell, 1989) and Indonesia (Potter, 1987; Peluso, 1993a) present a valid basis for looking into the field level intricacies of sustainable forms of development and thus initiate discussion on the matter. These cases, and previous attempts at exposing and articulating the contradictions of sustainable development, have essentially used 'place-based' or location-specific and 'bottom-up' analytical approaches (Blaikie, 1985). Place-based analysis provides a spatial and geo-physical context, while bottom-up analysis illustrates the specific socio-political and cultural dynamics.

It is in this light that the present chapter explores the application of sustainable development in the Philippines. The sustainable development paradigm and approach have become particularly prominent in the Philippines given the renewed political and economic climate it is now experiencing. Emboldened by the principles of the Rio meeting, the national agenda for economic development of the present government headed by President Ramos and entitled Philippines 2000, defines and articulates strategies that would result in the Philippines becoming a newly-industrialized country and at the same time realizing the sustainable development principles of the 1992 summit. For a country whose economic history had been loyal to the colonial or classical model, the result of which has left the country at the fringes of regional development processes, the Philippines saw the needed shift in development approach that would enable it to achieve economic growth and at the same time preserve its remaining and much depleted natural resource base. One of the government's programmes which expresses this shift and embodies this principle is the Community-Based Forest Management Programme (CBFMP) on which much of the discussion here will be focused. The chapter is divided into four parts. First, a national overview is presented, highlighting conditions that led to the implementation of a sustainable development approach. Then a brief discussion relating the concepts of sustainable development and community-based forest management is presented. This will be followed by a case study of a Community Forest Management Project site in Mati, Claveria, wherein a 'bottom-up' and 'place-based' approach of analysis is utilized to expose field-level learnings. Finally, the discussion concludes by relating these learnings as options for a broader clarification of sustainable development in the Philippines.

NATIONAL OVERVIEW

The historical experience of the Philippines that describes the severe exploitation of its natural resources, institutional disregard for social welfare, and the extreme greed of the privileged few made the task of economic recovery and achieving sustainable development an exceptional one. Exploitation of the Philippines' forests for commercial purposes has been taking place since the Spanish period (Boado, 1988) which left the country with 70–80 per cent forest cover at the end of the nineteenth century (Kummer, 1992).

During the American colonial period, the forest industry flourished as policies were enacted that encouraged further exploitation by large American companies which resulted in an annual deforestation rate of about 140,600 hectares from the 1920s to 1934 (Bautista, 1990) leaving almost 66 per cent of the country covered by forests by the end of the Second World War (Kummer, 1992).

It was, however, the post-war period that saw the rapid increase in forest exploitation activities peaking during the 1960s up to the mid-1970s. It was also the period that experienced immense economic growth as timber production significantly contributed to national revenue. During this period, log production contributed almost 58 per cent of the total share of exports (Bautista, 1990). By the mid-1970s, it was estimated that there were more than 400 timber licence agreements (TLAs) and 'special permits' that were conveniently awarded by the then President Marcos to his relatives, business partners, and military cronies (Vitug, 1993). With some concessions typically covering between 40,000 to 60,000 hectares (ibid.), annual timber production steadily increased and soared to almost 15 million cubic metres in the mid-1970s (Walpole *et al.*, 1993). By allowing these private enterprises to exploit the country's resources and build up the necessary structures, it was hoped by the national government that the revenues from the forest industry would be sourced back into the local economy and thus generate sustained economic growth. By capitalizing on its timber resources, the TLA system was expected to provide the impetus for the country's move towards industrialization (Guiang, 1993).

During its heyday, the Philippines was one of the world's premier single log producers and one of the region's fastest-developing countries. However, it was also the period that experienced great social unrest as the unbalanced distribution of wealth resulted in the glaring division of rich and poor. The economic benefits resulting from the timber industry enjoyed by the political élite resulted in increased migration from lowland to upland areas displacing many indigenous cultural communities (Ganapin, 1987b; Walpole *et al.*, 1993) and creating greater pressure on the forests. Indigenous cultural communities thereby lost their ancestral land to encroaching lowland migrants and to the timber companies simply because they were unable to produce the legal claims to their domain (Ganapin, 1987b). Furthermore, together with migrant upland farmers, they were identified by the government as the main cause of deforestation due to their unsustainable and destructive shifting cultivation or *kaingin* and illegal logging activities (Ganapin, 1987a; 1987b; Boado, 1988; Kummer, 1992). Threatened and abused, many of these people joined the Communist movement which at that time was the only means of coping with, and confronting, the prevailing social injustice. At the twilight of Marcos's political power in the early 1980s, the resources of the forests had been wantonly exploited and the impetus of economic development that was to be derived from them was gone (McDowell, 1989). Further burdened by an enormous international debt, the country lost its foothold on the global market economy and other countries in the region overtook it and achieved growth rates that have left the Philippines struggling on the bottom rung of the economic development ladder. By the end of the 1970s, only a little over 20 per cent of forest cover remained with less than 5 per cent primary old growth forests (Walpole *et al.*, 1993).

The administration of President Corazon Aquino, established in 1986 through a peaceful popular revolt forcing the dictator into exile, struggled to recover from the social, economic and political crisis that the Marcos regime had caused. Armed with popular charisma, a renewed international acceptance, and a sympathetic national agenda, the government was faced with the unenviable task of addressing issues of institutional political corruption, economic instability, absence of effective social services, military disenchantment, and communist and ethnic separatist insurgency. Economic stability was a strategic priority, recognizing the need for an economic agenda that would set the machines of development running again and thus hopefully enabling the government to respond to the prevailing situation of social and economic poverty and debt repayment. This was not to be an easy task as natural disasters that claimed thousands of lives and millions of pesos of destroyed property, damaged power-supplying infrastructure, and caused massive soil erosion, hence constraining agricultural and marine productivity (Ganapin, 1987a; Bautista, 1990), made the need to address environmental issues apparent and immediate. Local community movements by the people of San Fernando (1988) and Gabaldon (1989) exposed the growing concern and awareness of environmental issues. These forms of action, along with the constant power shortage experienced in recent years in urban areas, and the tragedy that took almost 2,000 lives in Ormoc city in 1991 (caused by heavy rains and floodwaters rushing down deforested slopes in nearby watersheds), and the massive landslides in Mindoro resulting from three strong typhoons in 1993 (Environmental Research Division, 1994), constantly reminded the government of the need to integrate socially committed environmental programmes with economic development.

It is in this context that the present community-based forest management approach came into existence as one of the country's landmark 'upland' (defined as land with 18 per cent or steeper slopes) development programmes. The concept embodied the shift from a privately owned production-oriented forest management approach to that of community stewardship and livelihood enhancement. The proper 'people-oriented' management of the environment, particularly the forests, was seen as an expression of the democratic principles of the government in responding to the sustainable development agenda. The Community-Based Forest Management Programme (CBFMP) of the Department of Environment and Natural Resources (DENR), the state agency responsible for the management and protection of the country's natural resources, served as the major element of the government's upland rural development agenda. The relevance of such an approach is emphasized by international concern for environmental issues articulated during the 1992 UNCED Conference. Moreover, there was a willingness among developed countries, through multi- and bi-lateral funding institutions, to support programmes such as the CBFMP in the Philippines that deal with issues such as deforestation and sustainable forest management (Korten, 1994).

PHILIPPINE SUSTAINABLE DEVELOPMENT AND COMMUNITY-BASED FOREST MANAGEMENT

The prominent status of the CBFMP and its linkage with the country's economic growth strategy are indicated by the extent to which it coincides with sustainable development principles, agreed in Rio, and which the Philippines is committed to implement as policy, and the financial support it is receiving from international funding institutions.

One of the steps the Philippine government took after the Rio Summit was to establish the Philippines Council for Sustainable Development (PCSD), one of the first country councils in the world. The council was primarily directed to monitor, conduct policy research, and draft legislation that initi-ates and encourages the Philippines to comply with commitments made at the 1992 UNCED Rio Summit. In order to establish a broad base of partic-ipation and support, the council is composed of representatives from all government agencies, private and business enterprises, civic organizations, and networks of peoples organizations (POs) and non-government organi-zations (NGOs). Through the PCSD, it is hoped that President Ramos's Philippines 2000 development agenda will be further integrated into the various programmes of government.

One of the main initiatives created by sustainable development in the Philippines is the participation and involvement of all sectors of civil society in the development process. To achieve this, decentralization of government functions was implemented and, through the local government code (LGC), devolved into the local government units. It is hoped that the devolution would encourage local government units to take the political initiative, strate-gize the development process, and implement an agenda that is contextually responsive and culturally dynamic. Hence, local government units have become politically powerful entities gaining political legitimacy in deter-mining local development policies and alternatives. Another feature of the sustainable development process is the involvement of the NGOs in many government-initiated programmes and projects. The collaboration between the government agencies and the NGOs reflects a sense of accountability of the government and the greater option for it actively to interact with the civil society (Teves and Lewis, 1993).

Although already existing prior to the establishment of the PCSD, the CBFMP was nevertheless regarded as integral to the development process.

The CBFMP thrust, which espouses a holistic and integrated sense of multiple land-use management and where the objectives of income generation, forest protection, and food production are high on the agenda, substantially applies sustainable development principles. Furthermore, the processes of the CBFMP seek to democratize access to forest resources by involving NGOs and POs and address the prevailing issue of upland rural poverty while at the same time protecting residual forest (Dugan, 1993). This immediately found justification within the sustainable development agenda.

The concept behind this scheme is not new. Known during the 1960s and 1970s as Social Forestry, the programme was implemented primarily as a deterrent to forest destruction caused by upland communities, generally known as *kaingeros*, who practised shifting cultivation and slash and burn farming, or *kaingin*. These people were labelled by the government as the main culprits of deforestation, and as such were identified as 'squatters', 'illegal residents', and 'trespassers' on government-owned land. Upland dwellers were thus looked upon as criminals and the main principle of the Social Forestry programme then was to control their illegal and destructive activities in the uplands and provide privileged security for resource utilization and exploitation over environmental considerations. However, the early 1980s saw the shift of the government's perception towards upland communities from culprits to partners in sustainable forest management. The establishment of programmes such as communal tree farming, family reforestation and forest occupancy management encouraged upland communities to integrate with protection efforts and promote sustainable upland agriculture and agroforestry (Gibbs *et al.*, 1990). Yet, structural constraints and institutional complacency allowed limited success. The state agency assigned to manage the forests, the Bureau of Forestry, still had a considerable number of personnel who believed in top-down 'scientific forestry', and consequently who had little faith in the progressive implications of community-based forest management. Moreover, it became a mechanism for siphoning the allocated reforestation funds for other self-serving interests and corruption (Vitug, 1993). It was during the term of President Aquino (1986 to 1992) that the political climate encouraged a broader and more progressive approach to sustainable forest management. Composed of a number of projects (Integrated Social Forestry Programme, the RP/German Integrated Rain Forest Project, the newly launched OECF-financed Programme, and the Community Forestry Programme), each with its own management strategy and tenurial instrument (Forest Land Management Agreement, the Community Forest Management Agreement, the Certificate of Community Forest Stewardship, and the Industrial Forest Management Agreement), the CBFMP is planned to cover 942,485 hectares for tree plantations and tree farms, 6.5 million hectares for protection, and 200,000 hectares of forests to be managed by the communities (Government of the Philippines: Department of Environment and Natural Resources, 1994). The CFP, launched in 1989, is probably the only upland development project that strongly embodies the basic principles of sustainable development (Vitug, 1993), such as multi-sectoral participation and community-based development initiatives.

The specific objectives of CFP are to provide financially attractive alternatives to swidden activities, to generate capital for investment through sustainable agriculture schemes appropriate for hilly lands, to develop and strengthen community capability to formulate rules while imposing sanctions that promote forest conservation, and to encourage an institutional context that promotes responsive policy development (Dugan, 1993). In a broader perspective, one of the basic assumptions that guide the conceptualization and implementation of the project is that forest destruction and the necessary protection strategy cannot be effective if the concerns of communities that reside in and adjacent to the forests are ignored.

Of the 3.4 million hectares of residual and degraded forests in the country, 1.5 million hectares of these forest lands are allotted for community forest management – in theory targeting the 10 to 18 million upland dwellers as recipients (Vitug, 1993). The programme's process includes the DENR selecting project sites and NGOs. The DENR also, together with the chosen NGOs, prepares and applies strategies that would lead to community organizing and training, continuous consultations, constant monitoring and evaluation, and process documentation. NGO help communities manage funds for 'pump-priming' activities as well as assist with alternative livelihood schemes such as reforestation, timber stand improvement, assisted natural regeneration, and agroforestry (Guiang, 1993). The emphasis and growing importance of NGOs as participating partners with the government in the country's development strategies highlight the continuing democratization process (Teves and Lewis, 1993). The programme has developed such that, today, there are fifty CFP pilot sites all over the country, each of which represents a variety of community forest management experiences. These experiences contribute to the greater understanding of community forest management and may provide valuable lessons integral to a greater understanding of sustainable development.

Another factor that raises hopes for the role of the CBFMP as a viable sustainable development strategy is the amount of support it is receiving from foreign funding institutions through loans and grants. The DENR was implementing loan- and grant-funded projects of United States Assistance for International Development (USAID), Asian Development Bank (ADB), United Nations Development Fund, United Nations Fund for Population Activities, Overseas Economic Contribution Fund (OECF), World Bank (WB), Ford Foundation, and those from the governments of Australia, Germany, New Zealand, and Canada. The DENR, one of the government's leading recipients of foreign assistance, between 1988 and 1992 alone, received a total of US$731 million through just five loans. This was estimated to be ten times the total amount of any previous forestry-related loan funds under the DENR's responsibility (Korten, 1994). In 1988, the forestry sector received about 36 per cent of the total allotted amount of US$240 million from ADB and the Japanese government in the hope that more than 300,000 hectares are reforested each year (Korten, 1994). USAID's contribution through the National Resource Management Programme (NRMP) hopes to enable government to develop sustainable management strategies for 100,000 hectares with each hectare estimated to cost around US$58. Thus, community forest management is a strategy that appeals to external support.

Yet despite the huge sums of money being placed into the programme, the popular political recognition it is getting, and its broad legal basis, disturbing contradictions have emerged putting into question not only some of the quantitative results but also the qualitative impact of the CBFMP, particularly the CFP. The enormous financial support from international funding agencies allowed the DENR fully to pursue its programme, but at the same time it also revealed its poor assessment of the existing environmental crisis, encouraged institutional corruption, and pursued ecologically damaging requirements all in exchange for the generation of foreign revenue (Korten, 1994). It has also been observed that, even though the legal framework of the stewardship programme is in place, there are much-needed adjustments within the structure and the process that are to implement them (Plantilla, 1993). There is also the structural conflict existing between government agencies with regards to programme 'territoriality'. For example, the DENR and the Department of Agrarian Reform (DAR) are still in the process of defining the terms 'upland' and 'lowland' and what activities must be implemented within these areas.

However, these observations reflect a 'top-down' analytical approach which consequently fails to point out the significant nuances in a field context where much of the qualitative lessons are revealed. By discussing the CFP project in Mati, Claveria, in the province of Misamis Oriental it is intended here that the case study will provide a basis of focusing on some of the 'bottom-up' and location-specific implications which have been overlooked. The lessons that emerge will then be analyzed in terms of community-based forest management as it relates to sustainable forms of development.

MATI, CLAVERIA

The municipality of Claveria is located in the province of Misamis Oriental, Mindanao (Figure 15.1). The *barangay* (the smallest geo-political administrative unit) of Mati in Claveria is a small town located 700 metres above sea level and is situated beside large areas of residual forests making it a suitable site for a community-based forest management project. The project site covers only 1,500 hectares of land which was formerly under logging concessions. After the initial processes, the Soil and Water Conservation Foundation Inc. (SWCFI) was chosen by the DENR as the assisting organization and

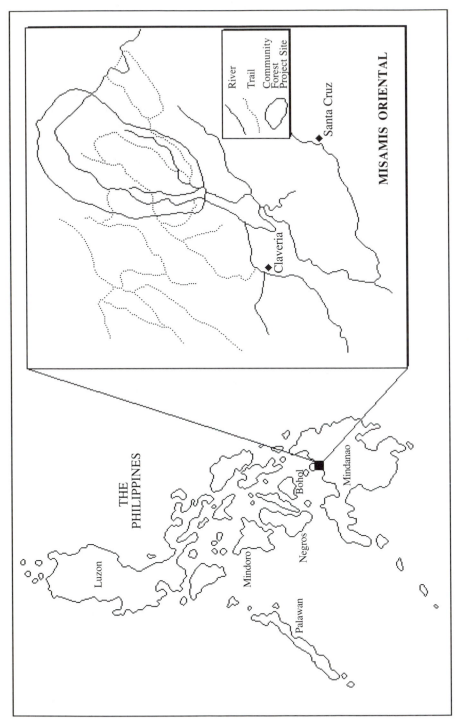

Figure 15.1 The Philippines, showing the municipality of Claveria

contractors of the project in Mati. The SWCFI is a Cebu-based NGO that was formed in 1985 as a response to the national and global challenges of ecological degradation through research, community organizing and environmental management. It was awarded the CFP project with an estimated US$88,000 budget. The project formally began in 1992 and the first activities of the SWCFI were to introduce themselves, establish relations, and discuss the project with the local government unit. The objective of the visit was mainly to inform the local government of their presence, since SWCFI is not known in the area, and gain the political legitimacy to conduct their activities and implement the project.

Rapid Rural Appraisals (RRA) were conducted to introduce the project and to gather information and establish a data base to be used in implementing the CFP. However, the visits to the different communities, the consultations, and the initial RRA results revealed a sense of resistance and indifference from the people. This was attributed to the significant presence of illegal loggers in the area and the community's frustrating experience in the past with government-supported projects. The research also showed that twenty groups on various occasions had pursued a variety of government projects in the area before SWCFI arrived, some of which still exist, such as an Integrated Social Forestry (ISF) project and a number of contract reforestation programmes. Attitudes towards the project changed dramatically after a series of videos about various environmental issues and a film on the 1991 Ormoc tragedy, when about 2,000 of its inhabitants were killed by floodwaters cascading down barren slopes, were shown.

After almost a year into the project's operation, the community finally agreed to participate. In July 1993, a co-operative was formed, the Mati Green Agroforestry Development Association Inc. (MAG-AGDA), which is basically a federation of existing people's organizations (POs) in the area. The co-operative is composed of 147 individuals which includes a number of Higao-nons, the area's indigenous cultural community who had lived in the area long before the influx of lowland migrants during the time of logging operations in the area. With the co-operative gaining local support, a perimeter survey or the delineation of the project site was undertaken. The Community Forest Management Agreement (CFMA), the tenure instrument, was then given and a resource inventory of the site was pursued. As soon as this was accomplished, the SWCFI would then assist the community in drawing up a Community Resource Management Development Plan (CRMDP) for the area. The procedures for a thorough resource inventory provided by the DENR, though, proved too technical for the SWCFI and the community and this inhibited them from developing the CRMDP.

Furthermore, it became evident that some of the community's resource use interests did not coincide with the project's objective. Some members of the co-operative wanted permits to harvest timber and wood to supply stakes for tomato plantations. The mayor of Claveria was interested in utilizing the water resource for irrigating the plantations and for a small-scale power supply. A few had expressed interest in extracting timber from the forests for their housing needs. Upland farmers were worried about the tenurial security of their land perceiving that it might be affected by the project since there was a dispute between the DENR and the DAR over land coverage. Finally, there was very little involvement from the area's indigenous cultural community, the Higao-nons, since the lowland migrants held most of the decision-making positions in the co-operative and since most of the activities centred on enhancing lowland livelihood activities such as agroforestry extension and livestock dispersal.

These local-led developments occurred against a backdrop of broader political change. By 1994, the CFP was five years old and many of the projects were being subjected to intense evaluation and scrutiny by the DENR and the funding agencies. It was also about this time that a strategy for implementation of the programme beyond the pilot level and on a national scale was being entertained by the government. Evaluation teams from both the government and funding institutions visited project sites, evaluated the activities, and developed summary reports. In September 1994, Mati was visited by the Philippine Working Group (PWG). As an alternative to the set and formal methods of evaluation being pursued by government agencies, the PWG, an informal and loose body of scientists, researchers, policy-makers, and government officials, undertook its own approach in assessing CFP with the objective of clarifying issues often overlooked, such as the use of indigenous management practices. It was recognized that to gain knowledge from the community forest management activities, research sites had to be identified that more or less represented the prospects of learning, hence a 'bottom-up' analysis. The framework for assessment is flexible with emphasis on areas of social conflict, organization, traditional forest management, indigenous knowledge, regeneration, cultural biodiversity, watershed perspective, process of spread, and the community's perceptions on policies. Based on this visit and report, general lessons from the experience are derived. Thus, the issues that have emerged concern the following:

1 Core community.
2 Regeneration and the practice of indigenous management knowledge.
3 Multi-sectoral 'interconnectivity'.
4 The influence of local government units.
5 Programme/project integration.
6 Aggregation.

These will be discussed below.

322

Core community

When the PWG discussed the project with the community and the SWCFI during the visit, one of the first concerns was to identify which sector of the community was benefiting the most from the project and if this result coincided with its objectives. Furthermore, the reasons behind the result were sought. From the visit, it was noted that the sector that benefited the most from the project was the lowland agricultural farmers mainly because they organized themselves into relatively influential groups within the community and held key positions in the co-operative. Hence, the project activities that catered to their interests, such as agroforestry extension and livestock and poultry dispersal, were especially prominent. Conversely, the Higao-nons, the cultural community whose history and tradition have long since been linked to the landscape, did not have any significant influence. Indeed, very little research was done to develop a cultural and historical understanding of the Higao-nons and to link them into the process and objectives of the project, notwithstanding the fact that this was one of the original aims of the programme. As a consequence, the Higao-nons have been marginalized in the process and have very little effect on the project's dynamics, and will most likely tend to avoid government-initiated projects that do not recognize their cultural claims on the landscape. This reveals what should be the basic concern for core communities or sectors within the larger community that hold the key to sustainable forest management. Clearly, the Mati case illustrates the need to focus on strategies that locate and identify core communities and utilize their knowledge and cultural management practices to achieve contextually relevant and sustainable forest management. The core community in this case were the Higao-nons, but influence from other, relatively more powerful sectors in the community have minimized their influence. The idea of a core community is emphasized because these communities have had a greater stake in the environment over the years, and their existence rests on the proper management and utilization of its resources. Strategically, they possess the knowledge and management schemes that are more contextually manageable. If the core communities are given greater emphasis and more effort is put into their empowerment, the process of development becomes culturally relevant and economically beneficial.

Regeneration and indigenous management knowledge

Another area of concern that the PWG tried to clarify was the community's strategy for managing the forest. From the discussions held, it was evident that agroforestry was widely implemented. Agroforestry activities basically involved contour farming, using legumes for controlling erosion and maintaining soil nutrients, and crop diversification. However, these activities were observed to have very little effect on the forests. Although there was an acknowledgement of the forest's significance in terms of providing water, wood, and protection from typhoons, very little activity has been planned in terms of its afforestation. Moreover, the forest was mainly recognized as serving an agricultural purpose which did not necessarily coincide with the programme's objectives. The unwillingness of the SWCFI and the co-operative to pursue a thorough resource inventory of the site indicates a lack of recognition of the forest's overall significance. Instead, what has been pursued are land management methods that are heavily based on capital and technical know-how. One of the management procedures for CFP is assisted natural forest regeneration. Very few CFP sites have pursued this strategy mainly because the method has neither been fully understood by the government nor has it been properly implemented. Regeneration is a complex management strategy that takes into consideration a very deep and cultural understanding of forest dynamics. The concept of regeneration, while integrated with traditional agroforestry techniques, has always been a feature of indigenous forest resource management but has not been fully documented and utilized as a popular approach. Thus, identifying factors and indicators that may affect forest growth and spread requires research that takes into account indigenous knowledge. It has been noted by the PWG that management through regeneration and the use of indigenous cultural knowledge may provide the key to a more sustainable, cost-saving, and culturally responsive form of management. This point reiterates the significance of the lack of input of the Higao-nons in the project.

Multi-sectoral interconnectivity

In trying to understand the social dynamics of the area, another concern that was brought up in the report was how the different sectors within the larger community relate to each other and to the project itself. Like many CFP sites, the larger community of Mati is composed of a number of sectors that have particular and not necessarily compatible interests in the utilization of the area's resources. Furniture makers, tomato growers, rice farmers, and the Higao-nons are some of these sectors. The experience of the larger community's initial reluctance to co-operate in the scheme, their indifference and unwillingness to conduct a resource inventory and proper forest management plan for the area, show the need to understand more fully multi-sectoral dynamics and processes that integrate disparate interests with that of the project. The concept of such relations was termed interconnectivity. In Mati, the semblance of interconnectivity was evident through the presence of the co-operative. However, as the PWG discussions progressed, other interests and concerns became evident and it became clear that simply organizing a community does not necessarily ensure shared objectives leading to sustainable forest management.

Local government units

The local government code enabled local governments to facilitate the development process by prioritising development strategies in ways they see as relevant to their context. For example, Mati's town mayor, in wanting to provide a sustainable source of municipal income at the same time as serving the agricultural sector, who comprise a major proportion of his community, and also preserving the nearby forests, wanted to tap the abundant supply of water and charge a minimal fee from those benefiting from it. These funds would then be used as capital for maintaining and protecting the water source, which are the forests around the area. Local projects therefore have to coincide with the local government unit's perception of what and how development strategies apply. The role of local government units was thus recognized by the PWG as essential in the process of implementing community forest management. In Mati, the legitimacy of SWCFI's presence and purpose and their ability to proceed with many of the project's activities reflect the local government's support for the project and the fact that it sees itself as integrative to the process and overall objective of community forest management.

Programme/project integration

The SWCFI was not the only group that wanted to implement a government project in the area and they soon found out that quite a number of projects already had either been implemented or were still being pursued in Mati. It was discovered that various government agencies and other rural development organizations have, at one time or another, implemented projects that catered to specific sectoral interests such as increased tomato production, cash crop diversification, and livestock and poultry raising. Essentially, all of these were designed to generate income and provide the residents with livelihoods. Some of these projects also involved the protection and reforestation of the nearby forests such as the Integrated Social Forestry project (ISF) and the community contract reforestation which were started a couple of years before SWCFI arrived. These projects, however, were discovered to be scattered, isolated and thus lacked coherence. Two things emerge from this observation: that the projects, being isolated, have produced an 'island' effect such as the ISF and the contract reforestation, and projects being implemented one on top of the other in the same area have produced a 'layered' effect such as the livelihood projects, the ISF, and the contract reforestation. Overall both the 'island' and the 'layered' effects in Mati have produced a kaleidoscope of environmental management and livelihood strategies that lacks cohesion and integration. If environmental and forest management is to be pursued on a more sustainable basis and on a broader scale while at the same time contributing to the greater effort on the localities' development processes, there should be efforts to prioritise and link these projects as a coherent whole if it is to achieve a greater and overall significance.

Aggregation

Finally, since the CFP in Mati is one of the many pilot cases being implemented all over the country, one of the concerns of the PWG is the prospect of linking these pilot-level projects into a coherent yet flexible whole. Is there a process which enables these pilot projects to spread themselves naturally, thereby creating a sense of an overall and national programme? Mati's forests are part of a larger forest that stretches into three adjoining provinces. It was realized that if these forests are to be managed sustainably, then each small project must incorporate concerns linked to the protection and management of the adjoining forests and the larger watershed. Thus there is a need to spread the scope and objective of the programme to other communities who live adjacent to or within these forests. The project in Mati is simply that which strictly implements the CFP which may only prove useful for documentation purposes. Discussions with the SWCFI and some of the community members show that there is very little consideration of going beyond the implementation procedures and establishing processes that would contribute to a framework of understanding and linking community forestry activities regionally. Hence, Mati is just one of the many pilot sites that does not see itself as part of a broader process of spread and aggregation. Unfortunately, this comes at a very crucial time when the DENR is already proposing a strategy to implement CFPs nationally but its assessment has remained focused on the macro-adjustments and there is hardly any emphasis on the micro-concerns that determine much of its field-level significance and overall impact. Hence there is a disjuncture between pilot schemes generally and the broader objectives of the programme.

CONCLUSION

After a period of environmental exploitation that enabled the Philippines to achieve a relative level of industrialization, the country is now at a cross-roads of sorts. Faced with vastly degraded forests and a depleted natural resource base, but with a renewed sense of motivation and purpose, the Philippines is struggling to achieve economic growth at the same time as it confronts its difficult socio-cultural, political and environmental issues. It is in this regard that the country is attempting to define its own form of sustainable development. There have been discussions in the Philippines on the relationship between environmental management and sustainable development wherein the former is emphasized as an integral component of the latter. For example, in a draft report assessing the CFP and which proposed a national implementation plan, the DENR highlights the developmental role of the programme in terms of its relations to the country's development agenda, termed Philippines 2000 (Government of the Philippines: Department of Environment and Natural Resources, 1994). This development agenda aimed at restoring a political, economic, and social climate that will encourage self-sustaining growth, thus enabling the government effectively and consistently to respond to national interests. This, and other such reports, however, understand and analyse the issues from a 'top-down' approach and have primarily dealt with macro-understandings. Hence, issues related to dislocation, marginalization, and the response of indigenous cultural communities such as that experienced in Palawan (Lopez, 1987) are typically neglected.

The objective of the chapter was to explore and identify the issues associated with attempts to pursue a 'place-based' and 'bottom-up' approach that would hopefully complement prevailing 'top-down' analytical approaches. Such a scheme enables further clarification and operationalization of community-based forest management as integral to a sustainable form of development. By focusing on a particular community-based forest management project site and pursuing a 'bottom-up' strategy that identifies and highlights field context realities, issues confronting community forest management were highlighted and clarified as alternative issues for local development frameworks. There is a need to focus on core communities such as the indigenous cultural community, the Higao-nons at Mati who have a great stake in the management of the environment but have been marginalized nevertheless in the process. Hence, indicators and mechanisms enabling participation and empowerment must be developed if marginalized communities are to gain from development endeavours. The indigenous and traditional skills and knowledge of those culturally linked to the natural forests or core communities, particularly regeneration, are alternative forms of management that may be more labour-intensive, cost-effective, culturally enriching and effective in the key objectives of promoting sustainable forest management. It is also important that these communities should be seen as relating to a larger community where interests and values are shared and incorporated into the objectives of the programme. Through the local government unit, these interests and values must further be given legal and political justification and worked into the broader framework. As these interests are put into concrete forms of activities, integration and a sense of cohesion among and between the projects must be established to avoid 'island' and 'layered' effects. Finally, a natural process of spread must be put in place to ensure effective forms of aggregating cultural forest management dynamics.

At present the CBFMP cover only about 5 per cent of the estimated needed number of engaged communities, 170 out of 3,250 targeted communities are recipients of some form of CBFMP, covering about 8 per cent of the targeted land area or only 80,000 hectares. Less than 2 per cent of the targeted area is covered under community management agreements

(Government of the Philippines: Department of Environment and Natural Resources, 1994). The concerns discussed here are results of preliminary inquiries which may reflect on the inadequacies of the CBFMP as a sustainable development and should therefore be regarded as requiring further clarification.

Emerging issues that may have to be stressed in future assessments are the implications of donor or funding agency co-ordination with regards to programme and project development, ancestral domain delineation in terms of achieving an integrated environmental management strategy, incorporating indigenous management knowledge systems, the condition and supply of water as an element of producing power, agriculture produce, health, and strategies for incorporating these lessons into the process and structures of policy development. Since 1990, there have been efforts to participate in policy dialogue with much of the emphasis on the issue of commercial timber extraction and log export ban. There has since been a clear effort to bring together community management experiences and the knowledge of forest regeneration. However, these two research concerns have not found a definite link in policy and the effort has been to bring about a development though dialogue. Initial efforts to acknowledge this have been initiated and sustained through the formation of groups such as the PWG. The orientation and purpose of the PWG were to identify issues of community forest management on a field-level and document processes as to how communities are achieving, or are in the process of trying to achieve, forest management. Although the findings of the PWG in assessing CFP would hopefully instigate more flexible and permanent forms of response from involved institutions, these findings also provide lessons that clarify community forest management as a development strategy. What may be significantly required is a methodology that enables government, NGOs and communities to be aware of and relate to the complex cultural and political nuances of the locality. In this regard, techniques such as community mapping hold the potential for stressing the social relations and traditional mechanisms of the community and thus enabling people and institutions properly to manage the forests. Community mapping is a research method being developed and utilized in the Philippines, notably by the Environmental Research Division (ERD) of the Manila Observatory, with other NGOs working in their own research sites. Although maps have been widely accepted as a form of legal document by the government, very little has been done as yet to develop the cultural aspects of mapping. What is being established of communities drawing their perceptions of the area and the forest is a cultural interpretation of their history and culture as represented on the physical landscape. The recent moves towards cultural communities claiming ancestral domain has placed greater significance on the concept and process of community mapping (Environmental Research Division, 1994). By emphasizing these points, it is hoped that options and alternatives are revealed and contribute to a more responsive form of sustainable development where directions are set and enforced by the field realities. It is only under such conditions that the poor and marginalized will be able to take the initiative in defining a sustainable form of development and create mechanisms that culturally determine the process and ensure that the goal of sustainable development in the forestry sector is met in the Philippines.

16

CONCLUSION

Towards sustainable development in South-East Asia?

Michael J.G. Parnwell and Raymond L. Bryant

The overarching impression given by the chapters in this volume is of a sharp imbalance between the forces of economic growth and those of environmental conservation in South-East Asia. As a result, the region has become equally renowned as one of the most dynamic zones of the global economy and as an area where the natural (and living) environment is being rapidly despoiled. The forests are being felled at an alarming rate, the seas have been extensively over-fished, massive fires sweep across the countryside with increasing frequency and intensity, industry and traffic have become a pollution menace, metropolitan areas have almost ground to a halt with congestion, and tourism is threatening to destroy the very resources upon which the industry has been built. Economic growth and environmental degradation, it appears, go hand in hand.

Such a simplistic view underplays considerably the complex and multifarious factors and forces which are contributing to this situation. We hope this volume has succeeded in identifying and unravelling the myriad political, social, cultural and economic ingredients of environmental change in South-East Asia. Environmentally destructive practices have been shown to involve more than the economic pressures of the capitalist system; they are encouraged, facilitated, allowed, tolerated and rued within a socio-political context. People make decisions against nature, not economic systems.

People also create the conditions within which sustainable development can function – not sustainable development as an end-state, for that is little more than an illusion; but sustainable development as an ideology, an initiating process (Drummond and Marsden, 1995, p. 62). We have seen in this volume that the process has been initiated, and the ideology adopted, in most South-East Asian countries. Many people no longer accept, if they ever did, that economic growth and environmental degradation must go hand in hand. Tolerance thresholds have been surpassed, or indeed have been reduced by a growing awareness of the externalities of environmental exploitation. People are protesting against environmental abuse, environmental pressure groups are mushrooming, schoolchildren are learning about the environment, the media is championing environmental causes, and practitioners are trying hard to right environmental wrongs. But – the problem continues and, by most accounts, appears to be growing. Why?

There are several possible answers to this question. The most obvious is that there is a wide discrepancy, and also a time-lag, between awareness and action (Ganapin, 1994, p. 77); between identifying a principle and putting it into practice. Education, the media and, sometimes, brutal first-hand experience (e.g. the devastating floods in southern Thailand in 1988, in which more than 300 people lost their lives) have brought environmental degradation quite sharply into focus. But, as Cochrane and others have shown in this volume, the growing espousal of the virtues of sustainable development typically emanates from people who have scrambled onto a thundering bandwagon, without any clear notion of where it is headed or, indeed, of how to steer it. Sustainability has also become an issue that politicians and parties are hurriedly writing into their speeches and manifestos. But none of this means very much if these words are empty; if the rallying call of 'sustainable development' is mere sloganism, the buzz word of the late twentieth century (Beckerman, 1995).

A second possibility is that not everyone has the awareness which was discussed above; that people are despoiling ecosystems and depleting resources out of ignorance. It is possible to find illustrations to support this notion – the squatter in Bangkok's Klong Toey slum who turfs refuse out of the window, only for it to become a health hazard over time – but in general it seems most implausible. For instance, in some of the peripheral parts of South-East Asia where environmental damage has a direct and lasting impact on people's lives and livelihoods, levels of formal education may be low, but traditional environmental knowledge is in contrast often very strong. Sea nomads, hunter–gatherers and, in many instances, shifting cultivators have evolved livelihood systems which are symbiotically interwoven with their environments. The coincidence of their exploitation practices and environmental degradation is more often than not a consequence of the intrusion of external actors and pressures which have upset the careful balance between people and their environment (Hurst, 1990, p. 270).

A more plausible explanation is that people are aware but do not care enough to act – people, such as the region's urban middle classes, who may espouse the virtues of environmental conservation but whose own deeds fail to match their noble words. At the same time, there are many people who are aware and do care, but are powerless to act. Tobin and White (1993, p. 51), citing the Asian environmental expert Stephen Yong, claim that 'poverty in Indonesia and the Philippines "forces people to exploit beyond sustainable limits the existing resources."' Seen together, these two groups of people probably make up the majority of the region's population.

If we look at the latter group first, we can begin to understand the importance of the 'development' component of the sustainability concept. One point must be made clear – sustainable development is not anti-development. According to Beckerman (1995, p. 2) 'society's proper objective should be to obtain the highest feasible welfare'. Thus, people cannot morally be condemned to live in poverty and destitution in the interests of preserving ecosystems. For example, the illegal encroachment of land-starved farmers into reserved forest – a widespread phenomenon in South-East Asia today (Potter, 1993) – is often (though not always) a desperate survival strategy: there may be no alternative. Encroachers may be acutely aware of the environmental impact of their actions, both on themselves and on others, but their individual short-term needs have to take priority over the longer-term consequences for society as a whole. These squatters and colonizers are routinely pilloried for their actions – indeed, they are often made scapegoats for rainforest degradation – but there are usually other forces at work which explain their predicament: e.g. the concentration of land-holdings, indebtedness, lack of land reform. Until these pressures are eased, the problem will persist.

Thus it has come to be widely accepted that sustainable development cannot effectively proceed without attention to the distribution of the spoils of growth. Pierce (1992, p. 317) has argued that 'the politics of ecology and redistribution are becoming inseparable'. This point is reiterated in the Brundtland Report (World Commission on Environment and Development, 1987, p. 8): '[a] world where poverty is endemic will always be prone to ecological and other catastrophes'. There is clearly a tension in South-East Asia between a requirement to prioritise the needs of the poor as a pre-requisite for sustainable development (Redclift, 1987, p. 36), and the prevailing orthodoxy of growth-driven development and an unhealthy reliance on the 'trickle-down effect'. As such, 'development' will continue to pass people by, and thus pressures will continue to be placed on the environment, however reluctantly and knowingly this may happen. The solution may not be easy, however: Tobin and White (1993, p. 55) ask, 'are there many meaningful policies that improve environmental quality without threatening the *immediate* legitimate economic needs of the poor in [South-East Asia]?'

If we switch now from the needy to the greedy, a very different set of circumstances prevails. In addition to the large number of people who are aware of environmental pressures but do not care enough to act (for example, by leaving their air-conditioned Mercedes at home and taking the bus to work), there are a much smaller number who often deliberately over-exploit resources for personal or corporate gain. These people, who are often drawn from the political, business and military élite, have been very much at the forefront of the analyses contained in this volume (see especially Lohmann, Eccleston and Potter). Their position is aided considerably by the vagueness or fuzziness of the sustainable development concept, and within it a lack of sharpness in separating 'good' from 'bad' ecological practices. People seek to rationalize the harmful environmental impact of their activities by emphasizing the economic and other benefits that these bring to society. By the same measure, and as we have seen in the context of so-called 'reafforestation' with commercial plantations, 'bad' can be called 'good' – the message being amplified through the media and other mechanisms of public relations.

This brings us back to the issue of overlapping or competing sustainabilities which has been referred to by several of the contributors to this volume (see e.g. Rigg and Jerndal, this volume). Ann Usher points to the contradiction involved in meeting the region's spiralling energy demand by promoting ostensibly 'green', renewable hydro-power schemes which, at the same time, undermine the sustainable livelihoods of those who must be resettled, or whose ecosystem will be radically transformed in the process. Colin Sage shows how resettlement in Indonesia offers the prospect of a fresh start for the destitute of Java, but their arrival in transmigration sites often significantly interferes with the livelihood strategies of native populations. Janet Cochrane's assessment of ecotourism is another case in point: the promotion of an environmentally more 'benign' form of tourism development in South-East Asia has, in turn, seriously destabilized local societies and economic systems. A final illustration of competing sustainabilities which is not discussed in this volume concerns the increasingly prevalent practice of fish- and shrimp-farming. At one stage it looked as though aquaculture and mariculture represented a panacea in face of the massive over-exploitation of marine and riverine resources, and a spiralling demand emanating from a booming population. However, it has quickly become apparent that aquaculture has brought with it a host of other ecological and economic problems which ultimately threaten the sustainability of this burgeoning industry and the ecosystems (especially mangrove forests) which have been transformed to support its rapid and largely uncontrolled development (Tobin and White, 1993).

A point which is overlooked in each of the above examples is that none of the outcomes to which we have referred is inevitable. Rather, it tends to happen that way because of the socio-political context within which these processes occur. If this is indeed the case, what are the prospects for a sustainable future? To attempt an answer to this question, we need to understand the dynamics of the nexus between society and environment (Redclift and Benton, 1994), and also the specific context within which their articulation occurs in South-East Asia. There are three scenarios which might prevail, which we shall loosely term 'worst-case', 'middle path' and 'ideal'. We shall explore these briefly below.

Worst-case scenario

Apocalyptic or worst-case scenarios have been suggested since at least the emergence of the 'Limits to Growth' hypothesis which was given substance by the Club of Rome in the early 1970s (Club of Rome, 1972). This hypothesis challenged the assumption of continued growth, and instead advocated moving towards a 'steady state economy'. Environment and development were seen as irreconcilable. Whilst antithetical to the notion of sustainable development because it is seen as anti-development, thereby disallowing the possibility for the poor of the South to play 'catch-up' with the industrialized North, the apocalyptic scenario has provided a convenient clarion call of the environmental movement (Pepper, 1993, p. 244). The ensuing alarmist discourse has tended to place local and national environmental changes within an international context. Thus global warming and the erosion of biodiversity threaten in the longer term to return to haunt the principal perpetrators of environmental degradation, even though in the short term there tends to be a marked separation between source and impact. Patchy evidence of this can already be seen in South-East Asia. Terry King's chapter on the problems and threats of drought and fire in archipelagic South-East Asia is a case in point. Scientific hypotheses and direct human experience suggest an increasing frequency and severity of the El Niño phenomenon, which is at least partly attributed to changes in the global climate to which deforestation allegedly makes a significant contribution.

In general such macro or global conceptions of environmental change have little relevance or meaning to the localized populations of South-East Asia with which this volume has been principally concerned – although, as an aside, it is interesting to report from work by one of the co-editors of this volume among up-river Iban communities in Sarawak (Parnwell, forthcoming(b)) that 'changes in the world's climate', 'wars overseas' (the survey took place shortly after Saddam Hussein had torched the Kuwaiti oil fields), and 'it's God's will' were among explanations proffered for the growing environmental pressures they were experiencing.

The worst-case scenario for South-East Asia would, arguably, involve the unchallenged continuation of present practices and trends. Such reaction and legislation that have occurred hitherto have clearly not been successful in inculcating the fundamental, even radical, changes in attitude and behaviour which are necessary to set South-East Asia on the path towards a sustainable future. There are at least four, closely inter-related grounds for making this seemingly value-laden statement. First, the 'pro-growth' school still has a great many followers. Here, growth is seen as essential to create the means with which to ameliorate the effects of human activity on the environment and on stocks of natural resources. Drummond and Marsden (1995, p. 53) summarize the neo-liberal position: 'the best way to provide for future generations is to exploit resources, not conserve them. Market forces and human ingenuity will always take care of shortages by providing solutions which will leave us better off than we were before.' This perspective puts faith in the market mechanism and the advancement and transfer of technology, and upon human reaction to environmental pressures to create the circumstances for change. The prevailing orthodoxy thus still places economic growth as the highest priority on the political agenda. However, the underlying philosophy in South-East Asia appears to be one of 'grow now, clean up later'. There is no compelling evidence yet of a shift in priorities towards a greater balance between growth and conservation, or a greater quality of growth. This view is reinforced by Pernia (1991, pp. 114–15; p. 130: our emphasis):

> Although perhaps existing in name, a real and effective environmental policy has been largely wanting in developing Asian countries in general. In Southeast Asia, environmental policy is just beginning to evolve in the official and popular consciousness and may take some time to have an impact, *unless a determined political will quickly evolves from the top leadership*. [. . .] Misunderstanding, lack of foresight and selfish interests are commonly at the root of the lack of resolve in, or opposition to, the formulation of appropriate environmental policies.

Second, and linked to the above, is the global view which many South-East Asian leaders (most notably the Malaysian Prime Minister, Dr Mahathir Mohammed) quite understandably take: that the world's environmental problems are principally attributable to over-consumption in the North, and that the South should not be prevented from mobilizing its resources in support of development by being made a scapegoat for the world's environmental ills (see the chapter by Eccleston and Potter). Whether this is a reasonable justification for a lack of speedy and decisive action within the region is hotly disputed by the pro-environment lobby both within and outside South-East Asia. The impacts of environmental abuse are local, national and regional just as much as they are global. Meanwhile, this debate only serves to slow the pace of action, and also provides a convenient smoke-screen for the continued rapacious exploitation of resources and the despoliation of the natural environment.

Third, there are few signs that the powerful and influential forces behind resource exploitation and environmental degradation are willing to yield ground – a point which is reinforced by Hurst (1990, p. 275) in the context of the native dwellers of South-East Asia's forests:

> The strongest opposition to tribal land rights has come from politicians and businessmen involved in the timber and mining industries. They frequently claim that the forests belong to the nation as a whole, not just those who live in them; . . . and that development for people in these remote areas can be provided only by exploiting the forests for timber.

As several of the contributors have shown (see especially Lohmann), there are formidable agents and agencies at work in South-East Asia – Thailand, Indonesia and Malaysia in particular – who use their power and authority to grant access to the region's resource base, and who in turn use the granting of such concessions as a means of underpinning their position of power (see e.g. for Thailand: Project for Ecological Recovery, 1992, p. xi). Until such power can be reduced or, as in Singapore, channelled directly towards policies of redistribution and conservation, the prospects for sustainable development are very slight indeed.

Finally, and also linked to the above, there is little evidence that the requirement of redistribution as a precursor to sustainable development is being met. This is partly because the powerful are reluctant to cede power and the full spoils of economic growth to the less privileged echelons of society; partly because of the inherent contradictions of market-led development; and partly because the socio-political barriers to mechanisms of redistribution in South-East Asia are so deeply entrenched. Thus it seems inevitable that the pressures of poverty, married to rapidly rising levels of expectation among the better-off, will continue to form a dual front of pressure on the natural environment, at least into the foreseeable future.

There are innumerable 'worst-case' scenarios for South-East Asia which we might consider, some clearly more fanciful than others, but because of constraints of space we will discuss just one at this juncture: the phenomenon of environmentally induced population movement. This, we argue, represents a potentially very significant future scenario (indeed, globally as well as regionally) which to date has largely been overlooked. The scenario is as follows: as environments become increasingly despoiled and stocks of resources depleted, growing numbers of people will be forced or will choose to move from their places of normal residence in order to preserve their means of livelihood. Movement will typically occur within nation states, but over time, pressures for international movement may also increase. Either way, environmentally induced population movement will almost inevitably contribute to mounting social and political tensions, the spreading rather than amelioration of poverty, and will at the same time contribute to the further intensification of environmental and resources pressures in the main receiving areas.

Lonergan and Parnwell (1995) argue that environmental degradation has a much greater influence on population displacement than is generally realized. Attention hitherto has mainly focused on 'environmental refugees' (e.g. El-Hinnawi, 1985; Westing, 1992; Bartelmus, 1994, p. 19): people displaced by environmental disasters, who globally may number more than 10 million where movements occur across international borders (Jacobsen, 1988), or more than 25 million if internally displaced persons are also included (Myers, 1993). Fortunately, this is still a relatively rare phenomenon in South-East Asia although, as has been shown in the case of the Mount Pinatubo eruption in 1991 (Brookfield and Byron, 1993), growing numbers of people are under threat from natural disasters, not so much because of the growing severity of such disasters but because land and other pressures are forcing increasing numbers of people to encroach into marginal and disaster-prone areas. The same also applies to areas which are prone to forest fires and sea-level rise.

There is also a growing realization that gradual, cumulative environmental changes are contributing to population displacement on a scale which may be several times greater than that caused by natural disasters (Ganapin, 1994). This is certainly the case in South-East Asia, where the degradation particularly of rainforest, coastal and upland ecosystems is causing the displacement of increasing numbers of people, even though they may appear to retain a modicum of choice about whether to leave or remain. Almost invisible in the literature hitherto has been the recognition that environmental degradation may also play an important contributory role in the ostensibly voluntary migration of people from their home areas (Lonergan and Parnwell, 1995). In other words, the presently huge and increasing volume of migration between rural areas and towards the principal cities of South-East Asia may, in many instances, have been partly influenced by environmental pressures. Whilst it clearly becomes extremely difficult to isolate environmental from other stimuli behind voluntary migration, there is little doubt that the livelihood consequences of soil erosion, resource depletion, climatic change, etc., contribute to the mobility decisions of a large and growing number of people in South-East Asia today.

As environmental pressures mount within the *laissez-faire* scenario, the pressures for population movement also rise considerably. In recent history this has not been too significant, as there has generally been somewhere for people to move to, however marginal. However, with the closure of the land frontier in most South-East Asian countries, alternative loci, particularly for displaced agriculturalists, are rapidly and increasingly becoming scarce. This gives rise to a variety of scenarios, some of which can already be observed in the region today. For instance, and as several chapters in this volume have shown (e.g. Parnwell and Taylor; Lang; Rigg and Jerndal), increasing economic and demographic pressures in the region's remaining forested zones are creating resource and environmental tensions which in turn are pushing lowland and upland peoples alike deeper into the interiors of their respective nation states (see also Ewers, 1994; Heyzer, 1995). In some instances the new arrivals have encroached into the territory of indigenous communities, leading simultaneously to social conflict (which may ultimately threaten internal security) and further environmental pressure, contributing to a cumulative process of resource depletion and population displacement. Where the scope for coping with these pressures within a nation's territory disappears, people may move across national boundaries in an effort to find land and resources to utilize. Such a form of movement is most likely between the peripheries of neighbouring South-East Asian states: e.g. from East Malaysia into Kalimantan and Brunei; Thailand into Burma and Cambodia; the Philippines into Sabah; and so on. This too could become an important source of future international tension in the region, in much the same way that the encroachment of national fishing fleets into the territorial waters of neighbours is already causing tensions between neighbouring states (e.g. between Thailand and both Cambodia and Malaysia).

A parallel scenario is that environmental pressures and their economic and social consequences will force or encourage increasing numbers of people to move towards the region's metropoli. Given that the cities are already showing severe signs of environmental stress as a result of their over-rapid and largely uncontrolled expansion (Pernia, 1991), it is difficult to envisage what the consequences of further rapid urbanization might be, environmentally, socially, economically and, especially, politically. Pressures for movement also emanate from within the city, as the urban middle classes fuel processes of extended metropolitanization (McGee and Greenberg, 1992) and counterurbanization as they seek to escape the polluted and congested urban environment. In the process, the negative environmental effects of urbanization are simply spread further afield.

We can see, therefore, that the costs and consequences of inaction (our worst-case scenario) may be enormous, even catastrophic. It is this which makes the scenario rather unlikely to prevail: as pressures mount and awareness increasingly dawns, people can be expected to react (if only by increasing their tolerance thresholds), and both preventative and remedial measures will be introduced. The 1988 floods in Southern Thailand, mentioned earlier, are a case in point: politicians and officials were suddenly awoken from their complacency by media images of floodwaters gushing through a southern Thai town, where the logging of the upper catchment had also removed the capacity of the natural vegetation to absorb the heavy rains. These shocking images, together with mounting political pressure from the environmental movement in Thailand at the time, combined to force a reaction from those in authority. Shortly afterwards a ban on logging activity was announced (Thailand: Project for Ecological Recovery, 1992).

Ideal scenario

Our 'ideal' (some would say idealistic) or best-case scenario is also rather unlikely, but for very different reasons (and not least because there will never be a consensus on the shape or condition of the ideal society). In this scenario, the antithesis of orthodox development prevails and society obtains the 'highest feasible welfare' without unacceptable cost to the environment and stock of resources. This matches the vision of 'Ecotopia' which is described by David Pepper (1984, p. 206):

'Ecotopia' . . . has decentralised production, where government control over the population is small, where the people at large decide on what technology will be developed and where workers own the means of production, and the direct investment of capital in an enterprise by absentees is not permitted.

This scenario also overlaps with the vision for sustainable development which has been described in the context of the Philippines (see Chapter 15), where local participation and local decision-making take place within a full democratic context where power and control are not usurped by others, where externalities are fully costed in the production process, where the needs of the poor and destitute are addressed and prioritized, and where altruism and concern prevail at the expense of greed and self-interest.

Thus, it should be clear that progress towards sustainable development requires more than the meshing of economic and ecological needs. As Cooper and Palmer show (1992, p. 185), a crucial ingredient of sustainable development is a change in values and expectations:

> If sustainable development is to be achieved, then the necessary fundamental changes in and modifications of agriculture, energy, forestry and other physical and industrial systems cannot stand alone. Alongside these changes must be a corresponding shift in attitudes and values – in the social, economic, political and moral aspects of human life. Development for a sustainable future must be as much about shifting values as it is about shifting practices.

It is this dual requirement of fundamental and superficial change which makes the best-case scenario very unrealistic: humankind does not have an impressive track-record of readily giving up what has been gained, or of putting the wider community's interests ahead of the individual's: capitalism engenders competition. Several chapters in this volume have highlighted how deep-seated are the political and social barriers in the way of a fundamental challenge to the *status quo*. Hurst (1990, p. 280) has shown how difficult it is for those groups who have sought to promote this challenge:

> Most Third World pressure groups must be cautious about any public statements they make regarding development issues; few governments will tolerate repeated criticism. Government authorities . . . accuse conservation groups in Asia of being anti-development and working against the national interest.

Gilbert Braganza's analysis has also shown that, even where the principle of popular participation is espoused from above, the impediments to its effective implementation are quite formidable. Furthermore, South-East Asian countries do not stand alone in the world: they are, to a greater or lesser extent, integral components of the global economy, and as such have compromised a certain degree of their autonomy in making decisions about development and environmental conservation within their national territories. Without a significant sea-change in practices and priorities at the global level, the room for manoeuvre within South-East Asia will continue to be restricted, however strong the political and popular will therewithin. Owen Cameron's view of Japanese economic and environmental policy, and his concluding remark that Japan's highest priority is to its own economic development, is symptomatic of this situation. Similarly, the way in which the Nordic countries were shown by Ann Usher, and also Jonathan Rigg and Randi Jerndal, to be using aid disbursements to further their own economic interests is a further case in point.

The middle path

Finally, we move briefly to consider the 'middle path' or compromise scenario, which we believe provides the most realistic foundation for sustainable development in South-East Asia. This scenario sees a gradual modification of directions and practices, where levels of environmental awareness steadily increase (through education and the media, and a growing social conscience); where legislation steadily improves and its implementation becomes more effective; where greater social responsibility is inculcated; where government, non-governmental organizations and business work in concert towards an agreed agenda rather than in opposition; where redistribution is given a higher priority; and where there is a genuine fusion of bottom-up and top-down initiatives. This is necessarily paralleled by shifting global priorities, attitudes and behaviour.

All of the above are already taking place albeit on a modest and variable scale in South-East Asia today, principally because awareness, concern and action are all increasing. The need now is for the momentum to build and for this mode of behaviour to be extended. An important barrier in this regard lies in the fact that the sustainable development message frequently falls on deaf ears. It is not in the interests of the perpetrators of environmental damage to listen. Meanwhile, the green message elsewhere commonly takes the form of 'preaching to the converted'. Those who are inclined to listen attentively to debates about the environment are also those who are most likely already to have changed their attitudes and behaviour. For this reason, education holds an important key, as Alan Dykes has stressed in Chapter 14, reiterating Schumacher's view (1974, p. 64) that 'education is the greatest resource'. This view is reinforced by David Pepper (1984, p. 215):

> [E]ducation can provide the panacea for our ills; whether we think that if only people knew the 'facts' of environmental abuse such abuse would stop, or whether we believe . . . that a complete change in *values* must accompany total reorganisation of society along more ecologically sound lines. Schumacher argued that education is the 'greatest resource' because it maintains and strengthens not only human dignity, initiative and constructive activity but also regard for nature.

We round off this discussion with a brief exploration of one of several possible models around which 'middle path' strategies for sustainable development in South-East Asia might be built: one which combines elements of both idealism and pragmatism. The illustration considers the potential role to be played by rural industrialization in addressing simultaneously the requirements of redistribution and ecologically tuned production. We focus on the case of Thailand because the Kingdom exhibits some of the region's most striking signs of spatial disparity and environmental stress (see Parnwell, 1994 for a fuller exploratory discussion of this notion).

Thailand is the epitome of uneven development. Bangkok is an example of the primate city *par excellence*, and there is also an overwhelming concentration of industrial activity in and around the capital city (see Parnwell, 1995). The consequences of over-urbanization are widely apparent: massive congestion, heavy pollution and a generally poor living environment. Meanwhile, Greater Bangkok's role as the country's economic dynamo does little to dissuade huge numbers of people from moving to the city is search of employment and other opportunities: a situation which is clearly not helped by the flip-side of uneven development, such as relative economic stagnation and a lack of opportunity in Thailand's countryside and periphery. As spatial inequalities widen through a lack of effective policy intervention, a cumulative process of migration, rural stagnation and urban concentration is set in train. The consequences of this for both the capital city and the countryside are potentially very serious. Bangkok is rapidly engulfing its hinterland region, creating urbanized corridors which stretch more than 100 kilometres from Rayong in the east to Ayutthaya and Nakhon Pathom to the north and west respectively. With the phenomenon of region-based urbanization (see McGee and Greenberg, 1992), the nuisance of industrial pollution and urban congestion is spread to a rapidly widening area. At the same time, the human resources upon which future rural development initiatives will depend are steadily being drained from the countryside by strongly perceived differentials in opportunities and prospects.

Rural industrialization offers a means whereby these urban and rural stresses might both potentially be alleviated. However, we are not proposing an idealistic notion which sees the city closed down and all forms of industrial activity moved to the countryside (see Friedmann and Douglass, 1978), or where the impressive momentum of economic growth is halted in the pursuit of distributive justice. Rather, we proffer a pragmatic approach which seeks to draw the countryside more fully and effectively into the economic frame, in the process helping to enhance comparative advantage, alleviate the pressures on the capital city and contribute to the development and diversification of the rural sector (see Parnwell and Khamanarong, 1990, for a fuller discussion). A parallel objective is to take more work to where it is needed, rather than relying excessively on people migrating to where the work is (or is perceived to be).

Rural industrialization entails the facilitation of industrial production within the rural setting, and may involve both the building up of certain forms of industrial activity in which rural areas have traditionally had a long involvement (e.g. handicrafts, textiles, ceramics), and also the importation of various forms of manufacturing activity (or, through the putting-out system, discreet parts of the production process) which are more typically associated with urban areas (in the Thai case, diamond- and gemstone cutting is a classic illustration: see Parnwell and Khamanarong, 1990). Encouraging a greater spatial spread in the pattern of industrial activity has been shown to enhance, not undermine, the cost competitiveness of industrialization. At the same time, it has brought considerable economic benefits to those rural areas which have become involved. Thus, the distributional requirement of the sustainable development principle is gradually satisfied by this model; what of the environment?

The prospects of a technological fix for the environmental problems associated with the heavy concentration of industrial activities in the Extended Bangkok Metropolitan Region are real, but may not realistically be expected to solve the problem of industrial pollution in the short to medium term. The socio-political and institutional barriers to 'quick-fix' solutions are immense, as we have seen throughout this volume. At the same time, a technological fix by itself would tend to act against prospects for a greater spread in the pattern of industrial activity. Whilst this may help to underpin economic efficiency, it would do little to reduce the continuing attraction of the city to further waves of migration, with all its attendant consequences.

The rural industrial sector, by contrast, has traditionally been host to forms of industrial production which in some, but not all, cases have a relatively benign environmental impact in terms of their use of raw materials and energy, and the forms of pollution which result. They are typically small in scale, employ labour-intensive production methods and often rudimentary levels of technology, and use locally available resources, many of which generate biodegradable wastes (see Kaplinsky, 1990). As such, they can be considered to be more in tune with the limits of the environment than many of the large-scale, capital-intensive and high-technology industries which have been sprouting up in the capital city. Research in Central and North-East Thailand has identified a wide variety of rural industries which fit the stylized description above, the development of which has also contributed significantly to local economic dynamism (see Parnwell, 1994). One illustration will have to suffice at this juncture.

The cushion-making industry, centred on the village of Ban Sri Than, Paa Tiw District in the Northeastern Thai province of Yasothon, provides evidence of an environmentally benign form of rural industrial activity, determined largely by the scale and technological simplicity of the manufacturing process. The production of traditional triangular cushions, which are in heavy demand from the tourist and handicraft export market as well as the domestic Thai market, uses locally available materials: cloth is woven locally, and the kapok which is used to stuff the cushions is also available from locally planted trees.

The cushions are almost entirely hand-made, and there is very little energy requirement. Waste is minimal and entirely biodegradable. Importantly, the industry has also been able to expand rapidly over the last fifteen years, in response to soaring market demand, without altering significantly the balance between production and environment. Local shortages of kapok have been met in innovative ways: straw, locally abundant after the rice harvest, has been used as a partial substitute for kapok in the stuffing of cushions; more intriguingly, some villagers now specialize in the recycling of kapok, visiting Bangkok and other urban centres and relieving people of their worn out or discarded mattresses and cushions, from which the kapok is extracted and brought back to the North-East. The modernization and development of this village industry have brought considerable benefits to the local population, and have also played a significant role in slowing the pace of out-migration towards Bangkok.

In line with our 'middle path' scenario, we thus emphasize the potential, and often overlooked, contribution that the rural industrial sector can make to sustainable development – but only up to a point. Notwithstanding the above example, the modernization of rural industrial enterprise brings the attendant risk of extending, and indeed intensifying resource and environmental pressures in areas which are much less well prepared to cope with them than urban areas, with their better provision of infrastructure, and greater availability of technological and managerial means of dealing with environmental externalities. Hence rural industrialization is not presented as a panacea, but as a pragmatic means of dealing simultaneously with the pressures for continued rapid economic growth and the need to match production with the needs and potential of the natural environment. As such, we argue that it provides a model, warranting closer attention than it has received hitherto, as a step along the middle path towards a sustainable future.

It is the fusion of idealism and pragmatism which provides the best foundation for sustainable development in South-East Asia. Without a clear vision towards which to aim, sustainable development will be nothing more than an illusion. Without a proper understanding of context and complexity, constraints and competing concerns, no amount of goodwill and good intention will challenge successfully the *status quo*.

BIBLIOGRAPHY

Abdoellah, O.S. (1987) 'Comparative "adaptive strategies" of transmigrants in Indonesia: a case study in Barambai, South Kalimantan', in T. Suzuki and R. Ohtsuka, (eds), *Human Ecology of Health and Survival in Asia and the South Pacific*, Tokyo: University of Tokyo Press.

Ackerman, E.A. (1953) *Japan's Natural Resources*, Chicago: University of Chicago Press.

Adams, P. (1991) *Odious Debts*, Toronto: Earthscan.

Adams, W.M. (1990) *Green Development: Environment and Sustainability in the Third World*, London: Routledge.

Adas, M. (1974) *The Burma Delta: Economic Development and Social Change on an Asian Rice Frontier, 1852–1941*, Madison: University of Wisconsin Press.

Adas, M. (1983) 'Colonization, commercial agriculture, and the destruction of the deltaic rainforests of British Burma in the late nineteenth century', in R.P. Tucker and J.F. Richards (eds), *Global Deforestation and the Nineteenth Century World Economy*, Durham: Duke University Press, pp. 95–110.

Agreement on Cooperation for the Sustainable Development of the Mekong River Basin, (1995) Chiang Rai.

Aiken, S.R. and Leigh, C.H. (1992) *Vanishing Rain Forests: The Ecological Transition in Malaysia*, Oxford: Clarendon Press.

Allen, J. (1992) 'Aspirations – and the reality', *Paper*, 4 February.

Allen, M.G. (1979) 'The Ulu Temburong expedition of 1978', *Brunei Museum Journal*, 4, 3, pp. 107–81.

Amad, W.Y.W. (1988) *Application of Landsat/SPOT Digital and Visual Analysis as a Tool for Forest Classification and Mapping in Lesong Forest Reserve, Peninsular Malaysia*, Kuala Lumpur: ASEAN Institute of Forest Management, Fellowship Report.

Anderson, B. (1990) *Language and Power: Exploring Political Cultures in Indonesia*, Ithaca: Cornell University Press.

Anderson, J. and Marsden, D. (1983) 'The Temburong renewable resources study', unpublished report, Bandar Seri Begawan: Forestry Department of Brunei Darussalam.

Anderson, J. and Marsden, D. (Forestry Consultants) Ltd. (1984) *Brunei Forest Resources and Strategic Planning Study*, unpublished report to the Government of Brunei Darussalam.

Andreasson, A. and Markgren, I. (1993) *Sustainable Use of the Laotian Forest Resources*, Gothenburg: Gothenburg University.

Annual Statistical Bulletin, Kuching, Department of Statistics, various issues.

Arase, D. (1994) 'Public–private sector interest coordination in Japan's ODA', *Pacific Affairs*, 67, 2, pp. 171–99.

Armitage, I. (1988) *Management of the Natural Forests, Technical Report No. 7, Forestry Sector Review Vietnam*, VIE/88/037, FAO/UNDP/MoF.

Armour, A.J.L. (1985) *Asia and Japan*, Dover, NH: The Athlone Press.

Arnold, J.E.M. (1993) 'Nonfarm employment in small-scale forest-based enterprises: policy and environmental issues', Working Paper No. 12, Oxford: Oxford Forestry Institute.

Ashton, P.S. (1964a) 'Ecological studies in the Mixed Dipterocarp forests of Brunei state', *Oxford Forestry Memoirs*, 25, Oxford: Clarendon Press.

Ashton, P.S. (1964b) *Manual of Dipterocarp Trees of Brunei State*, Oxford: Clarendon Press.

Asian Development Bank (ADB) (1993) *Indonesia: Private Sector Industrial Tree Plantations Program: Report of a Technical Assistance Consulting Team*, Manila: ADB.

Asian Development Bank (1994) *Summary Environmental Impact Assessment (SEIA) for the Proposed Theun Hinboun Power Project (Lao People's Democratic Republic)*, Manila: Asian Development Bank.

Asian Wall Street Journal, various issues.

Avé, J. and King, V.T. (1986) *Borneo: The People of the Weeping Forest*, Leiden: Rijksmuseum voor Volkenkunde.

Bailey, C., Dwiponggo, A. and Marahudin, F. (1987) 'Indonesian marine capture fisheries', ICLARM Studies and Reviews 10, Manila: International Centre for Living Aquatic Resource Management (ICLARM)/Jakarta: Directorate General of Fisheries and Marine Fisheries Research Institute, Department of Agriculture.

Baird, I. (1994) 'Another Franklin Dam in Laos?', unpublished paper, Vientiane.

Bangkok Bank Monthly Review (1994) 'Regional economic performance: the Northeast', 35, 2, pp. (19–27.

Bangkok Post (1994) 'Slash-and-burn', (Inside Indochina Supplement), 15 March.

Bangkok Post, various issues.

Barbier, E.B. (1993) 'Economic aspects of tropical deforestation in Southeast Asia', *Global Ecology and Biogeography Letters*, 3, pp. 215–34.

Barley, N. (1990) 'Tourist herds moving across the world', *The Independent*, 19 May 1990.

Barnett, A. (1992) *Deserts of Trees: The Environmental and Social Impacts of Large-Scale Tropical Reforestation in Response to Global Climate Change*, London: Friends of the Earth.

Barraclough, S. (1984) 'Political participation and its regulation in Malaysia', *Pacific Affairs*, 57, pp. 450–61.

Bartelmus, P. (1994) *Environment, Growth and Development: The Concepts and Strategies of Sustainability*, London: Routledge.

Bartu, F. (1993) *The Ugly Japanese: Nippon's Economic Empire in Asia*, Tokyo: Charles E. Tuttle.

Bautista, G.M. (1990) 'The forestry crisis in the Philippines: nature, causes and issues', *The Developing Economies*, 28, March, pp. 67–94.

Beaman, R.S., Beaman, J.H., Marsh, C.W. and Woods, P.V. (1985) 'Drought and forest fires in Sabah in 1983', *Sabah Society Journal, Kota Kinabulu*, 8, 1, pp. 10–29.

Beard, D. (1994) 'Keynote address to the International Commission on Large Dams (ICOLD)', Durban, South Africa, 9 November 1994.

Bebbington, A. (1994) 'Composing rural livelihoods: from farming systems to food systems', in I. Scoones and J. Thompson (eds), *Beyond Farmer First: Rural People's Knowledge, Agricultural Research and Extension Practice*, London: Intermediate Technology Publications.

Beckerman, W. (1995) *Small is Stupid: Blowing the Whistle on the Greens*, London: Duckworth.

de Beer, J.H. and McDermott, M.J. (1989) *The Economic Value of Non-Timber Forest Products in Southeast Asia (with emphasis on Indonesia, Malaysia and Thailand)*, Amsterdam: World Wildlife Fund.

Begley, S. (1989) 'The world's eco-outlaw?', *Newsweek*, 1 May, p. 70.

Belcher, M. and Gennino, A. (eds) (1993) *Southeast Asia Rainforests: A Resource Guide and Directory*, San Francisco: Rainforest Action Network in Cooperation with the World Rainforest Movement.

Beresford, M. and Fraser, L. (1992) 'Political economy of the environment in Vietnam', *Journal of Contemporary Asia*, 22, pp. 3–19.

Berkmüller, K. (1995) 'Hydropower Development and Protected Areas', unpublished paper, Vientiane.

Berma, M., forthcoming, 'The commercialisation of rural Iban handicrafts in Sarawak, Malaysia: problems and potential', PhD thesis, University of Hull, Centre for South-East Asian Studies.

Bernhard of the Netherlands, Prince (1991) 'WWF to assist travel industry promoting ecotourism', opening address to Pacific Asia Travel Association, April 1991 (*Jakarta Post*, 13.4.91).

Besant-Jones, J. (1994) 'A view of multilateral financing from a funding agency' in Financing Hydro Power Projects '94, sponsored by International Water Power and Dam Construction, 22–23 September 1994, Frankfurt.

Biodiversity Action Plan Planning Team (1993) *Biodiversity Prospectus of Vietnam*, Hanoi: Biodiversity Action Plan Planning Team, VIE/91/G31, October 1993.

Blaikie, P. (1985) *The Political Economy of Soil Erosion in Developing Countries*, London: Longman.

Blaikie, P. and Brookfield, H. (1987) *Land Degradation and Society*, London: Routledge.

Boado, E. (1988) 'Incentive policies and forest use in the Philippines', in P. Repetto and M. Gillis (eds) *Public Policies and the Misuse of Forest Resources*, Cambridge: Cambridge University Press, pp. 165–203.

Boo, E. (1990) *Ecotourism: The Potentials and Pitfalls*, Washington DC: WWF.

Brandon, K.E. and Wells, M. (1992) 'Planning for people and parks: design dilemmas', *World Development*, 20, 4, pp. 557–70.

Brazil, M. (1992) 'The wildlife of Japan', *Japan Quarterly*, 39, 3, pp. 328–38.

Bresnan, J. (1993) *Managing Indonesia: The Modern Political Economy*, New York: Columbia University Press.

Broad, R. with Cavanagh, J. (1993) *Plundering Paradise: The Struggle for the Environment in the Philippines*, Berkeley: University of California Press.

Brondijk, J.F. (1963) 'Reclassification of part of the Setap Shale Formation as the Temburong formation', *British Borneo Geological Survey Annual Report*, 1962, pp. 56–60.

Brookfield, H. and Byron, Y. (1990) 'Deforestation and timber extraction in Borneo and the Malay Peninsula – the record since 1965', *Global Environmental Change*, 1, 1, pp. 42–56.

Brookfield, H. and Byron, Y. (eds) (1993) *South-East Asia's Environmental Future: The Search for Sustainability*, Kuala Lumpur: Oxford University Press, with United Nations University Press.

Brosius, J.P. (1986) 'River, forest and mountain: the Penan Gang landscape', *Sarawak Museum Journal*, XXXVI, 57 n.s., pp. 173–84.

Brosius, J.P. (1991) 'Foraging in tropical rain forests: the case of the Penan of Sarawak, East Malaysia (Borneo)', *Human Ecology*, 19, 2, pp. 123–49.

Bruenig, E.F. (1969) 'On the seasonality of droughts in the lowlands of Sarawak (Borneo)', *Erdkunde*, 23, pp. 127–33.

Bruenig, E.F. (1993) 'Integrated and multi-sectoral approaches to achieve sustainability of ecosystem development: the Sarawak forestry case', *Global Ecology and Biogeography Letters*, 3, pp. 253–66.

Brunton, B.D. (1990) *Critique of the World Bank's TFAP Review for Papua New Guinea*, Working Paper no. 26, Law Reform Commission of PNG, March 1990.

Bruun, O. and Kalland, A. (1995a) 'Images of nature: an introduction to the study of man–environment relations in Asia', in O. Bruun and A. Kalland (eds), *Asian Perceptions of Nature: A Critical Approach*, London: Curzon Press, pp. 1–24.

Bruun, O. and Kalland, A. (eds) (1995b) *Asian Perceptions of Nature: A Critical Approach*, London: Curzon Press, pp. 189–214.

Bryant, R.L. (1992) 'Political ecology: an emerging research agenda in Third-World studies', *Political Geography*, 11, pp. 12–36.

Bryant, R.L. (1994) 'From laissez-faire to scientific forestry: forest management in early colonial Burma, 1826–85', *Forest and Conservation History*, 38, pp. 160–70.

Bryant, R.L. (1995) 'The politics of forestry in Burma', in P. Hirsch and C. Warren (eds), *Politics of Resources and Environment in Southeast Asia*.

Bryant, R.L., Rigg, J. and Stott, P. (eds) (1993) 'The political ecology of Southeast Asian forests: transdisciplinary discourses', *Global Ecology and Biogeography Letters*, Special issue, 3 (4–6), pp. 101–303.

Bullard, R.D. (ed.) (1993) *Confronting Environmental Racism*, Boston Massachusetts: South End Press.

Burgers, P.P.M. (1992) 'Rainforest and rural economy', Utrecht: Faculty of Geography, unpublished paper.

Burrough, P.A. (1986) *Principles of Geographical Information Systems for Land Resources Assessment*, Oxford: Clarendon Press.

Butler, R.W. (1991) 'Tourism, environment and sustainable development', *Environmental Conservation*, 18, 3, pp. 201–9.

Caldecott, J.O. (1988) *Hunting and Wildlife Management in Sarawak*, Cambridge University: IUCN Publications.

Caldecott, J.O. and Nyaoi, A. (1984) 'Sarawak wildlife: a resource to be taken seriously', *Sarawak Gazette*, 111, pp. 31–2.

Cameron, O. (1995) 'The emergence of environmental concern in post-oil crisis Japan', PhD thesis, forthcoming from the University of Cambridge, England.

Campbell, J.B. (1987) *Introduction to Remote Sensing*, London: Guildford Press.

Case, W. (1993) 'Semi-democracy in Malaysia', *Pacific Affairs*, 66, 2, pp. 183–205.

Cassettari, S. (1991) 'Introducing GIS into the National Curriculum for Geography', *Association for Geographic Information: Third National Conference and Exhibition*, ICC, Birmingham, 20–22 November 1991, pp. 341–4.

Cater, E. (1992) 'Profits from paradise', *Geographical Magazine*, March 1992, pp. 17–20.

Cater, E. (1995) 'Environmental contradictions in sustainable tourism', *The Geographical Journal*, 161, 1, pp. 21–8.

Cater, E. and Lowman, G. (eds) (1994) *Ecotourism: A Sustainable Option?*, Chichester: Wiley.

Chambers, R. (1983) *Rural Development: Putting the Last First*, London: Longman.

Chambers, R. (1993) *Challenging the Professions: Frontiers for Rural Development*, London: Intermediate Technology Publications.

Chambers, R. and Conway, G. (1992) 'Sustainable rural livelihoods: practical concepts for the 21st century', Discussion Paper 296, Institute of Development Studies, University of Sussex.

Chambers, R., Pacey, A. and Thrupp, L.A. (1989) *Farmer First: Farmer Innovation and Agricultural Research*, London: Intermediate Technology Publications.

Chan, S. and Clark, C. (1992) 'The rise of the East Asian NICs: Confucian capitalism, status mobility, and development legacy', in L. Clark and S. Chan (eds), *The Evolving Pacific Basin in the Global Political Economy: Domestic and International Linkages*, London: Lynne Rienner Publishers.

Chanda, N. (1982) 'Economic changes in Laos', in Martin Stuart-Fox, (ed.), *Contemporary Laos: Studies in the Politics and Society of the Lao People's Democratic Republic*, St Lucia: University of Queensland Press, pp. 116–28.

Chin, S.C. (1985) 'Agriculture and resource utilisation in a lowland rainforest Kenyah community', *Sarawak Museum Journal*, XXXV (56 n.s.), Special Monograph No. 4.

Chongkittavorn, K. (1989) 'Laos taking action to save forests', *The Nation*, 21 July, Bangkok.

Clark, D. (1994) 'Editorial', *Appita Journal*, 47, 3.

Clark, J. (1991) *Democratizing Development: The Role of Voluntary Organizations*, London: Earthscan.

Cleary, M. and Eaton, P. (1992) *Borneo: Change and Development*, Singapore: Oxford University Press.

Cleary, M. and Shaw, B.J. (1994) 'Ethnicity, development and the New Economic Policy: the experiences of Malaysia, 1971–1990', *Pacific Viewpoint*, 35, 1, pp. 83–107.

Club of Rome (1972) *The Limits to Growth: A Report for the Club of Rome's Project on the Predicament of Mankind*, Rome: Club of Rome.

Cochrane, J.E. (1991) *The Tourism Potential of Wasur National Park, Irian Jaya*, Jakarta: World Wide Fund for Nature.

Cochrane, J.E. (1992) *Management and Conservation of Tropical Forest Ecosystems and Biodiversity*, Jakarta: Asian Development Bank/Ministry of Forestry.

Cochrane, J.E. (1993a) *Gunung Leuser National Park Ecotourism Survey*, Medan: Leuser Development Programme.

Cochrane, J.E. (1993b) 'Tourism and conservation in Indonesia and Malaysia', in M.J. Hitchcock, V.T. King and M.J.G. Parnwell (eds), *Tourism in South-East Asia*, London: Routledge, pp. 317–26.

Cochrane, J.E., McKenzie, C. and Ratcliffe, J. (1994) *Microenterprise Establishment: Lore Lindu National Park*, Jakarta: The Nature Conservancy.

Colchester, M. (1989) *Pirates, Squatters and Poachers: The Political Ecology of Dispossession of the Native Peoples of Sarawak*, London: Survival International; Petaling Jaya: INSAN.

Colchester, M. and Lohmann, L. (eds) (1993) *The Struggle for Land and the Fate of the Forests*, London: Zed Books.

Colfer, C.J.P. (1983a and b) 'Change and indigenous agroforestry in East Kalimantan' (parts I and II), *Borneo Research Bulletin*, 15, pp. 3–21 and pp. 70–87.

Collins, M. (1991) 'Mapping the tropical forests', *Mapping Awareness and GIS Europe*, 5, 10, pp. 25–8.

Collins, N.M., Sayer, J.A. and Whitmore, T.C. (1991) *The Conservation Atlas of Tropical Forests: Asia and the Pacific*, London: Macmillan.

Cooper, D.E. and Palmer, J.E. (eds) (1992) *The Environment in Question: Ethics and Global Issues*, London: Routledge.

Cope, D. (1990) 'The rising sun in a green world', *Anglo-Japanese Journal*, 4, 2, pp. 14–17.

Costello, M.G. (1990) *The Economics of Protected Area Establishment and Management*, Jakarta: WWF/Asian Development Bank.

Cowen, D. (1988) 'GIS versus CAD versus DBMS: what are the differences?', *Photogrammetric Engineering and Remote Sensing*, 54, pp. 1551–5.

Craib, B.A. (1994) 'Japan as a development model: promise or problem?', *Japan Economic Institute Report*, Number 19A, 13 May, Washington: Japan Economic Institute.

Cramb, R.A. (1989) 'Shifting cultivation and resource degradation in Sarawak: perception and policies', *Borneo Research Bulletin*, 21, 1, pp. 22–48.

Cramb, R.A. (1990) 'Reply to John Palmer', *Borneo Research Bulletin*, 22, pp. 44–6.

Cramb, R.A. and Reece, R.W. (eds) (1988) *Development in Sarawak*, Melbourne: Monash University.

Cranbrook, Earl of (1992) 'Rainforest university', *The Geographical Magazine*, 64, 5 (May), pp. 26–30.

Cranbrook, Earl of (1993) 'Research and management of the Batu Apoi Forest Reserve, Temburong, Brunei: the Universiti Brunei Darussalam/Royal Geographical Society Rainforest Project 1991/92', in R.L. Bryant, J. Rigg and P. Stott (eds), *The Political Ecology of Southeast Asian Forests: Transdisciplinary Discourses. (Global Ecology and Biogeography Letters)*, Special Issue, 3, 4–6, 267–76.

Cranbrook, Earl of and Edwards, D.S. (1994) *Belalong: A Tropical Rainforest*, Singapore: Sun Tree Publishing.

Cross, P.A. (1991) 'GPS for GISs', *Mapping Awareness and GIS Europe*, 5, 10, pp. 30–4.

Crouch, H. (1992) 'Authoritarian trends, the UMNO split and the limits to state power', in J. S. Khan and F. L. Kok (eds), *Fragmented Vision: Culture and Politics and Contemporary Malaysia*, Sydney, NSW: Allen and Unwin.

Crouch, H. (1993) 'Malaysia: neither authoritarian nor democratic', in K. Hewson, R. Robinson and G. Rodan (eds), *Southeast Asia in the 1990s: Authoritarianism, Democracy and Capitalism*, St Leonards, NSW: Allen and Unwin.

Cutter, S.L. (1994) 'Environmental issues: green rage, social change and the new environmentalism', *Progress in Human Geography*, 18, pp. 217–26.

Daily Yomiuri (1994a) 'Asian nations see future powered by nuclear energy', 26 February 1994.

Daily Yomiuri (1994b) 'Conservationists say whale sanctuary within reach', 14 March 1994.

Dandot, W.B. (1992) 'Contributions of social science research and studies for development planning in Sarawak', *Jurnal Azam*, 8, pp. 12–39.

Dauvergne, P. (1994) 'The politics of deforestation in Indonesia', *Pacific Affairs*, 66, 4, pp. 497–518.

Davies, W.G. (1988) *Forest Industries, Technical Report No. 6, Forestry Sector Review Vietnam*, VIE/88/037, FAO/UNDP/MoF.

Development Today (Oslo), various issues.

Djohani, R.H. (1989) *Marine Conservation Development Programme in Indonesia*, Jakarta: WWF.

Djohani, R.H. (1995a) *The Combat of Dynamite and Cyanide Fishing in Indonesia; A Strategy to Decrease the Use of Destructive Fishing Methods in and around Komodo National Park*, Jakarta: The Nature Conservancy.

Djohani, R.H. (1995b) 'The sea is my home: the Bajau people of Bunaken Park', in R. Schefold (ed.), *Minahasa Past and Present: Tradition and Transition in an Outer Island Region of Indonesia*, Research School CNWS: Leiden, The Netherlands.

Djohani, R.H., Malik, R. and Muslimin (1993) 'Marine parks and the socio-economic implications for the Bajau people, particularly from the Togian Islands', paper prepared for the International Seminar on Bajau Communities, LIPI, Jakarta.

Douglass, I., Sutton, K., King, V.T., McMorrow, J.M., Parnwell, M. and Taylor, D. (1995) *Tropical Forests, Communities and Environmental Change in Borneo*, Final Report to Economic and Social Research Council.

Dove, M.R. (1981) 'Swidden systems and their potential role in agricultural development', *Prisma*, 21, pp. 81–104.

Dove, M.R. (1985) *Swidden Agriculture in Indonesia: The Subsistence Strategies of the Kalimantan Kantu*, Berlin, New York, Amsterdam: Mouton.

Dove, M.R. (1993) 'Smallholder rubber and swidden agriculture in Borneo: a sustainable adaptation to the ecology and economy of the tropical forest', *Economic Botany*, 47, 2, pp. 136–47.

Down to Earth, various issues.

Down to Earth (DTE) (1991) *Pulping the Rainforest*, London: Down to Earth.

Drummond, I. and Marsden, T. K. (1995) 'Regulating sustainable development', *Global Environmental Change*, 5, 1, pp. 51–63.

Dudley, N. (1992) *Forests in Trouble: A Review of the Status of Temperate Forests World-wide*, Gland: World Wide Fund for Nature.

Dugan, P. (1993) 'Community-based forest management', paper presented at Asian Development Bank, Manila.

Durrell Institute of Conservation and Ecology (1995) *Tourism, Conservation and Sustainable Development*, Canterbury: University of Kent.

Dykes, A.P. (1994) 'Landform processes', in Earl of Cranbrook and D.S. Edwards, *Belalong: A Tropical Rainforest*. Singapore: Sun Tree Publishing, pp. 31–47.

Dykes, A.P. (1995) 'Hydrological controls on shallow mass movements and characteristic slope forms in the tropical rainforest of Temburong District, Brunei', unpublished PhD thesis, University of London.

Dykes, A.P. in press, 'Analysis of factors contributing to the stability of steep hillslopes in the tropical rainforest of Temburong, Brunei', *Proceedings of the International Conference on Tropical Rainforest Research: Current Issues*. (Bandar Seri Begawan, Brunei, April 1993).

Eber, S. (ed.) (1992) *Beyond the Green Horizon*, Godalming: WWF UK.

Eccleston, B. (1995) 'Lessons from campaigning against the Bakun Dam project in Sarawak', unpublished paper, European Association for South-East Asian Studies Conference, University of Leiden, The Netherlands.

Eccleston, B., forthcoming, 'Does North-South collaboration enhance NGO influence on deforestation policies in Malaysia and Indonesia?', in *Journal of Commonwealth and Comparative Politics*, XXXIV, March 1996.

Ecologist (1986) 'Indonesia's transmigration programme: a special report in collaboration with Survival International and Tapol', *The Ecologist*, 16, 2/3.

Ecologist (1993) *Whose Common Future? Reclaiming the Commons*, London: Earthscan.

Ecologist/World Rainforest Movement (n.d.), 'TFAP: campaign dossier', *The Ecologist*/WRM.

Ecologist, various issues

Economist (1994) 'An Asian snail', 5 November, pp. 74–80.

Edington, J.M. and Edington, M. A. (1986) *Ecology, Recreation and Tourism*, Cambridge: Cambridge University Press.

Edwards, D. (1992) '16 months in Belalong', *Muhiba* (Royal Brunei Airlines), Jul./Aug. 1992, pp. 10–20.

Ehui, S.K. and Hertel, T.W. (1987) 'The impact of deforestation on agricultural production: an empirical investigation', *American Journal of Agricultural Economics*, 69, 5, p. 1094.

Einarsen, J. and Rodgers, K. (1993) *Whole Japan Indicator*, compilation of selections in Kyoto Journal.

BIBLIOGRAPHY

Ekachai, S. (1992) 'Man and the forest', *Bangkok Post*, 24 January.

Electrowatt Engineers and Consultants (1993) *Environmental and Financing Studies on the Yali Falls Hydropower Project (Basin Wide)*, Mekong Secretariat, May 1993.

El-Hinnawi (1985) *Environmental Refugees*, Nairobi: United Nations Environment Programme.

Elliot, J. A. (1994) *An Introduction to Sustainable Development*, London: Routledge.

Elmhirst, R. (1995) 'Transmigration and sustainable livelihoods in Indonesia: the role of women's environmental knowledge', paper prepared for a seminar at ICRAF, Bogor, April 1995.

English Tourist Board (1991) *Tourism and the Environment: Maintaining the Balance*, London: English Tourist Board.

Ensign, M. (1992) *Doing Good or Doing Well?: Japan's Foreign Aid Program*, New York: Columbia University Press.

Environmental Research Division (1994) *Upland Philippine Communities: Securing Cultural and Environmental Stability*, Research Report, June.

Ewart, B. (1993) 'Japan's businesslike approach to golf threatens the sport's simple pleasures', *Japan Times*, 24 February.

Ewers, K. (1994) 'Politics of biodiversity conservation in Thailand: global and local discourse', *Proceedings of the IIAS/NIAS Workshop on Environmental Movements in Asia*, Leiden, The Netherlands, 27–29 October 1994.

Exodus Discovery Holidays (1994/95).

Falconer, J. and Arnold, J.E.M. (1991) *Household Food Security and Forestry: An Analysis of Socio-Economic Issues*, Rome: FAO, Community Forestry Note No. 1.

Far Eastern Economic Review (1993) 'Seeds of growth [interview with Laos' Deputy Prime Minister]', 4 November, p. 32.

Far Eastern Economic Review, various issues.

Fernandez Carro, O. and Wilson, R. (1992) 'Quality management with fiber crops', *TAPPI Journal*, February, pp. 49–56.

Financial Times (London), various issues.

Finland National Board of Customs (1990–93) *Statistics of Foreign Trade*, Helsinki.

FIVAS (International Association for Water and Forest Studies) (1994) 'Kraft og Konflikter: norske vannkraftutbyggere i den tredje verden', Oslo: FIVAS.

Fletcher, H. (1988) 'The pulp and paper industry: a New Zealand perspective', in G. Schreuder (ed.), *Global Issues and Outlook in Pulp and Paper*, Seattle: University of Washington.

Food and Agriculture Organization (1982) *National Conservation Plan for Indonesia*, Vol. 1, Bogor.

Food and Agriculture Organization (1987) *The Tropical Forestry Action Plan*, FAO: Rome.

Food and Agriculture Organization (1991) *Shifting Cultivators: Local Technical Knowledge and Natural Resource Management in the Humid Tropics*, Rome: FAO, Community Forestry Note No. 8.

Food and Agriculture Organization (n. d.), *Shifting Cultivators of Indonesia: Marauders or Managers of the Forest? Rice Production and Forest Use among the Uma' Jalan of East Kalimantan*, Bangkok: FAO, Community Forestry Case Study Series 6.

Food and Agriculture Organization (n. d.), *The Impact of Social and Environmental Change on Forest Management: a case study from West Kalimantan, Indonesia*, Bangkok: FAO, Community Forestry Case Study Series No. 8.

Forrest, R.A. (1991) 'Japanese aid and the environment', *The Ecologist*, 21, 1, pp. 24–32.

Forum Keadilan (Jakarta), various issues.

Fowler, A. (1993) 'Non-governmental organizations as agents of democratization: an African perspective', *Journal of International Development*, 5, 3, pp. 325–39.

Fox, J.J. (1991) 'Managing the ecology of rice production in Indonesia', in J. Hardjono (ed.), *Indonesia: Resources, Ecology and Environment*, Oxford: Oxford University Press, pp. 61–84.

Freeman, J.D. (1970) *Report on the Iban*, London: Athlone Press.

Freeman, J.D. (1981) *Some Reflections on the Nature of Iban Society*, Occasional Paper, Australian National University: Advanced School of Pacific Studies.

Friedmann, J. and Douglass, M. (1978) 'Agropolitan development: towards a new strategy for regional planning in Asia', in UNCRD, *Growth Pole Strategy and Regional Development Planning in Asia*, Nagoya: UNCRD.

Fröberg, M., Jerndal, R. and Åkesson, A. (1990) *The Outlook for the Laotian Forest Sector: Domestic Needs Versus International Participation*, Gothenburg: Gothenburg University.

Fujisaka, S., Kirk, G. and Litsinger, L.A. *et al.*, (1991) 'Wild pigs, poor soils, and upland rice: a diagnostic survey of Sitiung, Sumatra, Indonesia', IRRI Research Paper Series No. 155, Manila: International Rice Research Institute.

Furnivall, J.S. (1956) *Colonial Policy and Practice: A Comparative Study of Burma and Netherlands India*, New York: New York University Press.

Ganapin, D. (1987a) 'Forest resource and timber trade in the Philippines' in *Forest Resources Crisis in the Third World*, Conference Proceedings, Sahabat Alam Malaysia, pp. 54–70.

Ganapin, D. (1987b) 'Philippine ethnic minorities: the continuing struggle for survival and self-determination', in *Forest Resources Crisis in the Third World*, Conference Proceedings, Sahabat Alam Malaysia, pp. 171–91.

Ganapin, D. (1994) 'Case study of the Philippines', in European Institute of South and South-East Asian Studies, *Population, Development and Environment in South and South-East Asia*, Proceedings of Seminar VIII, 23 September 1994, Brussels.

Geographical Magazine (1995) 'Wild cubs captured', March 1995.

Ghee, L.T. and Valencia, M.J. (eds) (1990) *Conflict over Natural Resources in South-East Asia and the Pacific*, Singapore: United Nations University Press.

Gibbs, C., Payuan E. and del Castillo, R. (1990) 'The growth of the Philippine social forestry program', in M. Poffenberger (ed.), *Keepers of the Forest*, Connecticut: Kumarian Press, pp. 253–65.

Global Network for Anti-Golf Course Action (GNAGCA) (1993) *'The World No Golf Day' Conference Proposal – April 26th-29th 1993 in Penang, Malaysia*, Chiba, Japan: GNAGCA, 5 March.

Global Resource Information Database (GRID) (1990) 'An analysis of deforestation and associated environmental hazards in northern Thailand: a joint Thailand-UNEP/GRID case study', *GRID Case Study Series*, 3, Bangkok: Global Environmental Monitoring System.

Goh K.C. (1992) 'Environmental management', *Singapore Journal of Tropical Geography*, 13, 1, pp. 14–24.

Goldsmith, E. and Hildyard, N. (1984) *The Social and Environmental Impact of Large Dams*, San Francisco: Sierra Club.

Gonseth, O. (1988) 'A look behind the tourism façade: some considerations on the development of tourism in the province of Ifugao (Philippines)', in P. Rossel (ed.), *Tourism: Manufacturing the Exotic*, Copenhagen: IWGIA, pp. 21–46.

Goodchild, M., Parks B. and Steyaert, L. (1993) *Environmental Modelling with GIS*, Oxford: Oxford University Press.

Goodfellow, D. (1993) 'Hosts versus guests: the desires and concerns of the passengers on society expeditions' ship "World Discoverer"', *Tjerutjuk (Bulletin of the Bali Bird Club)*, 2, 1, pp. 2–3.

Goodland, R. (1993) 'Ethical priorities in environmentally sustainable energy systems: the case of tropical hydro-power', paper delivered at an International Symposium,

'Energy Needs in the Year 2000 and Beyond: Ethical and Environmental Perspectives', Montreal, May 1993.

Goody, R.A. and Feaw, T.C. (1989) 'The profitability of smallholder rattan cultivation in Southern Borneo, Indonesia', *Human Ecology*, 17, 3, pp. 347–63.

Gough, K. (1990) *Political Economy in Vietnam*, Berkeley, California: Folklore Institute.

Government of Brunei Darussalam: Forestry Department (1993) *Sixty Years of Forestry in Brunei Darussalam: Towards Sustainable Management and Effective Biodiversity Conservation Through Sound Forestry Practice*, Bandar Seri Begawan: Forestry Department, Ministry of Industry and Primary Resources (Brunei Darussalam).

Government of Indonesia (1991) *Developmental Plan 1992–1996, Kutai National Park, East Kalimantan*, Jakarta; Directorate General of Forest Protection and Nature Conservation.

Government of Indonesia (1992) *Kumpulan Peraturan tentang Pungutan dan Iuran Bidang Pariwisata Alam serta Pungutan Masuk Kawasan Pariwisata Alam, (Regulations on fees for entry to national parks)*, Jakarta: Directorate General of Forest Protection and Nature Conservation.

Government of Indonesia: Ministry of State for Environment (1992) *Indonesian Country Study on Biological Diversity*, Jakarta: Ministry of State for Environment.

Government of Indonesia (1993) *Kantor Pengembangan Desa, Tinkat II, Lampung Utara*, Government of Indonesia: Lampung.

Government of Indonesia (1994) *The Operation of a Natural Tourism Business in the Utilization Zones of National Parks, Forest Parks and Parks for Tours in Nature*, Government Regulation No. 18/1994.

Government of the Lao PDR (1989) *Report on the Economic and Social Situation, Development Strategy, and Assistance Needs of the Lao PDR*, volume 1, Lao PDR: Geneva.

Government of the Lao PDR (1990) *Tropical Forestry Action Plan*, final draft, main report, Vientiane: Government of the Lao PDR.

Government of the Lao PDR (1992) *Forest Cover and Land Use in the Lao PDR: Final Report on the Nationwide Reconnaissance Survey*, Vientiane: Government of the Lao PDR, Ministry of Agriculture and Forestry, Department of Forestry, National Office of Forest Inventory and Planning, Lao-Swedish Forestry Co-Operation Programme, Forest Inventory Report No. 5, December 1992.

Government of the Lao PDR (1993a) Decree of the Prime Minister on the management and use of forests and forest land, Decree No. 169/PM, Vientiane, Laos.

Government of the Lao People's Democratic Republic: Directorate for Nature Management, (1993b) *Nam Theun 1/2 Hydropower Project, Laos: Assessment of Environmental Impacts*, Trondheim: Directorate for Nature Management.

Government of the Lao People's Democratic Republic: Ministry of Industry (n. d.) 'Hydro and thermal electric power projects under development plan up to 2000's', Vientiane.

Government of Malaysia: Department of Agriculture (1991) 'Remote sensing and land use mapping in Malaysia', *Malaysia Centre for Remote Sensing (MACRES) News*, 1, 2 (Dec.), pp. 8–9.

Government of Malaysia, Ministry of Primary Industries (1991) *Statistics on Commodities*, Kuala Lumpur: Ministry of Primary Industries.

Government of Myanmar: Forest Department (1994) *Greening Project for the Nine Critical Districts of the Arid Zone of Central Myanmar*, Yangon: Forest Department.

Government of the Philippines: Department of Environment and Natural Resources, (1994) *National CFP Implementation Plan: Background, Issues, Options and Planning Directives*, Manila: Department of Environment and Natural Resources, Draft, December.

Government of Vietnam (1988) *Main Report, Forestry Sector Review Vietnam*, VIE/88/037, FAO/UNDP/MoF.

Government of Vietnam (1990) *Tropical Forestry Action Plan Update No. 16*, 31 January 1990.

Government of Vietnam: Ministry of Forestry (1991) *Main Report: Vietnam Forestry Sector Review, Tropical Forestry Action Programme*, Hanoi, Ministry of Forestry, December.

Government of Vietnam (1993) *Vietnam: A Development Strategy*, Hanoi: Government of Vietnam, September 1993.

Government of Vietnam: Ministry of Forestry (1995) *Selected Government Decisions on Forestry during 1992, 1993 and 1994*, Renovation of Strategies for Forestry Development Project, Ministry of Forestry, Hanoi.

Graburn, N.H.H. (ed.) (1976) *Ethnic and Tourist Arts: Cultural Expressions from the Fourth World*, Berkeley, LA: University of California Press.

Grainger, A. (1993) *Controlling Tropical Deforestation*, London: Earthscan.

Grant, C.K. (1984) 'Geographical overview', in D.M.D. James (ed.), *The Geology and Hydrocarbon Resources of Negara Brunei Darussalam*, Bandar Seri Begawan: Muzium Brunei, pp. 11–33.

Gray, K. M. (1988) *Fuelwood Supply Analysis, Technical Report No. 9, Forestry Sector Review Vietnam*, VIE/88/037, FAO/UNDP/MoF.

Grimm, M. (1992) 'Japan and ASEAN: aspects of a new interdependence', *Japan Economic Institute Report*, Number 12A, 27 March, Washington: Japan Economic Institute.

Grimstad, P. (1994) 'Listen to countries in the South', Innsyn, Oslo: Norwegian Agency for Development Cooperation.

Gross, N. (1989) 'Charging Japan with crimes against the earth', *Business Week*, 9 October, p. 108.

Guardian (1994) 'Green groups to sue Suhatro', October 1994.

Guerreiro, A. (1988) 'Cash crop and subsistence strategies: towards a comparison of Kayan and Lahanan economies', *Sarawak Museum Journal*, XXXIX, 60 (n.s.), Dec. 1988, pp. 15–52.

Guiang, E.S. (1993) 'Community-based forest management: its evolution, emerging prototypes, and experiences from the Philippines', paper presented at the 4th Annual Common Property Conference of the International Association for the Study of Common Property, Manila, June.

Gunn, C.A. (1994) *Tourism Planning*, Washington D.C.: Taylor and Francis.

Gurung, C.P. and de Coursey, M. (1994) 'Nepal: pioneering sustainable tourism', *The Rural Extension Bulletin*, 5, University of Reading Agricultural Extension and Rural Development Department, pp. 17–21.

Hagler, R. (1993) 'Global forest', *Papermaker*, May: pp. 40–6.

Haines-Young, R., Green, D. and Cousins, S. (1994) *Landscape Ecology and Geographical Information Systems*, London: Taylor and Francis.

Håkangård, A. (1990) *Women in Shifting Cultivation, Luang Prabang Province, Lao PDR*, Development Studies Unit, Report No. 18, Department of Social Anthropology, Stockholm University.

Hall, J.C. (1993) 'Managing the tropical rain forest: swiddens, house gardens and trade in Central Kalimantan', unpublished PhD thesis, Oxford Brookes University.

Hanbury-Tenison, R. (1992) *Mulu: The Rain Forest*, London: Arrow (first publ. 1980), Weidenfeld and Nicolson.

Handley, P. (1993) 'Making connections', *Far Eastern Economic Review*, 4 November, pp. 28–31.

Handley, P. (1995) 'Trading on culture: Thailand takes a new tack with its neighbours', *Far Eastern Economic Review*, 30 March, pp. 28–9.

Hardjono, J. (1977) *Transmigration in Indonesia*, Kuala Lumpur: Oxford University Press.

Hardoy, J. *et al.* (1992) *Environmental Problems in Third World Cities*, London: Earthscan.

Harvey, D. (1993) 'The nature of environment: the dialectics of social and environmental change', in R. Miliband and L. Panitch (eds), *Real Problems, False Solutions: Socialist Register 1993*, London: Merlin Press, pp. 1–51.

Hasegawa, N. (1993) 'Logging in Sarawak, Malaysia and native customary rights', *Waseda Journal of Asian Studies*, 15, pp. 54–68.

Hatch, T. and Lim, C.P. (1978) *Shifting Cultivation in Sarawak*, Kuching: Department of Agriculture.

Healey, R.G. (1994) '"Tourist merchandise" as a means of generating local benefits from ecotourism', *Journal of Sustainable Tourism*, 2, 3, pp. 137–51.

Hecht, S. and Cockburn, A. (1990) *The Fate of the Forest: Developers, Destroyers and Defenders of the Amazon*, London: Penguin Books.

Helmsing, B. (1982) 'Agricultural production in the periphery: settlement schemes reconsidered', *Development and Change*, 13, 3, pp. 401–19.

Helsingen Sanomat (Helsinki), various issues.

Henley, D. (1994) 'Climatic variation in Indonesian history: a short review of recent research', *Indonesian Environmental History Newsletter*, 3, June, pp. 1–4.

Heppel, M. (1991) 'The manly pursuit of artistry: Iban woodcarving', in L. Chin and V. Mashman (eds), *Sarawak Cultural Legacy: A Living Tradition*, Kuching: Society Atelier Sarawak, pp. 33–43.

Heyzer, N. (1995) 'Gender, population and environment in the context of deforestation: a Malaysian case study', *IDS Bulletin*, 26, 1, pp. 40–6.

Hirsch, P. (1990) 'Forests, forest reserves and forest land in Thailand', *The Geographical Journal*, 156, 2, pp. 166–74.

Hirsch, P. forthcoming, 'Thailand and the new geopolitics of Southeast Asia: resource and environmental issues', in J. Rigg (ed.), *Counting the Costs: Economic Growth and Environmental Change in Thailand*, Singapore: ISEAS.

Hirsch, P. and Lohmann, L. (1989) 'Contemporary politics of the environment in Thailand', *Asian Survey*, 29, 4, pp. 439–51.

Hirsch, P., Bach Tan Sinh, Nguyen Nu Hoai Van, Do Thanh Huong, Nguyen Quoc Hung, Tran Ngoc Ngoan, Nguyen Viet Thinh and Vu Quyet Thang, (1992) *Social and Environmental Implications of Resource Development in Vietnam: The Case of Hoa Binh Reservoir*, University of Sydney, RIAP Occasional Paper No. 17.

Hitchcock, M. (1993) 'Dragon tourism in Komodo, Eastern Indonesia', in M.J. Hitchcock, V.T. King and M.J.G. Parnwell (eds), *Tourism in South-East Asia*, London: Routledge, pp. 303–16.

Hitchcock, M.J., King, V.T. and Parnwell, M.J.G. (eds) (1993) *Tourism in South-East Asia*, London: Routledge.

Hoffman, C.L. (1984) 'Punan foragers in the trading networks of Southeast Asia', in C. Shrire (ed.), *Past and Present in Hunter Gatherer Societies*, New York: Academic Press.

Holliman, J. (1987) 'Japan's trade in tropical timber with Southeast Asia', in Sahabat Alam Malaysia, *Forest Resources Crisis in the Third World*, Penang, Malaysia: Sahabat Alam Malaysia, pp. 250–9.

Hook, G.D. and Weiner, M.A. (eds) (1991) *The Internationalisation of Japan*, London: Routledge.

Hooker, M.B. (1976) *The Personal Laws of Malaysia*, Kuala Lumpur: Oxford University Press.

Hong, E. (1987) *Natives of Sarawak: Survival in Borneo's Vanishing Forests*, Penang: Institut Masyarakat.

Houghton, R. (1990) 'The future role of tropical forests in affecting the carbon dioxide concentration of the atmosphere', *Ambio*, 19, 4, pp. 204–9.

Houghton, R. (1991) 'Releases of carbon to the atmosphere from degradation of forests in tropical Asia', *Canadian Journal of Forest Research*, 21, 1, pp. 132–42.

Howard, M.C. (ed.) (1993) *Asia's Environmental Crisis*, Boulder, Colorado: Westview Press.

Hulme, D. (1987) 'State sponsored land settlement schemes: theory and practice', *Development and Change*, 18, 3, pp. 413–36.

Hurst, P. (1990) *Rainforest Politics: Ecological Destruction in South-East Asia*, London: Zed Books.

Hutomo, M., Uktolseya, H., Sloan, N.A., Abdullah, A., Djohani, R.H., Alder, J., Halim, M.H. and Sutardjo (1993) *Marine Conservation Areas in Indonesia: Two Case-Studies of Kepulauan Seribu, Java and Bunaken, Sulawesi*, Proceedings UNEP-COBSEA/NOSTE Workshop EAS 25, Penang.

Indonesian Biodiversity Country Study Standing Committee (IBCSSC) (1992) *Indonesian Country Study on Biological Diversity*, UNEP/Norway/Indonesia: Ministry of State for Population and Environment.

Innes-Brown, M. and Valencia, M.J. (1993) 'Thailand's resource diplomacy in Indochina and Myanmar', *Contemporary Southeast Asia*, 14, pp. 332–51.

Innsyn Special Edition (1994) Oslo: Norwegian Agency for Development Cooperation.

Inside Kvaerner Energy (1995) 28 February, Oslo: Kvaerner.

International Geosphere-Biosphere Programme (IGBP) (1992) 'Land use and land cover change', in *Report from the START Regional Meeting for Southeast Asia*, Stockholm: IGBP, pp. 51–3.

International Rivers Network (1995) *Technical Review of the Mekong Secretariat Report: 'Mekong Mainstream Run-of-River Hydropower'*, December 1994, San Francisco: International Rivers Network.

International Water Power and Dam Construction (London), various issues.

International Water Power and Dam Construction (1994a) *Financing Hydro Power '94*, proceedings of a conference sponsored by International Water Power and Dam Construction, Frankfurt, 22–23 September 1994.

International Water Power and Dam Construction (1994b) 'Wanted – private investors', International Water Power and Dam Construction, May 1994.

Ireson, C.J. (n. d.) 'The role of women in forestry in the Lao People's Democratic Republic', report prepared for the Swedish International Development Authority (SIDA), Vientiane.

Islam, S. (1991) *Yen for Development: Japanese Foreign Aid and the Politics of Burden-Sharing*, New York: Council on Foreign Relations Press.

Ison, R. (1990) 'Teaching threatens sustainable agriculture', Gatekeeper series, No. 21, International Institute for Environment and Development.

Ivarsson, S., Svensson, T. and Tönnesson, S. (1995) *The Quest for Balance in a Changing Laos: A Political Analysis*, Copenhagen, Nordic Institute of Asian Studies, NIAS Report No. 25.

Jaakko Poyry, (n. d.) *Jaakko Poyry Group in Brief*, Helsinki: Jaakko Poyry.

Jacobsen, J.L. (1988) *Environmental Refugees: A Yardstick of Habitability*, Worldwatch Paper No. 86, Washington: Worldwatch Institute.

Jakarta Post, various issues.

Japan Echo (1989) Special issue on 'Focus on foreign aid', Volume XVI, 1.

Japan Echo (1993) Special issue on 'Japanese views of Asia', Volume XX.

Japan Federation of Bar Associations, International Environmental Forum (1990) *Japan's Pollution Export and Environmental Destruction in Southeast Asia: A Look at Overseas Corporate Expansion and ODA*, Tokyo: JFBA Committee on Pollution Measures and Environmental Conservation.

Japan Federation of Bar Associations (1992) *Global Environmental Conservation Proposals for the Earth Summit*, Tokyo: Japan Federation of Bar Associations.

Japan Pulp and Paper, (n. d.) 'As conversion approaches 27 years of chip imports', 30, 1.

Japan Times, various issues.

Japan Times (1992) 'Broken environmental promises', 18 October, 1992.

Japan Times (1992) 'Conservationists moot timber protection act', 13 November 1992.

Japan Times (1994) 'Indonesia seeks eco-standards', 29 June 1994.

Jensen, Erik (1974) *The Iban and their Religion*, Oxford: Clarendon Press.

Jermy A. C. and Kavanagh, K.P. (eds) (1982; 1984) *Gunung Mulu National Park, Sarawak*. Kuching: Sarawak Museum Journal, Special Issue 2, 30, 51, I and II.

Jerndal, R. (1993) *The Geography of Laos*, Gothenburg: Gothenburg University.

Johannes, R.E. (1989) *Traditional Ecological Knowledge: A Collection of Essays*, IUCN.

Jones, L.W. (1962) *Sarawak: Report on the Census of Population taken on 15th June 1960*, Kuching: Government Printer.

Junus, I. (1994) *Indonesia Plywood Industry*, London: WI Carr, Banque Indosuez Group, June.

de Kadt, E. (1990) *Making the Alternative Sustainable: Lessons from Development for Tourism*, Discussion Paper 272, Institute of Development Studies, University of Sussex.

Kaplinsky, R. (1990) *The Economies of Small: Appropriate Technology in a Changing World*, London: Intermediate Technology.

Karnchanasutham, S and Wongwantanee, S. (1991) 'Provincial agricultural planning in Thailand', *GIS World*, 4, 3, pp. 98–100.

Kartawinata, K., Jessup, T.C. and Herwasono, S. (1982) *The Impact of Development on Interactions Between People and Forest in East Kalimantan*, Bogor: Man and Biosphere Project.

Kartawinata, K., Vayda, A.P. and Sambas Wirakusamah, R. (1978) 'East Kalimantan and the Man and Biosphere Program', *Borneo Research Bulletin*, 10, pp. 28–40.

Kathirithamby-Wells, J. (1995) 'Socio-political structures and the Southeast Asian ecosystem: an historical perspective up to the mid-nineteenth century', in O. Bruun and A. Kalland (eds), *Asian Perceptions of Nature: A Critical Approach*, London: Curzon Press, pp. 25–47.

Kedit, P.M. (1984) 'Bejalai is changing Iban society', *The Sarawak Gazette*, Vol. CX, No. 1487, pp. 24–6.

Kedit, P.M. (1989) 'An overview of Iban traditional cultural values and social norms, and their implications for contemporary Sarawak', *Sarawak Museum Journal*, special issue 4, part 4, 40, 61, pp. 1–14.

Kedit, P.M. (1991) '"Meanwhile, back home . . .": *Bejalai* and their effects on Iban men and women', in V.H. Sutlive (ed.), *Female and Male in Borneo: Contributions and Challenges to Gender Studies*, Williamsburg: Borneo Research Council Monograph Series, Vol. 1, pp. 295–316.

Keizai Koho Center (1992) *Japan's Industries Work for Conservation of Global Environment*, Tokyo, Japan: Keidanren Keizai Koho Center.

Kemp, K. (1991) 'GIS education around the world: year three of the NCGIA core curriculum project', *Association for Geographic Information: Third National Conference and Exhibition*, ICC, Birmingham 20–22 November 1991, pp. 381–3.

Kenchington, R. and Kim Looi, Ch'ng (1994) *Staff Training Materials for the Management of Marine Protected Areas*, Bangkok: UNEP.

Khamchoo, C. (1991) 'Japan's role in Southeast Asian security: "Plus ça change . . .", *Pacific Affairs*, 64, 1, pp. 7–22.

Khan, I.S. and Lim, R.P. (1991) 'Distribution of heavy metals in the Linggi River basin, Malaysia, with reference to pollution', *Australian Journal of Marine and Freshwater Research*, 42, pp. 435–48.

Khoo Kay Jim, (1992) 'The grand vision: Mahatir and modernisation', in J.S. Khan and F.L.K. Wah (eds), *Fragmented Vision: Culture and Politics and Contemporary Malaysia*, Sydney, NSW: Allen and Unwin.

King, V.T. (1986) 'Anthropology and rural development in Sarawak', *Sarawak Museum Journal*, 36, pp. 13–42.

King, V.T. (1993a) 'Tropical rainforests and indigenous peoples: symbiosis and exploitation', paper presented at the BAAS Festival, Keele, September 1993.

King, V.T. (1993b) *The Peoples of Borneo*, Oxford: Blackwell.

King, V.T. (1993c) 'Politik pembangunan: the political economy of rainforest exploitation and development in Sarawak, East Malaysia', *Global Ecology and Biogeography Letters*, 3, pp. 235–44.

King, V.T. (1993d) 'Tropical rainforests and indigenous peoples: symbiosis and exploitation', paper presented to the British Association Annual Science Festival, Keele, 1993 (forthcoming in a volume edited by I. Douglass).

King, V.T. (1993e) 'The Bintulu region of Sarawak, East Malaysia: the historical and socio-economic context', paper presented to the Workshop on Environmental Change in South-East Asia, University of Hull, September 1993.

King, V.T., Taylor, D.M. and Parnwell, M.J.G. forthcoming, 'Socio-economic responses to change among Iban communities in the Bintulu Region of Sarawak, Malaysia', *Jurnal Azam* (in press).

Kingsnorth, P. (1995) 'Golf war rages among paddyfields of Asia', *Independent on Sunday*, 30 April.

Kirby, E.S. (1980) 'Resource potentials of continental Asia and Japan's material needs', in S. Tsuru (ed.), *Economic Growth and Resources – Volume 5: Problems Related to Japan*, Proceedings of Session VI of the Fifth Congress of the International Economic Association held in Tokyo, Japan, London: Macmillan Press, pp. 257–84.

Kitazawa, Y. (1990) 'The Japanese economy and Southeast Asia: the examples of the Asahan aluminium and Kawasaki steel projects', in L.T. Ghee and M.J. Valencia (eds), *Conflict Over Natural Resources in South-East Asia and the Pacific*, Singapore: United Nations University Press, pp. 51–93.

Kitiprawat, S. (1992) 'Business opportunities in Laos today', *Bangkok Bank Monthly Review* 33, 11, pp. 15–22.

Klassekampen (Oslo), various issues.

Know-How Wire (Jaakko Poyry Client Magazine, Helsinki), various issues.

Ko, J.T.H. (1989) 'A socioeconomic study of the Iban to-day', *Sarawak Museum Journal*, special issue 4, part 4, 40, 61, pp. 79–96.

Kompas (Jakarta), various issues.

Korten, F. (1994) 'Questioning the call for environmental loans: a critical examination of forestry lending in the Philippines', *World Development*, 22, pp. 971–81.

Kubota, A. (1985) 'Foreign aid: giving with one hand?', *Japan Quarterly*, 32, 2, pp. 140–4.

Kuchera, K.P. (1986) 'Use of remote sensing for land resources mapping in Indonesia – overview and some practical examples of land resources mapping', in Mapping from Modern Imagery: Proceedings of the Symposium of the International Society for Photogrammetry and Remote Sensing and Remote Sensing Society, Edinburgh, Sept. 1986, *International Archives, Photogrammetry and Remote Sensing*, 26, 4, pp. 376–85.

Kummer, D. (1992) *Deforestation in the Philippines*, Manila: Ateneo de Manila University Press.

Kuroda, Y. and Nectoux, F. (1990) *Timber from the South Seas: An Analysis of Japan's Tropical Timber Trade and its Environmental Impact*, Gland, Switzerland: WWFN International.

Laili, N. and Darus, A. (1993) 'A case study in land use mapping using Landsat TM data in Malaysia', in *Proceedings of the South-East Asian Regional Conference on Education and Research in Remote Sensing*, Universiti Teknologi Malaysia, Johor Baru, 28–30 June 1993.

Laitalainen R. (1992) 'Letter to Witoon Permpongsacharoen', Bangkok: Project for Ecological Recovery, August.

Lam, Ying Fan (1992) 'A study on visual interpretation on SPOT and Landsat image to detect the change of land cover on two different date satellite data', paper presented at the SEAMEO-France Project Final Seminar and Steering Committee Meeting, Manila, Philippines, 17–20 October (1992.

Lauber, S. (1994) 'Japan and international environment law – economy over environment?', *Japanese Studies – Bulletin of the Japanese Studies Association of Australia*, 13, 3, pp. 37–54.

Leaman, D.J., Razali Yusuf and Harini Sangat-Roemantyo, (1991) *Dayak Forest Medicines, Kenyah Prospects for Development and Implications for Conservation*, Jakarta: World Wide Fund for Nature, Indonesia Programme.

Leighton, M. and Wirawan, N. (1986) 'Catastrophic drought and fire in Borneo tropical rain forest associated with the 1982–83 El Niño Southern Oscillation Event', in G.T. Prance (ed.), *Tropical Rain Forests and the World Atmosphere*, Boulder: Westview, pp. 75–102.

Leinbach, T. (1989) 'The Transmigration Programme in Indonesian national development strategy: current status and future prospects', *Habitat International*, 13, 3, pp. 81–93.

Leinbach, T., Watkins, J. and Bowen, J. (1992) 'Employment behaviour and the family in Indonesian transmigration', *Annals of the Association of American Geographers*, 82, 1, pp. 23–47.

Lele, S.M. (1991) 'Sustainable development: a critical review', *World Development*, 19, pp. 607–21.

Le Thac Can, Thuan, D.T. and Yem, T. (1993) *Environmental Impact Assessment of Bai Bang Pulp and Paper Factory*, EIA Unit, National Research Programme on Environment, Hanoi, January.

Leungaramsri, P. and Rajesh, N. (1992) *The Future of People and Forests in Thailand after the Logging Ban*, Bangkok: Project for Ecological Recovery.

Lewis, M.W. (1992) *Green Delusions*, Durham: Duke University Press.

Lian, F.J. (1993) 'On threatened peoples', in H. Brookfield and Y. Byron (eds), *South-East Asia's Environmental Future: The Search for Sustainability*, Kuala Lumpur: Oxford University Press, pp. 322–40.

Lindberg, K. and Hawkins, D.E. (1993) *Ecotourism: A Guide for Planners and Managers*, Vermont: The Ecotourism Society.

Lintner, B. (1994) 'Add water: Laos' hydroelectric plans seem overambitious', *Far Eastern Economic Review*, 13 October, pp. 70–1.

Lloyd-Parry, R. (1995) 'Food shortages in Northeast Asia', *Independent on Sunday*, 11 June.

Lohmann, L. (1990) 'Remaking the Mekong', *The Ecologist*, 20, 2 (March/April 1990), pp. 61–6.

Lohmann, L. (1991) 'Peasants, plantations and pulp', *Bulletin of Concerned Asian Scholars*, 23, 4, pp. 3–17.

Lohmann, L. (1992) 'Land, power and forest colonization in Thailand', in *Agrarian Reform and Environment in the Philippines and Southeast Asia*, London: Catholic Institute for International Relations, pp. 85–99.

Lohmann, L. (1995) 'No rules of engagement: the creative politics of 'environment' in Thailand', in J. Rigg (ed.), *Counting the Costs: Economic Growth and Environmental Change in Southeast Asia*, Singapore: ISEAS.

Lohmann, L. and Colchester, M. (1990) 'Paved with good intentions: TFAP's Road to oblivion', *The Ecologist*, 20, 3, May/June.

Lohmann, L., Wallgren, T. and Permpongsacharoen, W. forthcoming, 'Intercultural politics: the coevolution of a master plan and its critics', in S. Gudeman and S. Marglin, (eds), *People Count*.

Lonergan, S. and Parnwell, M.J.G. (1995) 'Assessing the linkages between environmental degradation and population displacement', paper presented to the session on 'Organizing Diversity: Migration Policy and Practice' at the Conference of the European Task Force on Canadian Studies, November 1995, Nijmegen, the Netherlands.

Lopez. M. E. (1987) 'The politics of lands at risk in a Philippine frontier', in P.D. Little and M.M. Horowitz (eds), *Lands at Risk in the Third World: Local Level Perspectives*, Boulder: Westview Press, pp. 230–48.

Lövgren, L. (1994) 'The dams debate in Sweden', in Swedish Society for Nature Conservation, *Nordic Dam-Building in the South*, Stockholm: Swedish Society for Nature Conservation.

McDowell, M. A. (1989) 'Development and the environment in ASEAN', *Pacific Affairs*, 62, pp. 307–29.

McGee, T.G. and Greenberg, C. (1992) 'The emergence of extended metropolitan regions in ASEAN: towards the year 2000', *ASEAN Economic Bulletin*, 9, 1, pp. 22–44.

McGregor, D.F.M. and Barker, D. (1991) 'Land degradation and hillside farming in the Fall River Basin, Jamaica', *Applied Geography*, 11, pp. 143–56.

Mackie, C. (1984) 'The lessons behind East Kalimantan's forest fires', *Borneo Research Bulletin*, 16, pp. 63–74.

MacKinnon, J. (1988) *The Nature Conservation System, National Parks and Protected Areas, Technical Report No. 3, Forestry Sector Review Vietnam*, VIE/88/037, FAO/UNDP/MoF.

MacKinnon, J., MacKinnon, K., Child, G. and Thorsell, J. (1986) *Managing Protected Areas in the Tropics*, Gland: IUCN.

McMorrow, J.M. and Douglass, I. (1995) 'Spatial and temporal patterns of forest and bush fires in Sabah', paper presented to Institute of British Geographers annual conference, Newcastle, January 1995.

McWilliam, F. (1995) 'Unnatural selection', *The Geographical*, February 1995, p. 20.

Mahmood, M. (1990) 'Political contestation in Malaysia', in N. Mahmood and Z. Haji Ahmad (eds), *Political Contestation: Case Studies from Asia*, Singapore: Heineman Asia.

Majid, A. *et al.* (1990) *Socio-Economic Development in Sarawak: Policies and Strategies for the 1990s*, Kuching: AZAM.

Malingreau, J.P. (1991) 'Remote sensing for tropical forest monitoring: an overview', in A.S. Belward and C.R. Valenzuela (eds), *Remote Sensing and Geographical Information Systems for Resource Management in Developing Countries*, ECSC/European Economic Community/EAEC, Brussels and Luxembourg, pp. 253–78.

Malingreau, J.P, Stephens, G. and Fellows, L. (1985) 'Remote sensing of forest fires: Kalimantan and North Borneo in 1982–83', *Ambio*, 14, 6, pp. 314–21.

Marchak, M. P. (1992) 'Latin America and the creation of a global forest industry', in H. Steen and R.P. Tucker (eds), *Changing Tropical Forests: Historical Perspectives on Today's Challenges in Central and South America*, New York: Forestry History Society.

Marsh, C.W. and Greer, A.G. (1992) 'Forest land-use in Sabah, Malaysia: an introduction to Danum Valley', *Philosophical Transactions of the Royal Society, London (B)*, 335, pp. 331–9.

Marsh, J.B. (ed.) (1992) *Resources and Environment in Asia's Marine Sector*, London: Taylor and Francis.

Marshall, A.G. (1992) 'The Royal Society's South-East Asian Rain Forest Research Programme: an introduction', in A.G. Marshall and M.D. Swaine (eds), *Tropical Rain Forest: Disturbance and Recovery*, London: Philosophical Transactions of the Royal Society of London, Series B, pp. 327–30.

Maull, H.W. (1992) 'Japan's global environmental policies', in A. Hurrell and B. Kingsbury, *The International Politics of the Environment*, Oxford: Clarendon Press, pp. 354–72.

Means, G.P. (1992) *Malaysian Politics: The Second Generation*, Singapore: Oxford University Press.

Mekong Secretariat (n.d.) *Response to the Review Made by the International Rivers Network on the Run-of-River Study*, Bangkok: Mekong Secretariat.

Middleton, N., O'Keefe, P. and Moyo, S. (1993) *The Tears of the Crocodile: From Rio to Reality in the Developing World*, London: Pluto.

Migdal, J. (1988) *Strong Societies and Weak States: State-Society Relations and State Capabilities in the Third World*, Princeton: Princeton University Press.

Miri-Bintulu Regional Planning Study (1974) London: Hunting Technical Services and Copenhagen: Hoff and Overgaard,.

Mongkhonvilay, S. (1991) 'Agriculture and environment under the New Economic Policy of the Lao People's Democratic Republic', in Nguyen Manh Hung, N.L. Jamieson and A.T. Rambo (eds), *Environment, Natural Resources, and the Future Development of Laos and Vietnam: Papers from a Seminar*, The Indochina Institute, George Mason University, Fairfax, Virginia, pp. 27–44.

Moore, E.H. (1984) 'Peter Williams-Hunt remembered', *Jurnal Persatuan Muzium Malaysia*, 3, pp. 98–103.

Moore, E.H. (1986) 'The moated Mu'ang of the Mun River Basin', unpublished PhD Thesis, University of London.

Morgan, R.P.C. (1986) *Soil Erosion and Conservation*, Longman: Harlow.

Morita, G. (1993) 'The global network for anti-golf course action', *Kyoto Journal*, 24, p. 13.

Morse, B. and Berger, T. (1992) *Sardar Sarovar: The Report of the Independent Commission*, Ottawa: Resource Futures International.

Moseley, H. (1994) 'Wary Laos erects border controls', *Phnom Penh Post*, 9–22 September, p. 16.

Mulla, Z. and Boothroyd, P. (1994) *Development-Oriented NGOs of Vietnam* (prepared through the co-operation of the Vietnam National Center for Social Sciences and Humanities), Vancouver: Centre for Human Settlements, University of British Columbia.

Mulyana, Y. (1995) 'Indonesia's national parks', *Conservation Indonesia*, 10, 4, January–March, Jakarta: WWF Indonesia Programme, p. 4.

Murdo, P. (1993) 'Cooperation, conflict in US–Japan environmental relations', *Japan Economic Institue Report*, Number 20A, 28 May, Washington: Japan Economic Institute.

Murdo, P. (1994) 'Japan's environmental foreign aid: what kind of edge?', *Japan Economic Institute Economic Report*, Number 31A, 12 August, Washington: Japan Economic Institute.

Mustapha, Z.H. (1991) 'In situ rural development in Sabah', in V.T. King and N.M. Jali (eds), *Issues in Rural Development in Malaysia*, University of Hull/Universiti Pertanian Malaysia, pp. 104–25.

Myers, N. (1993) 'Environmental refugees in a globally warmed world', *Bioscience*, 43, 11, pp. 752–61.

Narasimha M.P.A. (1990) 'The role of Japan', in J. Singh (ed.), *Developments in Asia-Pacific Region*, London: Tri-Service Press Limited, pp. 160–73.

Nation (Bangkok), various issues.

Nectoux, P. and Kuroda, Y. (1989) *Timber from the South Seas*, Gland: World Wide Fund for Nature.

Network First (1995) 'Vietnam – the last battle', ITV documentary by John Pilger, 25 April.

New Internationalist (1995) Special issue on 'Unmasked – the East Asian economic miracle', 263, January.

Ngidang, D. (1993) 'Media treatment of a land rights movement in Sarawak', *Media Asia*, 20, 2, pp. 93–9.

Nicholls, N. (1993) 'ENSO, drought and flooding rain in South-East Asia', in H. Brookfield and Y. Byron (eds), *South-East Asia's Environmental Future: The Search for Sustainability*, Kuala Lumpur: Oxford University Press, pp. 154–75.

Nishikawa, J. (1988) 'What is owed the debt-ridden Third World', *Japan Quarterly*, 35, 1, pp. 17–25.

NGO Forum Japan (1992) *People's Voice of Japan – Submission to UNCED 1992*, Tokyo: 92 NGO Forum.

Noakes, J.L. (1950) *Sarawak and Brunei: A Report on the 1947 Population Census*, Kuching: Government Printer.

Norpower, (1993) *Nam Theun 1/2 Hydropower Project Feasibility Study, Volume 1, Main Report*.

Norway: Directorate for Nature Management (1993) *Nam Theun 1/2 Hydroower Project, Laos: Assesment of Environmental Impacts*, Trondheim: Directorate for Nature Management, 16 August 1993.

Numata, M. (1995) 'Conservation education based on sustainability and biodiversity' *Natural History Research*, 3, 2, pp. 67–74.

Oberai, A.S. (1988) 'An overview of land settlement policies and programs', in A.S. Oberai (ed.), *Land Settlement Policies and Population Redistribution in Developing Countries: Achievements, Problems and Prospects*, New York: Praeger.

Ofreneo, R.E. (1993) 'Japan and the environmental degradation of the Philippines', in M. Howard (ed.), *Asia's Environmental Crisis*, Boulder, Colorado: Westview Press, pp. 201–20.

Ohlsson, B. (1988) *Forestry and Rural Development, Technical Report No.2, Forestry Sector Review Vietnam*, VIE/88/037, FAO/UNDP/MoF.

Olsen, L. (1970) *Japan in Post-War Asia*, New York: Frederick A. Praeger.

Orr, R.M. (1990) *The Emergence of Japan's Foreign Aid Power*, New York: Columbia University Press.

O'Tuathail, G. (1993) 'The new east-west conflict? Japan and the Bush administration's "New World Order" ', *Area*, 25, 2, pp. 127–35.

Overseas Development Administration (ODA) (1992) *British Overseas Aid Annual Review*, London: ODA.

Ovesen, Jan (1993) *Anthropological Reconnaissance in Central Laos: A Survey of Communities in a Hydropower Project Area*, Uppsala: Uppsala Research Reports in

Cultural Anthropology No. 13, Uppsala University.

Ovesen, J. (1995) *A Minority Enters the Nation State: A Case Study of a Hmong Community in Vientiane Province, Laos*, Uppsala: Uppsala Research Reports in Cultural Anthropology No. 14, Uppsala University.

Ozaki, R.S. and Arnold, W. (eds) (1985) *Japan's Foreign Relations: A Global Search for Economic Security*, Boulder, Colorado: Westview Press.

Pacific Affairs 1989–90 Special issue on 'Japanese aid', 62, 4.

Padoch, C. (1980) 'The environmental and demographic effects of alternative cash-producing activities among shifting cultivators in Sarawak', in J.I. Furtado (ed.), *Tropical Ecology and Development*, Kuala Lumpur: International Society of Tropical Ecology, pp. 475–81.

Padoch, C. (1982) 'Land use in new and old areas of Iban settlement', *Borneo Research Bulletin*, 14, pp. 3–14.

Padoch, C. (1988) 'Agriculture in interior Borneo: shifting cultivation and alternatives', *Expedition*, 30, 1, pp. 18–28.

Pain, M., Benoit, D., Levang, P. and Sevin, O. (1989) *Transmigration and Spontaneous Migrations in Indonesia*, Bondy, France: Orstom.

Paper and Packaging Analyst, various issues.

Paper Asia, various issues.

Park, C.C. (1992) *Tropical Rainforests*, London: Routledge.

Park, S.J. (1979) 'Foreign investment and new imperialism theories with special reference to Japanese foreign investment in East and Southeast Asia', in I. Nish and C. Dunn (eds), *European Studies on Japan*, Kent, England: Norbury, pp. 172–81.

Parnwell, M.J.G. (1993a) *Population Movements and the Third World*, London: Routledge (Introductions to Development series).

Parnwell, M.J.G. (1993b) 'Tourism and rural handicrafts in Thailand', in M. Hitchcock, V.T. King and M.J.G. Parnwell (eds), *Tourism in South-East Asia*, London: Routledge, pp. 234–57.

Parnwell, M.J.G., (1994) 'Rural industrialisation and sustainable development in Thailand', *TEI Quarterly Environment Journal*, Bangkok: Thailand Environment Institute.

Parnwell, M.J.G. (ed.), 1995 *Thailand: Uneven Development*, Aldershot Avebury.

Parnwell, M.J.G., forthcoming, 'Environmental change and Iban communities in Sarawak, East Malaysia: impact and response', in K. Ogino and E. Bruenig (eds), special edition on 'Socio-Sustainability Mechanisms in Tropical Rainforest Ecosystems' in the *Journal of the Japan Society of Tropical Ecology* (in press).

Parnwell, M.J.G. (n.d.) 'Environmental degradation, non-timber forest products and Iban communities in Bintulu Division, Sarawak: impact, response and future prospects', unpublished paper.

Parnwell, M.J.G. and Khamanarong, S. (1990) 'Rural industrialisation and development planning in Thailand', *Southeast Asian Journal of Social Science*, 18, 2, pp. 1–28.

Parnwell, M.J.G. and King, V.T. (1995) 'Environmental degradation, resource scarcity and population movement among the Iban of Sarawak, Malaysia', paper presented to a panel on 'Human-Environment Interaction in South-East Asia' at the First EUROSEAS Conference, Leiden, June 1995.

Parnwell, M.J.G. and Morrison, J. (1994) 'Deforestation and the response of Iban communities in East Malaysia: sustainable development or firefighting?', unpublished paper.

Pearce, D., Barbier, E. and Markandya, A. (1990) *Sustainable Development: Economics and Environment in the Third World*, London: Earthscan.

Pearce, F. (1991) *Green Warriors: The People and the Politics Behind the Environmental Revolution*, London: Bodley Head.

Pearce, F. (1992) *The Dammed: Rivers, Dams and the Coming World Water Crisis*, London: Bodley Head.

Pearce, F. (1994) 'Are Sarawak's forests sustainable?', *New Scientist*, 26 November, pp. 28–32.

Pearce, K.G., Amen, V.L. and Jok, S. (1987) 'An ethnobotanical study of an Iban community of the Pantu sub-district, Sri Aman, Division 2, Sarawak', *Sarawak Museum Journal*, XXXVII, 58 n.s., Dec., pp. 193–270.

Pearson, C.S. (ed.) (1987) *Multinational Corporations, the Environment and the Third World*, Durham: Duke University Press.

Peet, R. and Watts, M. (1993) 'Development theory and environment in an age of market triumphalism', *Economic Geography*, 69, pp. 227–53.

Peluso, N.L. (1991) 'Rattan industries in East Kalimantan, Indonesia', in Food and Agriculture Organization, *Case Studies in Forest-Based Small Scale Enterprises in Asia: Rattan, Matchmaking and Handicrafts*, Bangkok: FAO, Community Forestry Case Study Series No. 4, pp. 5–28.

Peluso, N.L. (1992) *Rich Forests, Poor People: Resource Control and Resistance in Java*, Berkeley: University of California Press.

Peluso, N. L. (1993a) 'Coercing conservation?', *Global Environmental Change*, 3, 2, June, pp. 199–217.

Peluso, N. (1993b) 'Coercing conservation: the politics of state resource control', in R. Lipschutz and K. Conca (eds), *The State and Social Power in Global Environmental Politics*, New York: Columbia University Press.

Pepper, D. (1984) *The Roots of Modern Environmentalism*, London: Routledge.

Pepper, D. (1993) *Eco-Socialism: From Deep Ecology to Social Justice*, London: Routledge.

Permpongsacharoen, W. (n.d.) *Tropical Forest Movements: Some Lessons from Thailand*, Bangkok: Project for Ecological Recovery.

Pernia, E.M. (1991) 'Aspects of urbanization and the environment in Southeast Asia', *Asian Development Review*, 9, 2, pp. 113–36.

Persoon, G.A. (1994) 'Vluchten of Veranderen: Processen van Veranderingen en Ontwikkelingen bij Tribale Groepen in Indonesia', unpublished PhD thesis, State University of Leiden.

Pham, C. (ed.) (1994) *Economic Development in Lao PDR*, Vientiane: Horizon 2000.

Phuu Jatkaan Raai Sapdaa (Bangkok), various issues.

Picard, M. (1993) 'Cultural tourism in Bali: national integration and regional differentiation', in M.J. Hitchcock, V.T. King and M.J.G. Parnwell (eds), *Tourism in South-East Asia*, London: Routledge, pp. 71–98.

Pierce, J.T. (1992) 'Progress and the biosphere: the dialectics of sustainable development', *The Canadian Geographer*, 36, 4, pp. 306–20.

Pigram, J. (1983) *Outdoor Recreation and Resources Management*, Kent: Croom Helm.

Ping, L.P. (1990) 'ASEAN and the Japanese role in Southeast Asia', in A. Broinowski (ed.), *ASEAN into the 1990's*, London: MacMillan.

Pitman, J. (1992) 'Japan seeks role as Earth's defender', *The Times*, 17 April.

Plantilla, J.R. (1993) 'Strengthening community stewardship agreements in the Philippines', in J. Fox (ed.), *Legal Frameworks for Forest Management in Asia*, Hawaii: East-West Center, pp. 73–87.

Pletcher, J. (1991) 'Regulation with growth: the political economy of palm oil in Malaysia', *World Development*, 19, 6, pp. 623–36.

Pleumarom, A. (1992) 'Golf tourism in Thailand', *The Ecologist*, 22, 3, pp. 104–10.

Poffenberger, M. (ed.) (1990) *Keepers of the Forest: Land Management Alternatives in Southeast Asia*, West Hartford: Kumarian Press.

Polunin, N.V.C. (1990) 'Marine regulated areas: an expanded approach for the tropics' in J.I. Furtado and K. Ruddle (eds), *Tropical Resources, Ecology and Development*, London: Wiley.

Potter, D.C. (1994) 'Assessing Japan's environmental aid policy', *Pacific Affairs*, 67, 2, pp. 200–15.

Potter, L. (1987) 'Degradation, innovation and social welfare in the Riam Kiwa Valley, Kalimantan, Indonesia', in P. Blaikie and H. Brookfield (eds) *Land Degradation and Society*, London and New York: Methuen, pp. 164–76.

Potter, L. (1993) 'The onslaught on the forests in South-East Asia', in H. Brookfield and Y. Byron (eds), *South-East Asia's Environmental Future: The Search for Sustainability*, Kuala Lumpur: Oxford University Press, pp. 103–23.

Prachachart Thurakit (Bangkok), various issues.

Primack, R.B. (1991) 'Logging, conservation and native rights in Sarawak forests', *Conservation Biology*, 5, pp. 126–30.

Pringle, R. (1970) *Rajahs and Rebels: the Ibans of Sarawak under Brooke Rule, 1841–1941*, London: Macmillan.

Project for Ecological Recovery and NGO Coordinating Committee for Rural Development (1993) *Fish, Forests and Food: Means of Livelihood in Mun River Village Communities*, Bangkok: Project of Ecological Recovery.

PR Watch (Madison, WI, USA), various issues.

Pulp and Paper International, various issues.

Puntasen, A., Siriprachai, S. and Punyasavatsut, C. (1992) 'Political economy of eucalyptus: business, bureaucracy and the Thai government', *Journal of Contemporary Asia*, 22, 2, pp. 187–206.

Purnomo, A. (1994) 'Policy influence for sustainable development: case studies from the environmental movement in Indonesia', unpublished MA Thesis, Tufts University.

Quy, V. and Hu Ninh, N. (eds) (1990) *Environment Newsletter*, Hanoi: Vietnam Centre for Natural Resources Management and Environment Studies (CRES).

Ratnapala, L. (1994) *Tourism in the Asia-Pacific Region and Opportunities for Indonesia*, Jakarta: Travel Indonesia.

Redclift, M. (1984) *Development and the Environmental Crisis: Red or Green Alternatives?*, London: Methuen.

Redclift, M. (1987) *Sustainable Development: Exploring the Contradictions*, London: Methuen.

Redclift, M. (1992) 'The meaning of sustainable development', *Geoforum*, 23, pp. 395–403.

Redclift, M. and Benton, T. (eds) (1994) *Social Theory and the Global Environment*, London: Routledge.

Remigio, A.A. (1993) 'Philippine resource policy in the Marcos and Aquino governments: a comparative assessment', *Global Ecology and Biogeography Letters*, 3, pp. 192–212.

Repetto, R. and Gillis, M. (eds) (1988) *Public Policies and the Misuse of Forest Resources*, Cambridge: Cambridge University Press.

Richards, J.F. and Tucker, R.P. (eds) (1988) *World Deforestation in the Twentieth Century*, Durham: Duke University Press.

Rigg, J.D. (1991) 'Grassroots development in Thailand: a lost cause?', *World Development*, 19, 2–3, pp. 199–211.

Rigg, J.D. (1995) '"In the fields there is dust": Thailand's water crisis', *Geography*, 346, 80, 1, pp. 23–32.

Rigg, J. forthcoming, '*Chin thanakaan mai*: new thinking in the Lao RDR', *Contemporary Southeast Asia*.

Rigg, J. and Stott, P. forthcoming, 'Forest tales: politics, environmental policies and their implementation in Thailand', in U. Desai (ed.), *Comparative Environmental Policy and Politics*, New York: SUNY Press.

Rix, G. (1993) *Japan's Foreign Aid Challenge: Policy Reform and Aid Leadership*, London: Routledge.

Robertson, B. (1988) *Marketing Perspectives on Tourism in National Parks*, Sydney, (proceedings of a seminar on National Parks and Tourism).

Robinson, A. and Sumardja, E. (1990) 'Indonesia', in C. Allin (ed.), *International Handbook of National Parks and Nature Reserves*, Greenwood Press, pp. 197–214.

Robison, R. (1993) 'Indonesia: tensions in state and regime', in K. Hewson, R. Robinson and G. Rodan (eds), *Southeast Asia in the 1990s: Authoritarianism, Democracy and Capitalism*, St Leonards NSW: Allen & Unwin.

Rudner, M. (1989) 'Japanese ODA to Southeast Asia', *Modern Asian Studies*, 23, 1, pp. 73–116.

Rudner, M. (1995) 'APEC: the challenges of Asia Pacific economic cooperation', *Modern Asian Studies*, 29, 2, pp. 403–37.

Rush, J. (1991) *The Last Tree: Reclaiming the Environment in Tropical Asia*, New York: Asia Society.

Ryder, G. (1995) 'Case Study: Pak Mun Dam in Thailand', paper presented to the Both Sides of the Dam Symposium, Delft.

Ryder, G. and Rothert, S. (1994) 'Rent-a-river, build a dam', *World Rivers Review*, 9, 4, p. 9.

Sabah Forest Department (1989) 'Forestry in Sabah', unpublished report, Sabah Forest Department, Sandakan.

Sabah Forest Department (n.d.) (c. 1992) 'Forestry and forest industry: fact sheet on Sabah', unpublished report, Sabah Forest Department, Sandakan.

Saccheri, I. and Walker, D. (1992) 'Subsistence and environment of a highland Kenyah community', *Sarawak Museum Journal*, 42, pp. 193–249.

Sahabat Alam Malaysia/World Rainforest Movement (1989) *The Battle for Sarawak's Forests*, Penang, Malaysia: Sahabat Alam Malaysia/World Rainforest Movement.

Sajogyo (1993) 'Agriculture and industrialization in rural development', in J-P. Dirkse, F. Husken and M. Rutten (eds), *Development and Social Welfare: Indonesia's Experiences Under the New Order*, Leiden: KITLV Press.

Sakurai, K. (1993) 'Japan's new government and Official Development Assistance', *INTEP Newsletter*, 3.

Salafsky, N. (1994) 'Drought in the rain forest: effects of the 1991 El Niño–Southern Oscillation Event on the rural economy in West Kalimantan, Indonesia', *Climatic Change*, 27, pp. 373–96.

Salim, E. (1988) 'Towards sustainable development of aquatic resources', in P.R. Burbridge, Koesoebiono, H. Dirschl and B. Patton (eds), *Coastal Zone Management in the Strait of Malacca*, 3–7, DESC/EMDI, School of Resource and Environmental Studies, Dalhousie University, Halifax, Canada.

Sanchez, P., Garrity, D. and Bandy, D. (1993) 'Sustainable alternatives to slash and burn agriculture and the reclamation of degraded land in the humid tropics', paper presented at the Global Forest Conference: Beyond UNCED – response to Agenda 21, Bandung, Indonesia, February 1993.

Sandin, B. (1980) *Iban Adat and Augury*, Penang: Penerebit Universiti Sains Malaysia.

Saramany, W. (1989) 'Status of wood-based industry in Laos', unpublished paper, Vientiane.

Sarawak Gazette, issues from 1899 to 1980.

Sargent, C. (1988) *Land Use Issues, Technical Report No. 1, Forestry Sector Review Vietnam*, VIE/88/037, FAO/UNDP/MoF.

Sargent, C. (1990) *The Khun Song Plantation Project*, London: IIED.

Sargent, C. and Bass, S. (1992) *Plantation Politics: Forest Plantations in Development*, London: Earthscan.

Sargent, C., Palmer, J. and Morrison, E. (1991) *Proceedings of the National Seminar on Setting Priorities for Research in the Land Use Continuum in Vietnam*, Hoa Binh, September 9–13.

Sattaur, O. (1991) 'Last chance for rainforest plan', *New Scientist*, 2 March 1991.

Sayakoummane, N. (1994) 'Damming the Theun River: a local perspective', in A.D. Usher (ed.), *Nordic Dam-Building in the South*, proceedings of an International Conference, Stockholm, 3–4 August 1994, Stockholm: Swedish Society for Nature Conservation, pp. 26–8.

Scholten, H.J. and de Lepper, M.J.C. (1991) 'The benefits of the application of Geographical Information Systems in public and environmental health', *World Health Statistics Quarterly*, 44, pp. 160–70.

Schramm, G. and Warford, J.J. (eds) (1987) *Environmental Management and Economic Development*, Baltimore: John Hopkins University Press.

Schreuder G. and Anderson, E. (1988) 'International wood chip trade: past developments and future trends, with emphasis on Japan', in G. Schreuder (ed.), *Global Issues and Outlook in Pulp and Paper*, Seattle: University of Washington.

Schumacher, E.F. (1974) *Small is Beautiful: A Study of Economics as if People Mattered*, London: Abacus.

Schwarz, A. (1989) 'Tourism in Indonesia', *South*, May 1989, pp. 63–6.

Scoones, I. and Thompson, J. (eds) (1994) *Beyond Farmer First: Rural People's Knowledge, Agricultural Research and Extension Practice*, London: Intermediate Technology Publications.

Scott, J.C. (1976) *The Moral Economy of the Peasant: Rebellion and Subsistence in Southeast Asia*, New Haven: Yale University Press.

Secrett, C. (1986) 'The environmental impact of transmigration', *The Ecologist*, 16, 2/3, pp. 77–88.

Sexton, S. (1992) 'Asia in the bunker', *The Ecologist*, 22, 3, p. 109.

Seymour, R. (1993) 'Yen and love for emotional drama to keep Japan paralysed', *Mainichi Daily News*, Readers Forum, 10 March.

Sharma, P.N. (1988) *Soil and Water Protection and Watershed Management, Field Document No. 2, Forestry Sector Review Vietnam*, VIE/88/037, FAO/UNDP/MoF.

Shell and WWF (1993) *Tree Plantation Review*, Study No. 3: Industrial Wood, London: Shell International Petroleum Company and World Wide Fund for Nature.

Sherman, C. (1995) *Thailand's Energy Tentacles: Power Plants, Dams and Disaster Fuelling 'Development' in Indochina*, Woollahra: Aid/Watch.

Shiraishi, S. and Shiraishi, T. (eds) (1993) *The Japanese in Colonial Southeast Asia*, Cornell University Southeast Asia Programme, Translation Series Number 3, Ithaca: Cornell University.

SIDA (Swedish International Development Agency) (1991) 'Agreed minutes on discussions between the Secratariat of the Mekong Committee and the Swedish International Development Agency on Development Cooperation', Stockholm: SIDA Travel Report, 5 May 1991.

Siew, K.Y. (1973a) *The Present Land Use of the Tawau Residency, Sabah, 1970*, Sabah Present Land Use Report No. 3, Department of Agriculture, Sabah.

Siew, K.Y. (1973b) *The Present Land Use of the Sandakan Residency, Sabah, 1970*, Sabah Present Land Use Report No. 4, Department of Agriculture, Sabah.

Silva, E. (1994) 'Thinking politically about sustainable development in the tropical forests of Latin America', *Development and Change*, 25, pp. 697–721.

Singh, K.D. (1994) 'The status of resources in the tropical forests based on FAO 1990 assessment', in *Multiple Resource Inventory and Monitoring of Tropical Forests*, proceedings of ASEAN Institute of Forest Management (AIFM) international conference, Seremban, Malaysia, 21–24 November 1994.

Sirindhorn, HRH Princess Maha Chakri, Silapacharanan, S. and Srisuksawadi, K. (1990) 'Land use/land cover map accuracy assessment of Landsat Thematic Mapper data using the DIMAPS image processing system for Narathiwat Province, Thailand', *Geocarta International*, 5, 1, pp. 15–24.

Sjöström, P. (1994) *Environmental Considerations Regarding the Theun Hinboun Hydropower Project*, Stockholm: Vattenfall.

Skanska Annual Report 1993, Stockholm.

SKEPHI (1992) *To Sink the Island: Logging Problems on Yamdena Island*, Jakarta: Network for Forest Conservation in Indonesia.

Sklar, L. and McCully, P. (1994) *Damming the Rivers: The World Bank's Lending for Large Dams*, San Francisco: International Rivers Network.

Sloan, N.A. and Sugandhy, A. (1994) 'An overview of Indonesian coastal environmental management', *Coastal Management*, 2, pp. 215–33.

Sluiter, L. (1992) *The Mekong Currency*, Bangkok: TERRA/Duang Kamol.

Smyth, I. (1991) 'The Indonesian family planning programme: a success story for women?', *Development and Change*, 22, pp. 781–805.

Soegiarto, A. (1993) 'A successful prediction using unconventional data', in H. Brookfield and Y. Byron (eds), *South-East Asia's Environmental Future: The Search for Sustainability*, Kuala Lumpur: Oxford University Press, pp. 176–8.

Soegiarto, A. and Polunin, N.V.C. (1981) *The Marine Environment of Indonesia*, Bogor, Indonesia: IUCN/WWF.

Soemarwoto, O. (1991) 'Human ecology in Indonesia: the search for sustainability in development', in J. Hardjono (ed.), *Indonesia: Resources, Ecology and Environment*, Oxford: Oxford University Press, pp. 212–35.

Soetikno, A. (1993) 'Appita's Asia profile', *Appita Journal*, 46, 2, pp. 90–2.

Sopher, D.E. (1977) *The Sea Nomads*, Singapore: National Museum of Singapore.

Sparks, S. (1988) 'Japanese banks and the Third World', *Multinational Monitor*, November, p. 24.

Stallings, B. and Sakurai, M. (1993) 'Development in the 1990's: U.S. and Japanese paradigms' in B. Stallings *et al.*, *Common Vision, Different Paths: The United States and Japan in the Developing World*, Overseas Development Council: Washington D.C.

Stankey, G.H. (1988) 'Tourism and national parks: perils and potential', in *National Parks and Tourism*, proceedings of a New South Wales National Parks and Wildlife Service seminar, Sydney.

State of Sarawak, Forest Department (1986) *et seq.*, *Annual Report*, Kuching: Sarawak Forest Department.

Stocking, M.J. and Elwell, H.A. (1973) 'Soil erosion hazard in Rhodesia', *Rhodesian Agriculture Journal*, 70, pp. 93–101.

Stocking, M.J. and Elwell, H.A. (1976) 'Rainfall erosivity over Rhodesia', *Transactions of the Institute of British Geographers*, NS1, pp. 231–46.

Stuart-Fox, M. (1986) *Laos: Politics, Economics and Society*, London: Pinter.

Stuart-Fox, M. (1995) 'Laos: towards sub-regional integration', *Southeast Asian Affairs 1995*, Singapore: ISEAS, pp. 177–95.

Suara Pembaruan (Jakarta), various issues.

Sudo, S. (1988) 'The road to becoming a regional leader: Japanese attempts in Southeast Asia, 1975–1980', *Pacific Affairs*, 61, 1, pp. 27–50.

Sugardjito, B.H. (1995) 'The development of national parks in Indonesia', *Conservation Indonesia*, 10, 4, Jakarta: WWF, pp. 18–19.

Sukpanich, T. (1990) 'Killing the land for money', *Bangkok Post*, 22 February.

Sumardja, E.A. (1995) 'National parks and foreign aid', *Conservation Indonesia*, 10, 4, Jakarta: WWF, pp. 14–15.

Sun, M. (1988) 'Japan prodded on the environment', *Science*, 241, 15 July, p. 284.

Susilawati, S. and Weir, M.J.C. (1990) 'GIS applications in forest land management in Indonesia', *ITC Journal*, 1990–3, pp. 236–44.

Sutlive, V.H. (1978) *The Iban of Sarawak*, Arlington Heights (Ill.): AHM Pub. Corp.

Sutlive, V.H. (1989) 'Ethnic identity and aspirations', *Sarawak Museum Journal*, XL, 61 (1), pp. 35–49.

Sutlive, V.H. (1992) *The Iban of Sarawak: Chronicle of a Vanishing World*, Kuala Lumpur: S. Abdul Majeed & Co.

Sutton, K. (1988) 'Land settlement in Sabah: from the Sabah Land Development Board to the Federal Land Development Authority', *Malaysian Journal of Tropical Geography*, 18, pp. 46–56.

Suzuki, M. (1993) 'Whirlpools in the mainstream', *Resurgence*, 137, pp. 24–5.

Sveriges Natur various issues.

Taillard, C. (1989) *Le Laos – stratégies d'un état-tampon*, Montpellier: Reclus.

Tanaka, S. (1986) *Post-War Japanese Resource Politics and Strategies: The Case of Southeast Asia*, Ithaca: Cornell China–Japan Program.

Tanjung, O. (1992) 'Gerakan Advoksi Lingkun', Medan: WIM, cyclostyled.

Tasker, R. and Ai, B. (1994) 'Virgin territory: Malaysia and Indonesia form huge forest reserve', *Far Eastern Economic Review*, 8 December.

Taylor, D.M., Hortin, D., Parnwell, M.J.G., King, V.T. and Marsden, T.K. (1994) 'Recent changes in tropical forest quality in Sarawak, East Malaysia and its implications for future management policies', *Geoforum*, 25, 3, pp. 351–69.

Tempo (Jakarta), various issues.

TERRA (1993) *Comment on the May 1993 Draft Fensibility Study by Norpower of the Nam Theun 1/2 Hydropower Project: Comments for discussion in Norway*, Bangkok: TERRA, 4 August 1993.

TERRA (1995) 'Nordic Hydropower Briefing', Appendix I, *First Nordic-Asian Development Bank BOT Venture in The Lao PDR: Nam Theun Hinboun Project*, Bangkok: TERRA Update, April 1995.

Teves, A. and Lewis, D. (1993) 'Overview', in J. Farrington and D. Lewis (eds), *Non-Government Organisations and the State in Asia*, London: Routledge, pp. 227–39.

Thailand: Project for Ecological Recovery (1992) *The Future of People and Forests in Thailand After the Logging Ban*, Bangkok: Project for Ecological Recovery.

Thai NGOs (1995) *Statement on Cooperation for the Sustainable Development of the Mekong River Basin*, Chiang Rai: Thai NGOs.

Thang, H.C. (1993) 'Monitoring forest cover with the use of satellite data in Malaysia', paper presented at the Regional Workshop on Tropical Forest Ecosystem Research, Conservation and Repatriation, Hanoi, Vietnam, 25 June–1 July 1993.

Thorbecke, E. and van der Pluijm, T. (1993) *Rural Indonesia: Socio-Economic Development in a Changing Environment*, New York: International Fund for Agricultural Development.

Ting, L.B. (1979) 'The effects of shifting cultivation on sustained yield management for Sarawak national forests', *Malaysian Forester*, 42, pp. 418–422.

Tjondronegoro, S. (1991) 'The utilization and management of land resources in Indonesia, 1970–1990', in J. Hardjono (ed.), *Indonesia: Resources, Ecology, and Environment*, Oxford: Oxford University Press.

Tobin, R.J. and White, A.T. (1993) 'Coastal resources management and sustainable development: a Southeast Asian perspective', *International Environmental Affairs*, 5, 1, pp. 50–65.

Totman, C. (1989) *The Green Archipelago: Forestry in Pre-Industrial Japan*, California: UCL Press.

Townsend, J. (1993a) 'Housewifisation and colonisation in the Colombian rain-forest', in J. Momsen and V. Kinnaird (eds), *Different Places, Different Voices: Gender and Development in Africa, Asia and Latin America*, London: Routledge.

Townsend, J. (1993b) 'Gender and the life course on the frontiers of settlement in Colombia', in C. Katz and J. Monk (eds), *Full Circles: Geographies of Women Over the Life Course*, London: Routledge.

Townsend, J. (1995) *Women's Voices from the Rainforest*, London: Routledge.

Traisawasdichai, M. (1994), 'Laotians "look likely losers" from foreign-funded forestry', *The Nation* (Bangkok), 11 July.

Traisawasdichai, M. (1995) 'Chasing the little white ball', *New Internationalist*, 263, pp. 16–17.

Trankell, I. (1993) *On the Road in Laos: An Anthropological Study of Road Construction and Rural Communities*, Uppsala: Uppsala Research Reports in Cultural Anthropology No. 12, Uppsala University.

Transmigration Advisory Group (1989) *Mid-Term Review*, Ministry of Transmigration, Directorate General Settlement Preparation, Jakarta.

Tsuruoka, D. (1994) 'Awakening giant – Sarawak sees its future as Asia's energy provider', 'In Sarawak's forests, there is more at risk than trees', *Far Eastern Economic Review*, 21 July, pp. 68–71.

Tucker, R.P. and Richards, J.F. (eds) (1983) *Global Deforestation and the Nineteenth Century World Economy*, Durham: Duke University Press.

Tudge, C. (1991) *Last Animals at the Zoo*, London: Michael Joseph.

Udol, A. (1993) 'The development of the pulp and paper industries in Thailand', *Appita Journal*, 46, 1, pp. 14–18.

Ui, J. (ed.) (1992) *Industrial Pollution in Japan*, Tokyo: United Nations University Press.

United Nations Development Programme (1990) *Development Co-operation: Lao People's Democratic Republic*, Vientiane, Laos: United Nations Development Programme.

United Nations Environment Programme (1993) *Environmental Data Report 1993–94*, Oxford: Basil Blackwell.

United States Department of the Treasury (USDT) (1993) *Treasury News*, 18 November.

Usher, A.D. (1990a) 'A forest policy sadly gone awry', *The Nation* (Bangkok), 10 May.

Usher, A.D. (1990b) 'Eucalyptus – widening the gap', *The Nation* (Bangkok), 14 June.

Usher, A.D. (1993) 'The Bai Bang legacy', *The Nation*, Bangkok, 23 June 1993.

Usher, A.D. (1994a) 'Dam-building in the south: The Nordic connection', *Sveriges Natur*, 3, pp. 30–34.

Usher, A.D. (ed.) (1994b) *Nordic Dam-Building in the South*, proceedings of an international conference in Stockholm, Stockholm: Swedish Society for Nature Conservation.

Vandergeest, P. and Peluso, N.L. (1993) 'Fixing property in national space: territorialization of the state in Siam/Thailand', New Haven: unpublished manuscript.

Vayda, A.P. (1979) 'Human ecology and economic development in Kalimantan and Sumatra', *Borneo Research Bulletin*, 11, pp. 23–32.

Vayda, A.P. (1981) 'Research in East Kalimantan on interactions between people and forests: a preliminary report', *Borneo Research Bulletin*, 13, pp. 3–15.

Vayda, A.P., Colfer, C.J.P. and Brotokusomo, M. (1980) 'Interactions between people and forests in East Kalimantan', *Impact of Science on Society*, 30, pp. 179–90.

Verstappen, H. (1956) 'The physiographic basis of pioneer settlement in Southern Sumatra', Kementerian Pertahanan, Jakarta.

Vitug, M. D. (1993) *Power from the Forest*, Manila: Philippine Center for Investigative Journalism.

de Vylder, S. and Sonnerup, B. (1994) *Lao PDR Energy Sector Review*, report commissioned by the Swedish International Development Authority, Stockholm/Malmö.

Wahana Lingkungan Hidup Indonesia (WALHI) and Yayasan Lembaga Bantuan Hukum Indonesia (YLBHI) (1992) *Mistaking Plantations for Indonesia's Tropical Forest*, Jakarta: WALHI and YLBHI.

Walker, R.B.J. (1993) *Inside/Outside: International Relations as Political Theory*, Cambridge: Cambridge University Press.

Wallace, C. (1992) 'Is Asia robbing rural poor to power the rich?', *Los Angeles Times*, 22 February 1992.

Walpole, P. *et al.* (1993) 'Upland Philippine communities: guardians of the final forest frontiers', in M. Poffenberger and B. McGean (eds), *Southeast Asia Sustainable Forest Management Network Research Report*, Number 4, August.

Walsh, R.P.O. (n.d.) 'Drought frequency changes in Sabah and adjacent parts of northern Borneo since the late nineteenth century and possible implications for tropical rain forest dynamics', unpublished paper.

Wapner, P. (1995) 'Politics beyond the state: environmental activism and world civic politics', *World Politics*, 47, pp. 311–40.

Weinstock, J.A. (1983) 'Rattan: ecological balance in a Borneo rainforest swidden', *Economic Botany*, 37, 1, pp. 58–68.

Western, A. (1979) *Small-Scale Papermaking*, Rugby: Intermediate Technology Industrial Services.

Westing, A.H. (1992) 'Environmental refugees: a growing category of displaced persons', *Environmental Conservation*, 19, 3, pp. 201–7.

Westoby, J. (1987) *The Purpose of Forests*, Oxford: Blackwell.

Wheat, S. (1994) 'Taming tourism', *Geographical Magazine*, April 1994, pp. 16–19.

Wheat, S. (1995) 'Tourism and human rights', *Geographical Magazine*, March, p. 10.

Wheeller, B. (1991) 'Tourism's troubled times: responsible tourism is not the answer', *Tourism Management*, June 1991, Oxford: Butterworth Heinemann.

Wheeller, B. (1993) 'Sustaining the ego', *Journal of Sustainable Tourism*, 1, 2, pp. 121–9.

White, A.T. and Palaganas (1991) 'Philippines Tubbataha Reef National Marine Park: status, management, issues and proposed plan', *Environmental Conservation*, 18, pp. 148–57.

Whitmore, T.C. (1975) *Tropical Rain Forests of the Far East*, Oxford: Clarendon Press.

Wilford, G.E. (1961) 'The geology and mineral resources of Brunei and adjacent parts of Sarawak with descriptions of Seria and Miri oilfields', Memoir 10, Geological Survey Department, British Territories in Borneo.

Williams, M. (1990) 'Forests', in B.L. Turner, W.C. Clark, R.W. Kates *et al.* (eds), *The Earth as Transformed by Human Action*, Cambridge: Cambridge University Press.

Wilson, E.O. (1992) *The Diversity of Life*, London: Penguin.

Winichakul, T. (1994) *Siam Mapped: A History of the Geo-body of a Nation*, Honolulu: University of Hawaii Press.

Winterbottom, R. (1990) *Taking Stock: TFAP After 5 Years*, London: WRI.

Wirawan, N. (1993) 'The hazard of fire', in H. Brookfield and Y. Byron(eds), *South-East Asia's Environmental Future: The Search for Sustainability*, Kuala Lumpur: Oxford University Press, pp. 242–60.

de Wok, W.J. and Pearce, J. (1991) *Japanese Bear Parks*, London: WSPA Report Series.

Wong, J.K.M. (1992) *Hill Logging in Sarawak,* Kuching: Sarawak Press.

Woon, W.C. and Lim, H.F. (1990) 'The non-government organisations and government policy on environmental issues in Malaysia', *Wallaceana*, 59 and 60, pp. 10–15.

World Bank (1988) *Indonesia: The Transmigration Program in Perspective*, Washington, DC: World Bank.

World Bank (1991) *The Pak Mun Hydropower Project Environmental Fact Sheet*, Washington, DC: World Bank.

World Bank (1993) *Lao People's Democratic Republic: Environmental Overview*, Washington, DC: World Bank.

World Bank (1994) *Indonesia Environment and Development: Challenges for the Future*, Washington, DC: World Bank.

World Commission on Environment and Development (1987) *Our Common Future*, Oxford: Oxford University Press.

World Rivers Review (San Francisco), various issues.

Worldwatch Institute (1995) *State of the World, 1995*, London: Earthscan/Worldwatch Institute.

Worldwide Fund for Nature (1995) 'Wasur National Park Project Report', *Conservation Indonesia*, 10, 4, January–March, Jakarta: WWF Indonesia Programme.

Wright, R. (1993) 'World pulp market: forecasts and prospects as at mid-1992', *Paper and Packaging Analyst*, 14, pp. 13–21.

Wright, R. (1994) 'New markets; new developments: Indonesia', presentation at *Financial Times* conference on World Pulp and Paper, London, May.

Wu, Y.L. (1977) *Japan's Search for Oil: A Case Study of Economic Nationalism and International Security*, Stanford: Hoover Institute Press.

Yamada, K. (1990) 'The triple evils of golf courses', *Japan Quarterly*, 37, 3, pp. 291–7.

Yapp, W., Ooi, S.T. and Gill, S.S. (1988) 'Approach to rural development in Sabah', in T.T. Chuan and W. Yee (eds), *Towards Modernising Smallholding Agriculture in Sabah*, Institute of Development Studies (Sabah), Kota Kinabalu, pp. 15–38.

Young, M. (1992) 'People and parks: factors for the success of community-based ecotourism in the conservation of tropical rainforest', paper presented to the IVth World Congress on National Parks and Protected Areas, Caracas, Venezuela.

Zainab Bakir, S. and Humaidi, M. (1991) 'Lampung: spontaneous transmigration', in H. Hill (ed.), *Unity and Diversity: Regional Economic Development in Indonesia since 1970*, Singapore: Oxford University Press.

Zasloff, J.J. (1991) 'Political constraints on development in Laos', in J.J. Zasloff and L. Unger (eds), *Laos: Beyond the Revolution*, Basingstoke: Macmillan, pp. 3–40.

Zasloff, J.J. and Unger, L. (eds) (1991) *Laos: Beyond the Revolution*, Basingstoke: Macmillan.

Zerner, C. (1992) 'Indigenous forest-dwelling communities in Indonesia's Outer Islands: livelihood, rights, and environmental management institutions in the era of industrial forest exploitation', unpublished report commissioned by the World Bank.

INDEX